ROBUST MODEL PREDICTIVE CONTROL OF PERMANENT MAGNET SYNCHRONOUS MOTOR UNDER PARAMETER MISMATCH

ROBUST MODEL PREDICTIVE CONTROL OF PERMANENT MAGNET SYNCHRONOUS MOTOR UNDER PARAMETER MISMATCH

GONGPING WU

College of Electrical and Information Engineering, Changsha University of Science and Technology, Changsha, China

SHENG HUANG

College of Electrical and Information Engineering, Hunan University, Changsha, China

QIUWEI WU

School of Electrical and Information Engineering, Tianjin University, Tianjin, China

VLADIMIR TERZIJA

School of Engineering, Newcastle University, Tyne and Wear, United Kingdom

ELSEVIER

Elsevier
Radarweg 29, PO Box 211, 1000 AE Amsterdam, Netherlands
125 London Wall, London EC2Y 5AS, United Kingdom
50 Hampshire Street, 5th Floor, Cambridge, MA 02139, United States

ISBN: 978-0-443-44209-4

For information on all Elsevier publications
visit our website at https://www.elsevier.com/books-and-journals

Publisher: Matthew Deans
Acquisitions Editor: Sophie Harrison
Editorial Project Manager: Siddharth Anand
Production Project Manager: Maria Bernard
Cover Designer: Bakyalakshmi S

Typeset by Integra Software Services

Contents

About the authors

Gongping Wu received a PhD in Electrical Engineering from the College of Electrical and Information Engineering, Hunan University, Changsha, China, in 2021. He studied in College of Electronic and Information Engineering, Tongji University, Shanghai, China, from September 2011 to September 2013. From 2019 to 2021, he was a visiting student at the Department of Electrical Engineering, Technical University of Denmark, Lyngby, Denmark. He has been working at the College of Electrical and Information Engineering, Changsha University of Science and Technology since July 2021 (assistant professor July 2021 to September 2024; associate professor since September 2024). His research interests include motor fault diagnosis, model predictive control, and permanent magnet synchronous motor systems.

Sheng Huang completed an MS and a PhD in Electrical Engineering from the College of Electrical and Information Engineering, Hunan University, Changsha, China, in 2012 and 2016, respectively. Since 2023, he has been a professor at the College of Electrical and Information Engineering, Hunan University. His research interests include renewable energy generation, modeling and integration study of wind power, and permanent magnet synchronous motor systems.

Qiuwei Wu received a PhD in Electrical Engineering from Nanyang Technological University, Singapore, in 2009. He is a chair professor at the School of Electrical and Information Engineering, Tianjin University, Tianjin, China. His research interests are distributed optimal operation and control of low-carbon power and energy systems, including distributed optimal control of wind power, optimal operation of active distribution networks, and optimal operation and planning of integrated energy systems.

Vladimir Terzija completed Dipl-Ing., MSc, and PhD in Electrical Engineering from the University of Belgrade, Belgrade, Serbia, in 1988, 1993, and 1997, respectively. Since 2023, he has been a full professor at the School of Engineering, Newcastle University, the UK. From 2021 to February 2023, he was a full professor and the head of the Laboratory of Modern Energy Systems at Skoltech, Russian Federation. During 2006–2020, he was the EPSRC Chair Professor in Power System Engineering at the Department of Electrical and Electronic Engineering, the University of Manchester, Manchester, UK. From 1997 to 1999, he was an assistant professor at the University of Belgrade, Belgrade, Serbia. From 2000 to 2006, he was a senior specialist for switchgear and distribution automation at ABB, Ratingen, Germany. His research interests include smart grid applications, wide-area monitoring, protection and control, multienergy systems, switchgear and transient processes, ICT, data analytics, and digital signal processing applications in power systems. He was the recipient of the prestigious Alexander von Humboldt Fellow. He is the editor-in-chief of the *International Journal of Electrical Power and Energy Systems*.

CHAPTER 1

Classic control strategies and practical application of permanent magnet synchronous motor

1.1 Introduction

The motor system is an indispensable and important power foundation in modern production, and it has become a key piece of basic equipment supporting national defense construction and economic development. According to statistics from the International Energy Agency, motor systems consume over 65% of global industrial electricity, and their total capacity is still increasing year by year. As the core carrier and key execution device of rail transit equipment, aerospace equipment, energy-saving and new energy vehicles, robots, and other equipment, the operation quality of the motor system has become the core technical support for promoting the development of high-end equipment. In the current industrial transmission field, motor systems are mainly divided into DC motor systems and AC motor systems according to the power supply mode. With the innovative development of power electronics and power devices, AC motor systems and drive technology have shown significant advantages in the industrial drive field. In AC motor servo systems, permanent magnet synchronous motors (PMSMs) have received widespread attention due to their significant advantages, such as high power factor, good dynamic performance, wide speed range, and compact structure. In the 1970s, the emergence of high-performance rare earth permanent magnet materials promoted the large-scale application of PMSMs in the industrial field. PMSMs generate electromagnetic torque through the interaction between the excitation magnetic field of the permanent magnet and the magnetic field of the armature, which can achieve the conversion of electrical and mechanical energy. According to the different installation positions of permanent

Robust Model Predictive Control of Permanent Magnet Synchronous Motor Under Parameter Mismatch
https://doi.org/10.1016/B978-0-443-44209-4.00016-6

magnets, PMSMs are mainly divided into two categories: surface-mounted permanent magnet synchronous motors (SPMSMs) and internal permanent magnet synchronous motors (IPMSMs). The IPMSMs have a higher torque density and can be widely used in transportation fields that require high motor torque, such as new energy vehicles and ship electric propulsion. SPMSMs are widely used in industrial equipment such as robots and precision machine tools due to their advantages of simple control and excellent operational performance.

The three-phase PMSM system is currently the mainstream of AC drives in the industrial field. With the continuous expansion of application fields, traditional three-phase PMSMs are still unable to meet the demands of high-end technical equipment for the quality of motor system operation. Long-term service in extreme natural environments such as the deep sea, deep earth, and deep space has become the new norm for modern motor systems. Compared with traditional three-phase PMSM systems, multiphase PMSM systems have received widespread attention and favor in special application fields such as aerospace equipment, advanced rail transit equipment, and military engineering equipment due to their stronger fault tolerance, smoother output torque characteristics, and higher reliability. As early as the 1960s, many scholars showed a strong interest in studying multiphase motor systems and conducted in-depth research on multiphase motors and their control. E.E. Ward studied five-phase induction motors and their inverter drive systems in 1969, laying the foundation for the development of subsequent multiphase motor drive systems [1]. Compared with three-phase motor systems, multiphase motor systems have more control degrees of freedom and flexibility, so almost all existing analysis methods and control algorithms can be applied to the control of multiphase motors. However, compared with three-phase motor systems, the model and coupling relationship of multiphase motor systems are more complex, so the analysis and control of multiphase motor systems are significantly different from those of traditional three-phase motor systems.

1.2 Advantages and applications of multiphase motor system

Before the 1980s, motor drive technology severely constrained the research and application of multiphase motor systems. However, over the past two to three decades, with the rapid development of power electronics technology and modern motor control theory, motor systems driven by power electronic converters are no longer limited by the number of phases supplied by the power grid. The application scope of high-performance multiphase motor drive systems has rapidly expanded. Research institutes and universities have taken high-performance multiphase motor systems as their research focus and have led the development of advanced technological equipment such as high-power multiphase direct-drive permanent magnet wind power generation systems, high-power comprehensive all-electric propulsion systems, and high-power rail transit drive systems. The reason multiphase motor systems are highly valued by researchers is that, compared with traditional three-phase motor systems, they offer significant advantages such as strong fault tolerance and high reliability. The advantages of multiphase motor systems are as follows:

1. Multiphase motor systems have the advantages of strong fault tolerance and high reliability. The long-term operation of modern motor systems has become the new normal under complex working conditions, which places higher demands on the reliable operation of motor systems in the event of faults. If one or more phases of a multiphase motor experience an open-circuit fault, the multiphase motor system only needs to reduce the load without stopping to maintain normal operation. This significantly improves the reliability and fault tolerance of the motor system. The unique advantages of multiphase motor systems make them highly suitable for applications such as rail transit, aerospace, military equipment, and other fields that require high reliability and do not allow downtime.
2. A multiphase motor system is an effective way to achieve low-voltage, high-power technology solutions. Multiphase motors are particularly suitable for low-voltage, high-power systems due to their advantages, such as a large number of windings and low stator current harmonics. When the motor power and phase current are constant, increasing the number of windings in the motor system will lower the power supply voltage. The use of low-voltage equipment can avoid the problems of voltage and current sharing caused by traditional centralized converters. The advantages of multiphase motor systems make them highly suitable for applications such as ship electric propulsion, ship integrated systems, electric locomotives, and other scenarios where the supply voltage level is strictly limited.
3. Multiphase motor systems have the advantages of low torque ripple and good dynamic and static performance. In traditional three-phase motor systems, the interaction between the fundamental wave and harmonics

of the air-gap magnetic potential can generate harmonic torque ripple, which can affect the dynamic and static performance of the motor system. By increasing the number of phases in the motor, magnetic potential harmonics can be canceled out, thereby significantly reducing the content of magnetic potential harmonics in the motor winding. This will significantly reduce torque ripple and improve the dynamic and static performance of the motor system. The low torque ripple and high dynamic performance of multiphase motor systems can meet the demands of high-end machine tools and robots for dynamic qualities such as high speed, precision, and stability.

4. Multiphase motor systems have the advantages of high efficiency and high controllability. As the number of phases in the motor increases, the fundamental current of the multiphase motor system becomes smaller under the same torque, the stator copper loss is significantly reduced, and the system efficiency is significantly improved. In addition, as the number of phases in the motor increases, the available voltage vector of the multiphase motor systems also doubles under an inverter power supply, which greatly improves the controllability of the motor system. The combination of multiphase motor systems and multilevel converters is beneficial for expanding wind power systems toward larger capacity, higher voltage levels, and higher efficiency. It can effectively solve the problem of safe and efficient operation of high-voltage and high-power wind power systems.

Although multiphase motor systems have many advantages, as mentioned above, further exploration is needed for key fundamental theories such as optimization design and high-performance control of multiphase motor systems of different types and categories. Therefore researchers have conducted in-depth studies on the design methods and efficient control theories of different types and categories of multiphase motor systems, based on the need to improve efficiency and optimize the control performance of these systems. The purpose is to address a series of fundamental and common bottleneck technical problems in the design of the ontology structure, analysis of the operating mechanism, establishment of mathematical models, optimization of control strategies, and efficient operation of multiphase motor systems. Research scholars' theoretical studies on multiphase motor systems mainly focus on two types: multiphase induction motors and multiphase permanent magnet motors.

Regarding the research on multiphase induction motor systems, scholars have conducted extensive studies from different perspectives in various application fields, mainly focusing on five-phase induction motor systems, six-phase induction motor systems, and nine-phase induction motor systems. Professor Thomas A. Lipo and his research team have conducted extensive and detailed research on the precise establishment of mathematical models for multiphase induction motor systems, the improvement of motor torque density, and space vector PWM control since the 1980s. Zhao et al. [2] presented a companion vector space decomposition control technique for dual three-phase induction machines, which can be extended to the control of induction machines with an arbitrary number of phases. Zhao et al. [3] proposed a magnetic field orientation control technology for multiphase induction motors with an unbalanced stator winding structure. This technology is developed from the balanced winding structure and is also applicable to multiphase induction motors with unbalanced stator winding structures. Lyra et al. [4] presented a technique for injecting third harmonic zero-sequence current components into six-phase induction motors, which can solve the problem of uneven distribution of winding current and improve the torque density of the machine. Munoz et al. [5] proposed a novel dual stator winding asynchronous motor driver that can eliminate the negative effects of stator resistance during low-speed operation. This greatly simplifies the implementation of sensorless control schemes for multiphase induction motors. Jones et al. [6] presented a modified current control scheme for multiphase induction motor drives, which can simultaneously compensate for any imbalance of the fundamental wave and unwanted low-order stator current harmonics. Barrero et al. [7] proposed an enhanced predictive current control method for asymmetrical six-phase motor drives, which can achieve fast torque and current response while facilitating stator current harmonic suppression. Pandit et al. [8] presented an improved direct torque control method for asymmetrical six-phase induction motor systems, which retains the advantages of a simple structure and good transient response while being able to operate at higher average switching frequencies. Rubino et al. [9] proposed a deadbeat model predictive direct flux vector control for multiphase induction machines configured with multiple three-phase structures, which can achieve excellent dynamic control performance while maintaining the sinusoidal current of the multiphase motor.

Regarding the design of nine-phase induction motors, Umesh et al. [10] proposed a finite-element method model of a nine-phase squirrel-cage induction motor, which can improve the performance of the drive under high-pole operation. Titus et al. [11] proposed a new asymmetric nine-phase induction motor-based load-commutated inverter-fed drive configuration, which can provide an alternative rugged, low-cost solution in various applications. Bastos et al. [12] discussed the issue of replacing traditional three-phase motor drives with nine-phase motor drives and presented experimental results on the impact of faults on torque and current waveforms in multiphase drives. González-Prieto et al. [13] proposed a constrained model predictive control strategy for nine-phase induction motor drives using a three-step method, which includes the initial descent of switch states, dynamic selection of voltage vectors using hard constraints, and improved performance through the inclusion of soft constraints. Kong et al. [14] proposed a novel harmonic current injection method for multiphase induction motors, which can optimize the air-gap and yoke flux density distribution of multiphase variable-speed drive systems. Reddy et al. [15] presented a fault-tolerant multiphase multilevel inverter configuration for a pole–phase modulated nine-phase induction motor drive, which can improve the performance of multiphase induction motor drives in both normal and switch failure conditions. Liu et al. [16] proposed a novel rotor-flux-oriented control with fuzzy logic current controllers for a nine-phase induction motor drive system, which can effectively enhance fault-tolerant capability under various faulty conditions. Abdel-Khalik et al. [17] designed the postfault control of a nine-phase six-terminal induction machine during a single open-line fault, which can be used in high-power medium-voltage applications as an alternative to conventional six-phase induction machines.

In the theoretical research of multiphase PMSM drive systems, researchers have conducted in-depth exploration of basic scientific issues such as operation mechanism analysis, accurate modeling, coordinate transformation, harmonic current suppression, and fault-tolerant operation of multiphase PMSM. However, the main focus is on two types: five-phase PMSM drive systems and six-phase PMSM drive systems. Wang et al. [18] proposed a new decoupled modulation scheme based on the xy-space control voltage reference, which can effectively suppress the harmonic currents of five-phase PMSM. Yang et al. [19] presented a voltage sensorless scheme for the five-phase PMSM drives by using the extended state observer, which can suppress the parallel resonance of filter capacitors and winding inductors by estimating the feedback extended state. Wang et al. [20] presented an improved finite control set model predictive current control strategy with virtual voltage vectors and an adaptive control set, which can improve the steady and transient performance of a five-phase PMSM. Xu et al. [21] proposed a unified open-phase fault diagnosis method for five-phase PMSM systems under normal operation and fault-tolerant operation modes, which can achieve the diagnosis of open-phase faults and enhance the robustness against speed and load disturbances. Rajanikanth et al. [22] proposed an improved model predictive current control of five-phase PMSM drives by using preselection of voltage vectors, which can reduce magnetic flux pulsation, torque pulsation, and computational burden. Huang et al. [23] proposed a generalized fault-tolerant model predictive current control method for five-phase PMSM drives, which can effectively address open-switch faults by maintaining the torque production capacity and ensuring an undisturbed magnetic motive force. Li et al. [24] presented a fault-tolerant current vector control strategy based on a five-phase six-leg inverter for five-phase SPMSM, which can eliminate specific torque harmonics and achieve decoupling control under fault conditions. Rajanikanth et al. [25] proposed an enhanced model predictive current control scheme for a five-phase PMSM based on voltage vector preselection, which can effectively minimize the total harmonic distortion of the current by introducing additional voltage vectors. Saeed et al. [26] proposed a low-complexity model predictive current control for five-phase PMSM with constant switching frequency, which can improve steady-state performance and reduce computational burden. Wu et al. [27] proposed an improved model predictive torque control method without weighting factors for five-phase PMSM drives, which can improve torque performance and eliminate weighting factors in the cost function. Saeed et al. [28] presented a generalized low-complexity deadbeat solution for model predictive control schemes of five-phase PMSM drives, which can achieve lower computational burden, improved steady-state performance, and simple generalized execution. Xiong et al. [29] proposed a constant switching frequency multiple-vector-based finite-control-set model predictive current control scheme for five-phase PMSM drives, which can enhance the robustness of the multiphase motor system against parameter mismatch. Du et al. [30] proposed a fault-tolerant model predictive current control method for five-phase PMSM drives with open-circuit faults, which can improve the fault-tolerant ability of the multiphase motor system under single and multiple open-circuit fault operations. Wang et al. [31] proposed an online full-range minimum copper loss method based on fault factors for five-phase PMSM, which can achieve high torque output and low copper loss under single-phase short-circuit faults. Wang et al. [32] proposed a new multivector model predictive control with voltage error tracking for a five-phase

PMSM, which can achieve satisfactory fault tolerance and enhance steady-state performance under short-circuit faults. Saeed et al. [33] presented an improved finite control set model predictive control scheme based on dual vectors for five-phase PMSM drives, which can improve tracking accuracy and DC link voltage utilization. Huang et al. [34] presented a novel model predictive torque control combined with space vector modulation for five-phase PMSM drives, which can enhance the fault-tolerant performance of the multiphase motor system under open-circuit faults.

Regarding the research on six-phase PMSM, Lin et al. [35,36] proposed an intelligent complementary sliding-mode control method for six-phase PMSM drive systems, which can improve control performance and maintain stability under fault conditions. Luo et al. [37] proposed a direct torque control method for an asymmetric six-phase PMSM based on feedback linearization, which can achieve smaller torque and flux ripple in the primary subspace and less harmonic current in the secondary subspace. Zhou et al. [38] proposed a deadbeat direct current control strategy for six-phase PMSM by using dynamic time programming, which can solve the optimization problem of adjustment time under multiobjective complex constraints. Luo et al. [39,40] proposed a novel model predictive torque control strategy with discrete duty ratio optimization for a six-phase PMSM, which can eliminate the influence of weighting factor variation and achieve minimal torque ripple. Das et al. [36,40] proposed a voltage injection-based current harmonic minimization method for wide-bandgap device-based six-phase PMSM, which can avoid phase error problems and improve the steady-state performance of harmonic compensation. Tang et al. [41] proposed a fault-tolerant control strategy suitable for the asymmetric six-phase PMSM, which can enhance fault-tolerant performance and capability in multiple open-phase fault modes. Hu et al. [42] proposed a flux-weakening control for six-phase PMSM based on vector space decomposition, which is superior to traditional control methods in reducing current imbalance and harmonic currents. Song et al. [43] proposed a novel carrier-based modulation strategy for this special asymmetrical six-phase series-winding PMSM, which can achieve higher speed and suppress current distortion within the nonlinear modulation range. Lu et al. [44] proposed a new fault-tolerant PWM predictive control of six-phase PMSM drives based on pulsewidth modulation, which can achieve less current harmonics and optimum current control performance. Gonçalves et al. [45] proposed a novel fault-tolerant predictive current control strategy for six-phase PMSM drives, which can guarantee motor drive operation under post-fault conditions with ripple-free torque and low current harmonic distortion.

In summary, the common basic scientific problems of multiphase motor systems have been systematically solved. However, the advantage of multiphase motor systems lies in enhancing the fault tolerance of the entire system by increasing the number of phases in the motor. The increase in the number of phases in a multiphase motor makes the control system strategy increasingly complex, and vector control requires more and more voltage vectors. This significantly increases the design difficulty and computational burden of the control system, thereby reducing the reliability and practicality of the multiphase motor system. In addition, more stringent fault conditions need to be considered in some high-speed and high-torque motor applications, such as the fault removal of the entire unit motor. The traditional structure of multiphase motors can no longer meet the requirements for fault-tolerant operation of the system. Therefore, to further expand the application of PMSMs in harsh environments and extreme working conditions, research on common and fundamental key technologies, such as new modular multiphase motor design and high-performance control methods, has significant engineering value and practical significance.

1.3 Research status and significance of modular multiphase PMSM

The multiphase PMSM systems can achieve higher power density and efficiency by leveraging the advantages of both multiphase motors and PMSMs. With the continuous improvement in the operational quality of motor systems in fields such as rail transit, ship propulsion, electric vehicles, and wind power equipment, multiphase permanent magnet motor systems have demonstrated significant competitive advantages in these areas due to their high power density, high efficiency, and strong fault tolerance. The high-power-density multiphase PMSM systems have significant research value and importance for space-constrained mobile platforms such as all-electric ships, all-electric tanks, and all-electric aircraft. According to the different structures of multiphase PMSMs, they can be divided into two categories: conventional multiphase structure and new modular multiwinding structure. Taking the nine-phase PMSM as an example, the schematic diagrams of the conventional multiphase structure and the new modular multiwinding structure are shown in Figs. 1.1 and 1.2, respectively. The conventional multiphase structure PMSM can ensure the safe and stable

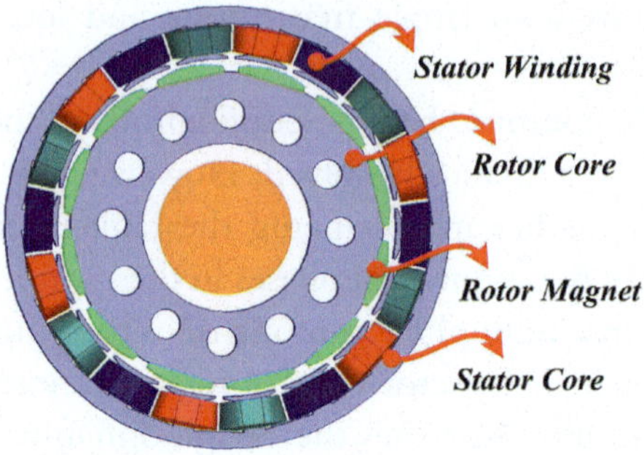

Fig. 1.1 Conventional multiphase structure.

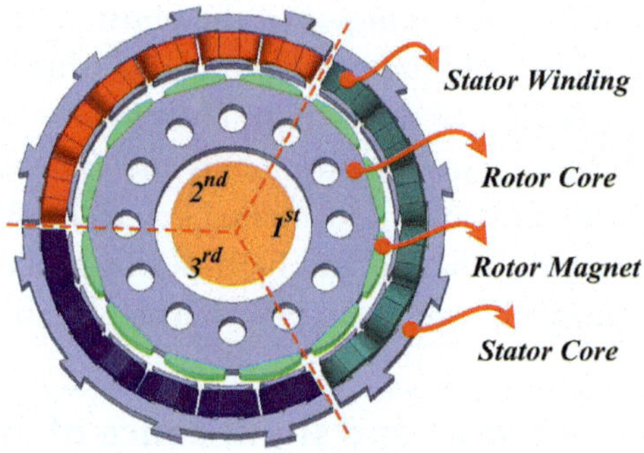

Fig. 1.2 New modular multiwinding structure.

operation of the motor system under faults by increasing the number of phases, but there are challenges such as complex control strategies and difficult controller design. In comparison, the new modular multiwinding structure PMSM offers the following advantages.

1. With the continuous improvement of motor manufacturing technology, the motor units in modular multiwinding PMSMs can achieve physical isolation, including electrical, magnetic, and thermal isolation. When the multiphase winding experiences extreme faults such as an open circuit or short circuit, the entire faulty motor unit can be cut off, and the remaining normal motor units can be used to maintain the normal fault-tolerant operation of the motor system under the fault. Compared with conventional multiphase structure PMSMs, the new modular structure PMSMs can further enhance the fault tolerance and reliability of motor systems.

2. Due to the modular structure of PMSMs with isolated motor units in modular multiwinding motors, each motor unit can be controlled using conventional three-phase motor control strategies. Therefore the controller for modular structure PMSMs can use existing three-phase motor controllers. Compared with conventional multiphase structure PMSMs, it greatly simplifies the complexity of the control system and is more conducive to the integrated design of the motor system.
3. The modular structure PMSM combines the mature, conventional three-phase motor system with a modular topology structure that has strong fault tolerance. Each motor unit is controlled by an independent controller, which can allocate high-power and high-torque requirements to each motor unit. Compared with conventional multiphase structure PMSMs, it effectively reduces the current and power levels borne by the windings and drive units in the modular structure PMSM.

In summary, compared with conventional multiphase structure PMSMs, modular structure PMSMs have many advantages, such as simple topology, simple controller design, and better fault-tolerant control performance, which can better meet the operational quality requirements of high torque, wide speed range, and high fault tolerance. In addition, due to the unique body structure design of modular structure PMSMs, each motor unit has an independent set of three-phase stator windings. When a fault occurs in a certain motor unit, it does not affect the normal operation of other motor units, and it can effectively avoid the circulating current caused by parallel power devices in the context of high-power and high-torque applications.

The modular structure PMSM drive system is essentially an electrical transmission system composed of a modular multiwinding PMSM, an inverter, and a controller. Due to the emerging research on modular structure PMSM drive systems, current research on modular structure PMSM mainly focuses on motor body design, topology optimization, and other aspects. Levi et al. [46] designed a novel modular multiwinding PMSM consisting of eight motor units, which fully embodies the advantages of strong fault tolerance and high reliability of modular multiwinding motors in electric vehicles. Cengelci et al. [47] proposed a new modular motor and modular inverter concept in high-performance medium-voltage adjustable speed drive systems, where each group is powered by a separate three-phase pulse-width modulation inverter that can continue to operate at a lower power level in the event of inverter and/or motor failure. Tang et al. [48] introduced fault-tolerant strategies for modular motor systems and modular permanent magnet synchronous wheel hub motors under fault conditions, which can significantly reduce the maximum temperature of the windings and make the temperature distribution of the windings more uniform. Han et al. [49] proposed the concept of modular stator design for fractional slot concentrated winding PMSMs and derived the prerequisites for motors to meet stator modularization, which can suppress motor losses without reducing output performance. Tang et al. [50] proposed a novel modular tubular permanent magnet synchronous linear motor, in which the primary component of the unit adopts a large slot structure containing multiple coils. The principle of three-phase balance of the entire motor magnetic flux was explained, and the end tooth design further suppressed the braking force and improved the thrust density. Tang et al. [51] proposed a modular permanent magnet for wheeled motors and designed a permanent magnet capable of achieving high fault-tolerant performance and low magnetic potential spatial harmonics. Based on the working mechanism and main fault types of modular fault-tolerant motors, corresponding fault-tolerant control strategies were proposed. Yao et al. [52] proposed a novel modular stator tubular PMSM with a modular complementary stator structure that can decouple adjacent phase windings and provide the required fault tolerance. The proposed motor has good fault tolerance and improved force performance and is suitable for electromagnetic active vehicle suspension. Mohamed et al. [53] proposed and analyzed a converter topology that can be reconfigured to operate in both conventional and modular driving modes. In the modular driving method, each stator coil is driven by a separate converter module. The modular driving method provides a wider speed range and better fault tolerance at rated torque. Rubino et al. [54] proposed a vector control scheme for modular three-phase PMSM, which uses modular modeling methods to independently decouple torque control for each three-phase unit, allowing for torque-sharing strategies to be implemented between the three-phase windings of the modular motor. Wang et al. [55] proposed a novel axial flux modular-double-stator hybrid excitation switched reluctance motor by applying a modular stator and hybrid excitation structure and analyzed the effects of permanent magnet thickness and rotor pole number on the torque of the proposed modular motor. Davarpanah et al. [56] presented a new modular hybrid excited switched reluctance motor with permanent magnets embedded between the stator teeth. The modular switched reluctance motor with four teeth per stator pole and two teeth per stator pole has better dynamic performance and higher average steady-state electromagnetic torque. Mohamed et al. [57] proposed a modular improved integrated topology for switched reluctance motors, which integrates modular power converters into the motor to achieve highly compact, fault-tolerant, lightweight, and high-power-density motor drivers. Rubino et al. [58] proposed a new decoupling transformation for decoupling torque control of modular three-phase induction motor drivers, which can reliably operate in deep weak magnetic fields under healthy and faulty operation as well as Maximum Torque Per Voltage (MTPV) operation. The optimization design of modular multiwinding PMSM has been extensively and deeply explored, but the controller of modular multiwinding PMSM still uses traditional proportional-integral controllers. The parameter settings of the proportional-integral controller only correspond to a specific operating range of the motor, while the modular multiwinding PMSM is a nonlinear system with time-varying parameters, making it difficult to achieve satisfactory dynamic and static performance throughout the entire operating range. The advantages of modern advanced control methods have not been fully utilized in the field of modular multiwinding PMSM control, and there is a lack of research on high-performance control theory and methods for modular multiwinding PMSM, resulting in unsatisfactory operational performance and control effects of modular multiwinding PMSM. Therefore studying high-performance control strategies and methods for modular multimotor systems has important scientific significance and engineering application value for the development of their control technology, which is in line with the development trend of modern motor systems. This book is based on the demand for improving control performance and enhancing the robustness of modular multiwinding PMSM systems. It aims to address common and fundamental key scientific issues faced by modular multiwinding PMSM systems, such as operating characteristics, model establishment, parameter mismatch sensitivity analysis, and high-performance robust predictive control under parameter mismatch.

1.4 Classic control strategy of motor drive system

The hysteresis current control strategy is a simple and real-time method for regulating motor drive systems, as shown in Fig. 1.3. Its working principle is to act as a fast-response, nonlinear controller by comparing the measured stator current with its sinusoidal reference command. If the current error exceeds a preset upper or lower limit, the inverter's power switches are directly commanded to apply voltage that forces the current back within the band. This creates a self-oscillating pattern where the

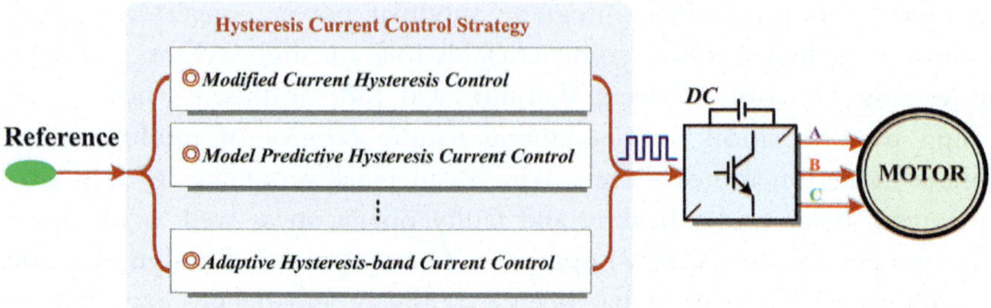

Fig. 1.3 Block diagram of the hysteresis current control strategy for the motor drive system.

actual current tightly tracks the reference. While offering excellent dynamic response and parameter insensitivity, its main drawback is a variable inverter switching frequency, which depends on operating conditions and the chosen bandwidth, balancing tracking accuracy against switching losses. Jiang et al. [59] proposed a multimode current hysteresis control for brushless DC motors in motor and generator states, which can complete commutation successfully and reduce commutation torque ripple. In addition, a novel transition control method based on the background of current hysteresis control with fast response is presented to carry out commutation and the transition processes simultaneously. Kadjoudj et al. [60] presented a hybrid intelligent current controller that combines a hysteresis current controller and a ramp comparator simultaneously, which can significantly reduce the distortion component of switching frequency and has better performance and a simple structure. Liu et al. [61] proposed a multiprocessor-based fully digital control architecture for PMSM drives using the hysteresis current control concept, which can execute current-loop and position-loop algorithms in floating-point formats and in parallel processing mode. Wang et al. [62] proposed a phase-shifting fault-tolerant control method based on hysteresis current control for a primary permanent-magnet linear motor traction system, which can improve transient and steady-state performance without being affected by changes in motor parameters. Xue et al. [63] presented a novel double-vector-based model predictive hysteresis current control for asynchronous motors using a two-level VSI, which has advanced steady-state control performance and lower sampling frequency. Bose et al. [64] proposed an adaptive hysteresis-band current control PWM technique for machine drive systems, which is easy to implement through dedicated hardware, has fast response speed, and can automatically limit the peak current of the equipment. Chang et al. [65] proposed a novel vector control hysteresis current controller for induction motor drives, which can take full advantage of adding the zero voltage vector and greatly reduce the inverter switching frequency. Rebeiro et al. [66] presented a novel interior PMSM drive that combines a fuzzy logic controller-based tuned PI speed controller and an adaptive hysteresis current controller, which exhibits better dynamic response in torque and speed over a wide speed range. Dong et al. [67] presented an improved zero-sequence current hysteresis control-based space vector modulation for open-end winding PMSM drives, which can provide a satisfactory zero-sequence current suppression effect under different speed ranges. Shawier et al. [68] presented a new hysteresis current control technique based on vector-space decomposition and restrained voltage vectors, which can enhance the performance of hysteresis current control for six-phase induction motors. Mamdouh et al. [69,70] proposed a current gradient-based modified hysteresis controller for asymmetrical six-phase induction motors, which has the advantages of fast dynamic response, simple structure, and robustness against machine parameter variations.

Proportional-Integral (PI) control serves as a cornerstone linear compensation technique in motor drives, commonly applied to regulate speed, torque, or current, as shown in Fig. 1.4. Its operation is based on the error signal, which is the deviation between a commanded setpoint and the measured feedback. This error is processed through two parallel paths: the proportional (P) term delivers an instantaneous corrective action proportional to the error, governing dynamic response speed; the integral (I) term integrates historical error to eliminate steady-state offset. By precisely adjusting the proportional and integral gains, the PI controller can achieve stable and accurate regulation under steady-state conditions, with the steady-state error reaching zero. Shin et al. [71] proposed a new antiwindup PI controller for variable-speed motor drives, which can improve speed control performance and achieve small overshoot and

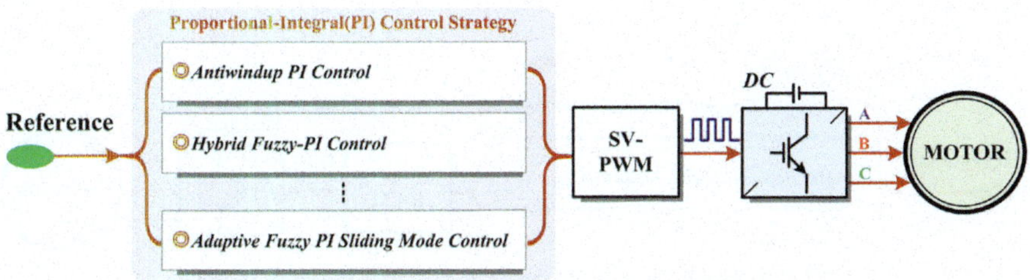

Fig. 1.4 Block diagram of proportional-integral control strategy for the motor drive system.

fast settling time. Seok et al. [72] presented an automatic P/PI speed controller switching scheme for industrial servo drives, which can improve the quality of motor dynamic characteristics and reduce the engineering effort for machine builders. Shahnazi et al. [73] proposed a novel adaptive fuzzy PI sliding mode control (SMC) method for position control of induction and DC servomotors, which can significantly improve the robustness of control performance in motor systems with unknown parameters and uncertainty in load conditions. Sant et al. [74] proposed a hybrid fuzzy-PI speed controller based on a combination of a fuzzy logic controller and a PI controller, which can improve the performance of the fuzzy logic controller under transient conditions for PMSM drives. Cheng et al. [75] proposed a new self-tuning fuzzy PI controller with conditional integral for doubly salient permanent-magnet motor drives, which has better adaptability and steady-state dynamic performance than ordinary linear PI control. Errouissi et al. [76] presented a novel control design of a PI speed controller for AC motor drives with consideration of input saturation, which can maintain the required transient performance even under external disturbances such as model uncertainty and load torque changes. Moallem et al. [77] proposed a novel multiobjective optimization method based on a genetic-fuzzy algorithm, which can optimize the gains of PI controllers in the indirect field-oriented control of induction motor drives.

SMC is a robust nonlinear control strategy for motor drives, as shown in Fig. 1.5. This principle involves defining a sliding surface that is typically a function of velocity or current tracking error to represent the desired system dynamics. Then, the controller applies discontinuous control laws aimed at forcing the system's state trajectory to reach and subsequently remain on that surface. The sliding mode controller enters the sliding surface, and the system exhibits strong robustness to parameter changes and

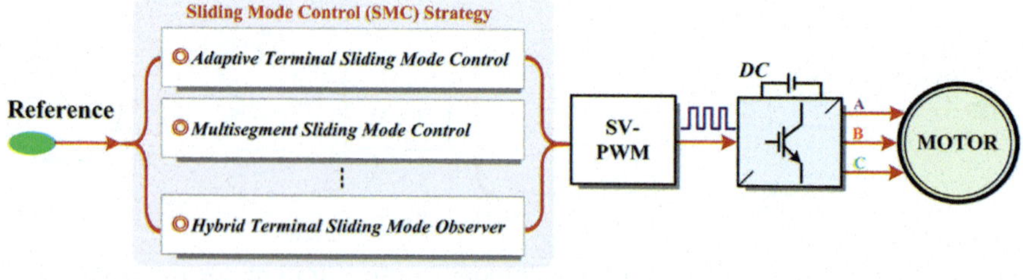

Fig. 1.5 Block diagram of the sliding mode control strategy for the motor drive system.

external disturbances, as its dynamics are only controlled by predefined surface equations. Çavuş et al. [78] proposed a new adaptive terminal SMC method for induction motors with direct torque control, which can improve the control performance of induction motors and overcome the shortcomings of flux control based on model predictive control. Shyu et al. [79] proposed a multisegment SMC based on a trapezoidal velocity profile for synchronous reluctance motors, which can effectively meet the expected speed and acceleration of the motor in incremental motion speed control mode. Yan et al. [80] proposed a novel speed- and flux-sensorless sliding-mode control for an induction motor, which can achieve decoupling of stator flux and torque/speed control loops and has strong robustness. Feng et al. [81] presented a hybrid terminal sliding-mode observer based on the nonsingular terminal sliding-mode and the high-order sliding-mode, which can estimate the rotor position/speed and achieve sensorless control of PMSMs. Foo et al. [82] proposed a nonlinear sliding-mode speed-control scheme by incorporating the maximum-torque-per-ampere trajectory, which can achieve high dynamic and steady-state performance of PMSMs over a wide speed range. Nguyen et al. [83] proposed an adaptive sliding-mode control strategy for PMSMs based on a modified reduced-order proportional integral observer, which can achieve better dynamic response and estimate uncertain parameters/external load torque. Hou et al. [84] proposed a continuous terminal SMC algorithm for servo motor systems, which can be extended to nonlinear systems with high-order dimensions and mismatched uncertainties. Wang et al. [85] presented a passivity-based sliding-mode position control method for induction motor drives based on an adaptive load torque estimator, which can achieve good position tracking response and strong robustness to mechanical parameters/load torque disturbances. Zhang et al. [86] proposed an adaptive sliding mode reaching law based on the exponential function and existing sliding mode reaching law, which can effectively accelerate the SMC approaching process and improve the speed control performance of PMSMs. Baik et al. [87] proposed a robust nonlinear speed control method for permanent magnet synchronous motors based on the integral SMC technique, which can effectively ensure robustness in the presence of parameter changes and speed measurement errors. Vu et al. [88] proposed a new sliding-mode control method based on the Takagi-Sugeno fuzzy model for PMSM drive systems, which has faster dynamic response and stronger robustness than the conventional SMC scheme under parameter uncertainties and unknown external disturbances. Du et al. [89] presented a new discrete-time fast terminal SMC law for position control of permanent magnet linear motors, which has the advantages of the current discrete-time fast terminal SMC approach over the PID control method. Wang et al. [90] proposed a new sensorless control strategy for PMSM drive systems based on an adaptive nonsingular fast terminal sliding mode observer, which can achieve better control performance, stronger robustness, and better system stability.

Neural network control is an intelligent and data-driven strategy that does not require a clear mathematical model of the motor drive system, as shown in Fig. 1.6. After training, the artificial neural network acts as a controller to generate commands for the motor drive system to achieve precise tracking. The advantage of the neural network control strategy lies in its excellent adaptability and robustness to parameter changes and load disturbances. Moreover, it can effectively compensate for the uncertainty of traditional linear controllers. Fu et al. [91] proposed a new neural network control strategy for induction motor drives based on vector control, which can achieve lower switching frequencies or sampling rates compared to conventional vector control. Yin et al. [92] presented a plant-physics-guided backpropagation neural network learning method for PMSM drive systems, which can effectively enhance the online learning speed of motor controller parameters. Alanis et al. [93] presented a new real-time discrete backstepping neural control technique for induction motor drive systems, which can solve the tracking problem for a class of multiple-input–multiple-output nonlinear systems. Chen et al. [94] proposed a novel robust speed control method based on a dual-layer neural network device estimator and a dual-layer neural network PI controller, which can improve the performance of induction motor drives and reduce their sensitivity to load disturbances and parameter variations. Shyu et al. [95] proposed a model reference adaptive speed control scheme for induction motor drives using neural networks, which is not only simple and practical to implement but also robust to parameters and disturbances. Yu et al. [96] proposed a new

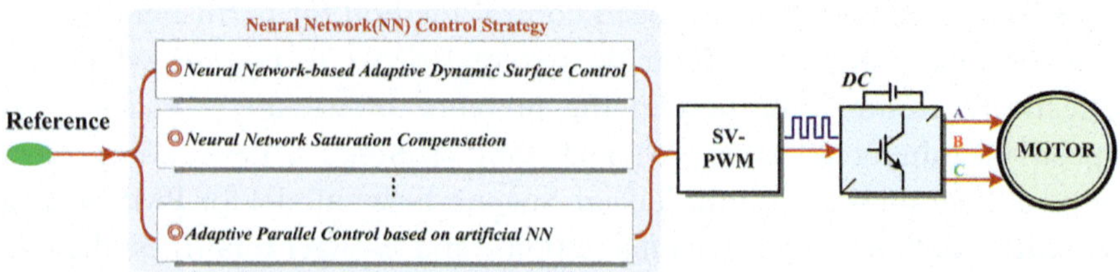

Fig. 1.6 Block diagram of the neural network control strategy for the motor drive system.

neural network-based adaptive dynamic surface control method for PMSM drives, which can effectively enhance robustness against load disturbances and parameter uncertainties. Kung et al. [97] presented an adaptive speed control method for induction motor drives using the neural network concept, which can obtain perfect tracking and regulating control characteristics under wide operating ranges. Hudson et al. [98] proposed a novel sensorless control technique for single-switch-based switched reluctance motors using neural networks, which can reduce the number of calculations required to estimate rotor position from the input set. Wlas et al. [99] proposed artificial neural networks for online identification of induction motor drives, which can significantly improve the speed estimation process and make the whole control system operate better. Jang et al. [100] proposed a neural network saturation compensation strategy for DC motor drive systems, which can effectively compensate for saturation nonlinearity in the presence of system uncertainty. Sheu et al. [101] proposed a new self-tuning PI speed controller for an induction motor based on neural network identification, which can eliminate the influence of load disturbances and system parameter uncertainties on the speed response. Senjyu et al. [102] proposed a new speed-control scheme for ultrasonic motors using a neural network, which can compensate for changes in motor characteristics and approximate the nonlinear input-output mapping of ultrasonic motors. Sousy et al. [103] presented an intelligent control system for PMSM servo drives based on a recurrent wavelet-based Elman neural network, which can achieve high-precision tracking performance and handle the presence of parameter uncertainties. Rubaai et al. [104] proposed an adaptive parallel control architecture based on artificial neural networks, which can effectively solve the optimal tracking control problem of highly nonlinear multiinput and multioutput dynamics in induction motor drives.

Model predictive control is an advanced control strategy for motor drives that utilizes a dynamic model of the system to predict its future behavior over a finite time horizon, as shown in Fig. 1.7. In each control cycle, the algorithm simulates the effects of all possible control actions (e.g., inverter voltage vectors) on future states like currents and speed. These predictions are evaluated against a predefined cost function, which quantifies control objectives such as tracking accuracy and switching losses. The control action that minimizes this cost function is selected and applied to the inverter. This "receding horizon" process is repeated every sampling period, providing optimal, constrained multivariable control that inherently handles system nonlinearities and limits. However, its implementation requires

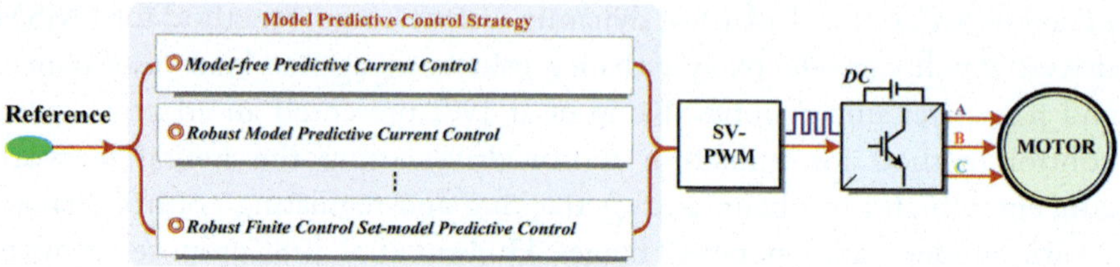

Fig. 1.7 Block diagram of the model predictive control strategy for the motor drive system.

significant real-time computational power for solving the optimization problem. Rubaai et al. [105] proposed the model predictive control with a cross-coupling control structure algorithm for a dual-motor system, which can improve the dynamic performance and reduce the online execution time. Han et al. [106] proposed the current difference-based model-free predictive current control strategy for switched reluctance motors, which can achieve better dynamic performance and reduce the computation cost. Zhang et al. [107] presented a new model predictive current control of PMSM drive systems based on the nonparametric prediction model, which can eliminate the dependence on the accuracy of the motor model parameters. Zhu et al. [108] presented a model-free predictive current control with an extended control set for PMSM drive systems, which can significantly reduce the computational burden and torque ripple. Lin et al. [109] presented a model-free predictive current controller based on an ultra-local model employing a fixed-time observer, which can effectively reduce the current fluctuations of the PMSM drive system in the steady state. Zhang et al. [110] proposed a simple motor-parameter-free model predictive current control method for PMSM drive systems, which can eliminate the influence of resistance and flux-linkage parameters in the predictive model. Li et al. [111] proposed a model-free predictive current control with adaptive resonance for PMSM drive systems based on an ultra-local model, which can suppress periodic and non-periodic disturbances in the system. Ismail et al. [112] proposed an improved adaptive discrete linear plant model of PMSM drive systems under different speed regions, which can reduce the speed drop and torque ripple compared with those of conventional model predictive control strategies. Zhang et al. [113] proposed a robust model predictive current control method for PMSM drive systems based on a nonlinear extended state observer, which can obtain lower harmonic content and better dynamic performance. Zhou et al. [114] proposed a new continuous voltage vector model-free predictive current control method based on the uncoupling feature, which can improve the robustness and steady-state responses of PMSM drive systems. Mousavi et al. [115] proposed a new model-free predictive current control strategy with parallel cost functions for the open-end winding synchronous reluctance motor, which can reduce the computational burden while maintaining the high performance of standard implementation. Wei et al. [116] proposed a model-free predictive current control method for PMSM drive systems based on adaptive ultralocalized time-series, which can improve the model flexibility and accuracy of the current operating state of the system. Fobi et al. [117] proposed a new and simple modulated model-free predictive current control method for PMSM drive systems based on constrained optimization, which can achieve excellent performance in current tracking, harmonic content suppression, and current ripple reduction. Niu et al. [118] proposed a robust finite control set-model predictive control method for PMSM drive systems based on an indirect reference vector, which can improve robustness to parameter changes and higher current control accuracy. Sun et al. [119] proposed an improved continuous-control-set model predictive control method for dual three-phase PMSMs, which can improve the robustness of the system while retaining the advantages of fast response speed and small overshoot. Ahmed et al. [120] presented a new model-free predictive current control method for synchronous reluctance motors based on real-time current estimation, which can obtain excellent dynamic performance while achieving better control performance and tracking error. An et al. [121] proposed an adjustable model predictive control strategy for PMSM drives based on online parameter identification, which can suppress the torque ripple and eliminate the current distortion caused by inductance parameter mismatch. Petkar et al. [122] proposed a modified model predictive current control method for PMSM drive systems, which can exhibit better response and enhance robustness to parameter mismatch. Wei et al. [123] proposed a new model-free predictive current control strategy for PMSM drive systems based on extended affine ultra-local models, which can improve current quality and robustness compared to conventional model predictive control.

1.5 Conclusion

In this chapter, the evolution and current situation of motor drive systems are described, and the development trend of motors is explained from the perspectives of motor body design and motor system control.

The pursuit of higher power, reliability, and fault tolerance has driven the evolution from traditional three-phase motor systems to multiphase motor topologies. The advantage of multiphase motors is that they can continue to operate even after phase faults, making them widely used in fields such as electric ships, aerospace, and advanced electric vehicles. Modular multiphase PMSMs have the advantages of high power density and strong reliability, and the application of modular stator concepts simplifies manufacturing and maintenance while enhancing system robustness. Then, the classic control theories of motor systems are described in detail, such as hysteresis current control, proportional-integral (PI) control, SMC, neural network control, and model predictive control. This chapter provides a systematic overview of the evolution of motor topology and the development of control strategies, offering new research perspectives for scholars in the field of motors and drives.

References

[1] W.E. E, H. Harer, et al., Preliminary Investigation of an Inverter-fed 5-phase Induction Motor, Proc. Inst. Electr. Eng. 116 (6) (1969) 980–984.
[2] Y. Zhao, T.A. Lipo, et al., Space vector PWM control of dual three-phase induction machine using vector space decomposition, IEEE Trans. Ind. Appl. 31 (05) (1995) 1100–1109.
[3] Y. Zhao, T.A. Lipo, et al., Modeling and control of a multi-phase induction machine with structural unbalance, IEEE Trans. Energy Convers. 11 (03) (1996) 578–584.
[4] R.O.C. Lyra, T.A. Lipo, et al., Torque density improvement in a six-phase induction motor with third harmonic current injection, IEEE Trans. Ind. Appl. 38 (05) (2002) 1351–1360.
[5] A.R. Munoz, T.A. Lipo, et al., Dual stator winding induction machine drive, IEEE Trans. Ind. Appl. 36 (05) (2000) 1369–1379.
[6] M. Jones, S.N. Vukosavic, et al., A synchronous current control scheme for multiphase induction motor drives, IEEE Trans. Energy Convers. 24 (04) (2009) 860–868.
[7] F. Barrero, J. Prieto, et al., An Enhanced Predictive Current Control Method for Asymmetrical Six-Phase Motor Drives, IEEE Trans. Ind. Electron. 58 (08) (2011) 3242–3252.
[8] J.K. Pandit, M.V. Aware, et al., Direct Torque Control Scheme for a Six-Phase Induction Motor With Reduced Torque Ripple, IEEE Trans. Power Electron. 32 (09) (2017) 7118–7129.
[9] S. Rubino, R. Bojoi, et al., Model Predictive Direct Flux Vector Control of Multi-three-Phase Induction Motor Drives, IEEE Trans. Ind. Appl. 54 (05) (2019) 4394–4404.
[10] B.S. Umesh, K. Sivakumar, et al., Multilevel Inverter Scheme for Performance Improvement of Pole-Phase-Modulated Multiphase Induction Motor Drive, IEEE Trans. Ind. Electron. 63 (04) (2016) 2036–2043.
[11] J. Titus, K. Hatua, et al., An Asymmetric Nine-Phase Induction Motor for LCI-Fed Medium Voltage Drive Applications, IEEE Trans. Power Electron. 35 (05) (2020) 5047–5056.
[12] R. Rodrigues, T.S. Souza, et al., Assessment of a Nine-Phase Induction Motor Drive for Metal Industry Applications, IEEE Trans. Ind. Appl. 56 (06) (2020) 7217–7226.

[13] I. González-Prieto, I. Zoric, et al., Constrained Model Predictive Control in Nine-Phase Induction Motor Drives, IEEE Trans. Energy Convers. 34 (04) (2019) 1881–1889.
[14] W. Kong, R. Qu, et al., Air-Gap and Yoke Flux Density Optimization for Multiphase Induction Motor Based on Novel Harmonic Current Injection Method, IEEE Trans. Ind. Appl. 53 (03) (2017) 2140–2148.
[15] B.P. Reddy, A Fault-Tolerant Multilevel Inverter for Improving the Performance of a Pole–Phase Modulated Nine-Phase Induction Motor Drive, IEEE Trans. Ind. Electron. 65 (02) (2018) 1107–1116.
[16] Z. Liu, Z. Zheng, et al., Enhancing Fault-Tolerant Ability of a Nine-Phase Induction Motor Drive System Using Fuzzy Logic Current Controllers, IEEE Trans. Energy Convers. 32 (02) (2017) 759–769.
[17] A.S. Abdel-Khalik, M.S. Hamad, et al., Postfault Operation of a Nine-Phase Six-Terminal Induction Machine Under Single Open-Line Fault, IEEE Trans. Ind. Electron. 02 (2018) 1084–1096.
[18] W. Wang, Z. Song, et al., Decoupled Modulation Scheme for Harmonic Current Suppression in Five-Phase PMSM, IEEE Trans. Power Electron. 37 (08) (2022) 8795–8799.
[19] S. Yang, P. Zheng, et al., Carrier-Based PWM Technique and Extended State Observer-Based Current Control Scheme for Current-Source Inverter-Fed Five-Phase PMSM Drives, IEEE Trans. Ind. Electron. 71 (07) (2024) 6635–6646.
[20] W. Wang, Y. Fan, et al., Finite control set model predictive current control of a five-phase PMSM with virtual voltage vectors and adaptive control set, CES Trans. Electr. Mach. Syst. 2 (02) (2018) 136–141.
[21] X. Xu, J. Hang, et al., Unified Open-Phase Fault Diagnosis of Five-Phase PMSM System in Normal Operation and Fault-Tolerant Operation Modes, IEEE Trans. Transp. Electrific. 11 (05) (2025) 12063–12075.
[22] P. Rajanikanth, V.K. Thippiripati, et al., An Improved Model Predictive Current Control of Five-Phase PMSM Drive for Torque and Flux Ripple Alleviation, IEEE J. Emerg. Sel. Top. Ind. Electron. 5 (03) (2024) 1273–1282.
[23] W. Huang, X. Zhu, et al., Generalized Fault-Tolerant Model Predictive Control of Five-Phase PMSM Drives Under Single/Two Open-Switch Faults, IEEE Trans. Ind. Electron. 70 (08) (2023) 7569–7579.
[24] G. Li, Q. Zhu, et al., Open-Circuit Fault-Tolerant Vector Control for Five-Phase SPMSM Based on Five-Phase Six-Leg Inverter, IEEE Trans. Energy Convers. 38 (01) (2023) 404–416.
[25] P. Rajanikanth, M.L. Parvathy, et al., Enhanced Model Predictive Current Control-Based Five-Phase PMSM Drive, IEEE J. Emerg. Sel. Top. Power Electron. 21 (01) (2024) 838–848.
[26] M. Saeed, W. Song, et al., Low-Complexity Deadbeat Model Predictive Current Control With Duty Ratio for Five-Phase PMSM Drives, IEEE Trans. Power Electron. 35 (11) (2020) 12085–12099.
[27] X. Wu, W. Song, et al., Low-Complexity Model Predictive Torque Control Method Without Weighting Factor for Five-Phase PMSM Based on Hysteresis Comparators, IEEE J. Emerg. Sel. Top. Power Electron. 6 (04) (2018) 1650–1661.
[28] M. Saeed, W. Song, et al., Generalized Deadbeat Solution for Model Predictive Control of Five-Phase PMSM Drives, IEEE Trans. Power Electron. 38 (04) (2023) 5178–5191.
[29] C. Xiong, H. XuA, et al., Constant Switching Frequency Multiple-Vector-Based Model Predictive Current Control of Five-Phase PMSM With Nonsinusoidal Back EMF, IEEE Trans. Ind. Electron. 67 (03) (2020) 1695–1707.
[30] J. Du, W. Huang, et al., Torque Ripple Reduction of Model Predictive Controlled Five-Phase PMSM Drives With Open-Circuit Faults, IEEE Trans. Transp. Electrific. 9 (02) (2023) 2228–2237.

[31] H. Wang, C. Gu, et al., Online Full Range Copper Loss Minimization for Single-Phase Short-Circuit Fault Tolerant Control in Five-Phase PMSM, IEEE Trans. Transp. Electrific. 10 (02) (2024) 2777–2788.
[32] X. Wang, G. Liu, et al., Multivectors Model Predictive Control With Voltage Error Tracking for Five-Phase PMSM Short-Circuit Fault-Tolerant Operation, IEEE Trans. Transp. Electrific. 8 (01) (2022) 675–687.
[33] M.S.R. Saeed, W. Song, et al., Double-Vector-Based Finite Control Set Model Predictive Control for Five-Phase PMSMs With High Tracking Accuracy and DC-Link Voltage Utilization, IEEE Trans. Power Electron. 37 (12) (2022) 15234–15244.
[34] W. Huang, W. Hua, et al., Model Predictive Torque Control With SVM for Five-Phase PMSM Under Open-Circuit Fault Condition, IEEE Trans. Power Electron. 35 (05) (2020) 5531–55407.
[35] F. Lin, Y. Hung, et al., Fault-Tolerant Control for Six-Phase PMSM Drive System via Intelligent Complementary Sliding-Mode Control Using TSKFNN-AMF, IEEE Trans. Ind. Electron. 60 (12) (2013) 5747–5762.
[36] F. Lin, I. Sun, et al., Recurrent Fuzzy Neural Cerebellar Model Articulation Network Fault-Tolerant Control of Six-Phase Permanent Magnet Synchronous Motor Position Servo Drive, IEEE Trans. Fuzzy Syst. 24 (01) (2016) 153–167.
[37] Z. Zhou, S. Yao, et al., Deadbeat Direct Current Control Using Dynamic Time Programming for Six-Phase PMSM Drives With Open-Phase Faults, IEEE Trans. Ind. Electron. 71 (09) (2024) 10064–10074.
[38] Y. Luo, C. Liu, et al., Model Predictive Control for a Six-Phase PMSM With High Robustness Against Weighting Factor Variation, IEEE Trans. Ind. Appl. 55 (93) (2019) 2781–2791.
[39] Y. Luo, C. Liu, et al., Elimination of Harmonic Currents Using a Reference Voltage Vector Based-Model Predictive Control for a Six-Phase PMSM Motor, IEEE Trans. Power Electron. 34 (07) (2019) 6960–6972.
[40] P.P. Das, S. Satpathy, et al., A Voltage Injection-Based Current Harmonics Suppression Strategy for Six-Phase PMSM With Nonsinusoidal Back EMF, IEEE J. Emerg. Sel. Top. Ind. Electron. 5 (01) (2024) 285–297.
[41] H. Tang, Z. Dong, et al., Fault-Tolerant Control for the Asymmetric Six-Phase Permanent Magnet Synchronous Motor With Open-Phase Faults, IEEE Trans. Transp. Electrific. 11 (02) (2025) 5793–5808.
[42] Y. Hu, Y. Li, et al., Flux-Weakening Control of Dual Three-Phase PMSM Based on Vector Space Decomposition Control, IEEE Trans. Power Electron. 36 (07) (2021) 8428–8438.
[43] Z. Song, Z. Dong, et al., A Novel Modulation Strategy for Asymmetrical Six-Phase Series-Winding PMSM Based on Predictive Controller, IEEE Trans. Ind. Electron. 70 (06) (2023) 5592–5603.
[44] H. Lu, J. Li, et al., Fault-Tolerant Predictive Control of Six-Phase PMSM Drives Based on Pulsewidth Modulation, IEEE Trans. Ind. Electron. 66 (07) (2019) 4992–5003.
[45] P. Gonçalves, S. Cruz, et al., Fault-Tolerant Predictive Current Control of Six-Phase PMSMs With Minimal Reconfiguration Requirements, IEEE J. Emerg. Sel. Top. Power Electron. 11 (02) (2023) 2084–2093.
[46] E. Levi, M. Jones, et al., A novel concept of a multiphase, multimotor vector controlled drive system supplied from a single voltage source inverter, IEEE Trans. Power Electron. 19 (02) (2004) 320–335.
[47] E. Cengelci, P.N. Enjeti, et al., A new modular motor-modular inverter concept for medium-voltage adjustable-speed-drive systems, IEEE Trans. Ind. Appl. 36 (03) (2000) 786–796.
[48] Y. Tang, F. Chai, et al., Investigation of Open-Circuit Fault-Tolerant Strategy in a Modular Permanent Magnet Synchronous In-Wheel Motor Based on Electromagnetic–Thermal Analysis, IEEE Trans. Transp. Electrific. 8 (01) (2022) 1085–1093.

[49] X. Zhu, M. Zhao, et al., Harmonic Loss Reduction of Fractional Slot Concentrated Winding Permanent-Magnet Vernier Motor With Different Stator-Modular Designs, IEEE Trans. Ind. Appl. 61 (01) (2025) 56–66.

[50] Q. Tan, X. Huang, et al., Analysis of Flux Linkage and Detent Force for a Modular Tubular Permanent Magnet Synchronous Linear Motor With Large Slots, IEEE Trans. Energy Convers. 34 (03) (2019) 1532–1541.

[51] Y. Tang, F. Chai, et al., Design of a Modular In-Wheel Motor With High Fault-Tolerant Performance and Low MMF Space Harmonic, IEEE Trans. Ind. Electron. 70 (07) (2023) 6574–6584.

[52] T. Yao, W. Zhao, et al., Design and Analysis of a Novel Modular-Stator Tubular Permanent-Magnet Vernier Motor, IEEE Trans. Appl. Supercond. 28 (03) (2018) 1–6.

[53] A.H. Mohamed, H. Vansompel, et al., Reconfigurable Modular Fault-Tolerant Converter Topology for Switched Reluctance Motors, IEEE J. Emerg. Sel. Top. Power Electron. 10 (03) (2022) 2890–2902.

[54] S. Rubino, O. Dordevic, et al., Modular Vector Control of Multi-Three-Phase Permanent Magnet Synchronous Motors, IEEE Trans. Ind. Electron. 68 (10) (2021) 9136–9147.

[55] G. Wang, H. Chen, et al., Design, Analysis, and Comparison of a Novel Axial Flux Modular-Double-Stator Hybrid Excitation SRM, IEEE Trans. Transp. Electrific. 10 (03) (2024) 6100–6114.

[56] G. Davarpanah, J. Faiz, et al., A Modular Hybrid Excited Switched Reluctance Motor With Two Groups of Permanent Magnets to Enhance the Performance of the Motor, IEEE Trans. Energy Convers. 39 (03) (2024) 1686–1698.

[57] A.H. Mohamed, H. Vansompel, et al., Polygon-Retrofitted Integrated Modular Motor Drive for Switched Reluctance Machines, IEEE Trans. Ind. Electron. 69 (12) (2022) 12469–12479.

[58] S. Rubino, R. Bojoi, et al., Decoupled and Modular Torque Control of Multi-Three-Phase Induction Motor Drives, IEEE Trans. Ind. Appl. 56 (04) (2020) 3831–3845.

[59] M. Kadjoudj, M.E.H. Benbouzid, et al., A robust hybrid current control for permanent-magnet synchronous motor drive, IEEE Trans. Energy Convers. 19 (01) (2004) 109–115.

[60] J. Weidong, W. Peixia, et al., Multimode Current Hysteresis Control for Brushless DC Motor in Motor and Generator State With Commutation Torque Ripple Reduction, IEEE Trans. Ind. Electron. 65 (04) (2017) 2975–2985.

[61] T.H. Liu, C.H. Liu, A multiprocessor-based fully digital control architecture for permanent magnet synchronous motor drives, IEEE Trans. Power Electron. 5 (04) (1990) 413–423.

[62] W. Wei, Z. Xinlu, et al., Phase-Shifting Fault-Tolerant Control of Permanent-Magnet Linear Motors With Single-Phase Current Sensor, IEEE Trans. Ind. Electron. 69 (03) (2022) 2414–2425.

[63] X. Yaru, M. Dongyi, et al., Vector-Based Model Predictive Hysteresis Current Control for Asynchronous Motor, IEEE Trans. Ind. Electron. 66 (11) (2019) 8703–8712.

[64] B.K. Bose, An adaptive hysteresis-band current control technique of a voltage-fed PWM inverter for machine drive system, IEEE Trans. Ind. Electron. 37 (05) (1990) 402–408.

[65] C. TingYu, L. KuieLin, et al., A novel vector control hysteresis current controller for induction motor drives, IEEE Trans. Energy Convers. 9 (02) (1994) 297–303.

[66] S. Ronald, M. Nasir, Performance Analysis of an FLC-Based Online Adaptation of Both Hysteresis and PI Controllers for IPMSM Drive, IEEE Trans. Ind. Appl. 48 (01) (2012) 12–19.

[67] D. Zhiping, S. Zaixin, et al., Improved Zero-Sequence Current Hysteresis Control-Based Space Vector Modulation for Open-End Winding PMSM Drives With Common DC Bus, IEEE Trans. Ind. Electron. 70 (10) (2023) 10755–10760.

[68] S. Abdullah, E. Wessam, et al., Hysteresis Current Control for Six-Phase Induction Motor Drives With Reduced Ripple and Improved Tracking Based on Subspace Decomposition and Restrained Voltage Vectors, IEEE Trans. Ind. Electron. 71 (07) (2024) 6534–6545.

[69] M. Mohamed, A. Mohammad, Current Gradient Based Modified Hysteresis Controller for Asymmetrical Six-Phase Induction Motor, IEEE J. Emerg. Sel. Top. Power Electron. 12 (05) (2024) 5164–5175.

[70] M. Mohamed, A. Mohammad, Space Vector-Based Hysteresis Current Controller of Asymmetrical Six-Phase Induction Motor, IEEE Trans. Ind. Electron. 72 (03) (2025) 2424–2434.

[71] S. Hwi-Beon, New antiwindup PI controller for variable-speed motor drives, IEEE Trans. Ind. Electron. 45 (03) (1998) 445–450.

[72] S. Jul-Ki, K. Kyung-Tae, et al., Automatic Mode Switching of P/PI Speed Control for Industry Servo Drives Using Online Spectrum Analysis of Torque Command, IEEE Trans. Ind. Electron. 54 (05) (2007) 2642–2647.

[73] R. Shahnazi, H.M. Shanechi, et al., Position Control of Induction and DC Servomotors: a Novel Adaptive Fuzzy PI Sliding Mode Control.IEEE, Trans. Energy Convers. 23 (01) (2008) 138–147.

[74] A.V. Sant, K.R. Rajagopal, et al., Permanent Magnet Synchronous Motor Drive Using Hybrid PI Speed Controller With Inherent and Noninherent Switching Functions, IEEE Trans. Magn. 47 (10) (2011) 4088–4091.

[75] C. Ming, S. Qiang, et al., New self-tuning fuzzy PI control of a novel doubly salient permanent-magnet motor drive, IEEE Trans. Ind. Electron. 53 (03) (2006) 814–821.

[76] E. Rachid, A. Ahmed, et al., Experimental Validation of a Novel PI Speed Controller for AC Motor Drives With Improved Transient Performances, IEEE Trans. Control Syst. Technol. 26 (04) (2017) 1414–1421.

[77] M. Moallem, B. Mirzaeian, Multi-objective genetic-fuzzy optimal design of PI controller in the indirect field oriented control of an induction motor, IEEE Trans. Magn. 37 (01) (2001) 3608–3612.

[78] B. Çavuş, M. Aktaş, A New Adaptive Terminal Sliding Mode Speed Control in Flux Weakening Region for DTC Controlled Induction Motor Drive, IEEE Trans. Power Electron. 39 (01) (2024) 449–458.

[79] S. Kuo-Kai, L. Chiu-Keng, Incremental motion control of synchronous reluctance motor via multisegment sliding mode control method, IEEE Trans. Control Syst. Technol. 10 (02) (2002) 169–179.

[80] Y. Zhang, J. Changxi, et al., Sensorless sliding-mode control of induction motors, IEEE Trans. Ind. Electron. 47 (06) (2000) 1286–1297.

[81] F. Yong, Z. Jianfei, et al., Hybrid Terminal Sliding-Mode Observer Design Method for a Permanent-Magnet Synchronous Motor Control System, IEEE Trans. Ind. Electron. 56 (09) (2009) 3424–3431.

[82] M.F.R. Gilbert Foo, Sensorless Sliding-Mode MTPA Control of an IPM Synchronous Motor Drive Using a Sliding-Mode Observer and HF Signal Injection, IEEE Trans. Ind. Electron. 57 (04) (2010) 1270–1278.

[83] N. Ton Hoang, N. Ty Trung, et al., An Adaptive Sliding-Mode Controller With a Modified Reduced-Order Proportional Integral Observer for Speed Regulation of a Permanent Magnet Synchronous Motor, IEEE Trans. Ind. Electron. 69 (07) (2021) 7181–7191.

[84] H. Huazhou, Y. Xinghuo, et al., Finite-Time Continuous Terminal Sliding Mode Control of Servo Motor Systems, IEEE Trans. Ind. Electron. 67 (07) (2019) 5647–5656.

[85] W. Wen-Jieh, C. Jenn-Yih, Passivity-based sliding mode position control for induction motor drives, IEEE Trans. Energy Convers. 20 (02) (2005) 316–321.

[86] Z. Zhang, Y. Xiaodong, et al., Enhanced Sliding Mode Control for PMSM Speed Drive Systems Using a Novel Adaptive Sliding Mode Reaching Law Based on Exponential Function, IEEE Trans. Ind. Electron. 71 (10) (2024) 11978–11988.
[87] B. In-Cheol, K. Kyeong-Hwa, et al., Robust nonlinear speed control of PM synchronous motor using boundary layer integral sliding mode control technique, IEEE Trans. Control Syst. Technol. 8 (01) (2000) 47–54.
[88] N.T.-T. V, Y. Dong-Young, et al., T–S Fuzzy-Model-Based Sliding-Mode Control for Surface-Mounted Permanent-Magnet Synchronous Motors Considering Uncertainties, IEEE Trans. Ind. Electron. 60 (10) (2012) 4281–4291.
[89] D. Haibo, C. Xiuping, et al., Discrete-Time Fast Terminal Sliding Mode Control for Permanent Magnet Linear Motor, IEEE Trans. Ind. Electron. 65 (12) (2018) 9916–9927.
[90] W. Sen, W. Haiyang, et al., Sensorless Control Strategy for Permanent Magnet Synchronous Motor Based on Adaptive Non-Singular Fast Terminal Sliding Mode Observer, IEEE Trans. Appl. Supercond. 34 (08) (2024) 5208905.
[91] F. Xingang, L. Shuhui, A Novel Neural Network Vector Control Technique for Induction Motor Drive, IEEE Trans. Energy Convers. 30 (04) (2015) 1428–1437.
[92] Y. Zhenxiao, C. Xu, et al., Plant-Physics-Guided Neural Network Control for Permanent Magnet Synchronous Motors, IEEE J. Sel. Top. Signal Process. 19 (01) (2025) 74–87.
[93] A.Y. Alanis, E.N. Sanchez, et al., Real-time Discrete Backstepping Neural Control for Induction Motors, IEEE Trans. Control Syst. Technol. 19 (02) (2011) 359–366.
[94] C. Tien-Chi, S. Tsong-Terng, Model reference neural network controller for induction motor speed control, IEEE Trans. Energy Convers. 17 (02) (2002) 157–163.
[95] S. Kuo-Kai, S. Hsin-Jang, et al., Model reference adaptive speed control for induction motor drive using neural networks, IEEE Trans. Ind. Electron. 45 (01) (1998) 180–182.
[96] Y. Jinpeng, S. Peng, et al., Neural Network-Based Adaptive Dynamic Surface Control for Permanent Magnet Synchronous Motors, IEEE Trans. Neural Netw. Learn. Syst. 26 (03) (2015) 640–645.
[97] Y.S. Kung, C.M. Liaw, et al., Adaptive speed control for induction motor drives using neural networks, IEEE Trans. Ind. Electron. 42 (01) (1995) 25–32.
[98] C.A. H, N.S. Lobo, et al., Sensorless Control of Single Switch-Based Switched Reluctance Motor Drive Using Neural Network, IEEE Trans. Ind. Electron. 55 (01) (2008) 321–329.
[99] W. Miroslaw, K. Zbigniew, et al., Neural-Network-Based Parameter Estimations of Induction Motors, IEEE Trans. Ind. Electron. 55 (04) (2008) 1783–1794.
[100] J.O. Jang, Neural Network Saturation Compensation for DC Motor Systems, IEEE Trans. Ind. Electron. 54 (03) (2007) 1763–1767.
[101] S. Tsong-Terng, C. Tien-Chi, Self-tuning control of induction motor drive using neural network identifier, IEEE Trans. Energy Convers. 14 (04) (1999) 881–886.
[102] T. Senjyu, H. Miyazato, et al., Speed control of ultrasonic motors using neural network, IEEE Trans. Power Electron. 13 (03) (1998) 381–387.
[103] F.F.M. El-Sousy, Intelligent Optimal Recurrent Wavelet Elman Neural Network Control System for Permanent-Magnet Synchronous Motor Servo Drive, IEEE Trans. Ind. Inform. 9 (04) (2013) 1986–2003.
[104] A. Rubaai, R. Kotaru, et al., Online training of parallel neural network estimators for control of induction motors, IEEE Trans. Ind. Appl. 37 (05) (2001) 1512–1521.
[105] Z. Zhanqing, J. Wang, et al., Speed Synchronization Control Strategy of Dual-Motor System With Explicit Model Predictive Control, IEEE J. Emerg. Sel. Top. Power Electron. 12 (03) (2024) 2787–2798.
[106] H. Guoqiang, Z. Huimin, et al., Model-Free Current Predictive Control Method for Switched Reluctance Motors, IEEE Trans. Ind. Electron. 71 (10) (2024) 12041–12050.

[107] Z. Xiaoguang, Z. Liu, et al., Model Predictive Current Control for PMSM Drives Based on Nonparametric Prediction Model, IEEE Trans. Transportation Electrification 10 (1) (2024) 711–719.
[108] Z. Zhihao, W. Xile, et al., Torque-Ripple-Mitigated Model-Free Predictive Current Control With Extended Control Set for PMSMs, IEEE Trans. Magn. 60 (09) (2024) 8203206.
[109] L. Xinpo, L. Jianxing, et al., Model-Free Current Predictive Control for PMSMs With Ultralocal Model Employing Fixed-Time Observer and Extremum-Seeking Method, IEEE Trans. Power Electron. 40 (08) (2025) 10682–10693.
[110] Z. Xiaoguang, C. Yuanhang, et al., A Simple Motor-Parameter-Free Model Predictive Current Control for PMSM Drive, IEEE Trans. Ind. Electron. 72 (04) (2023) 3292–3302.
[111] T. Li, S. Xiaodong, et al. Model-Free Predictive Control for Harmonic Suppression of PMSMs Based on Adaptive Resonant Controller. IEEE Trans. Energy Convers. 2025, 40(04): 3104–3114.
[112] M.M. Ismail, X. Wei, et al., Adaptive Linear Predictive Model of an Improved Predictive Control of Permanent Magnet Synchronous Motor Over Different Speed Regions, IEEE Trans. Power Electron. 37 (12) (2022) 15338–15355.
[113] Z. Zhenrui, L. Yancheng, et al., Robust Model Predictive Current Control of PMSM Based on Nonlinear Extended State Observer, IEEE J. Emerg. Sel. Top. Power Electron. 11 (01) (2023) 862–873.
[114] Z. Yanan, L. Hongmei, et al., Continuous Voltage Vector Model-Free Predictive Current Control of Surface Mounted Permanent Magnet Synchronous Motor, IEEE Trans. Energy Convers. 34 (02) (2019) 899–908.
[115] M.S. M, N. Behnam, et al., Applying Model-Free Predictive Current Control With Parallel Cost Functions for Open-End Winding Synchronous Reluctance Motor, IEEE Trans. Power Electron. 40 (05) (2025) 6801–6812.
[116] M.S. M, N. Behnam, et al., Adaptive Ultralocalized Time-Series for Improved Model-Free Predictive Current Control on PMSM Drives, IEEE Trans. Power Electron. 40 (05) (2025) 6801–6812.
[117] S.O. F, A. Sodiq, et al., Modulated Model-Free Predictive Current Control With Constrained Optimization for Permanent Magnet Synchronous Motor Drives, IEEE J. Emerg. Sel. Top. Power Electron. 13 (03) (2025) 3378–3387.
[118] Z. Zhenrui, L. Yancheng, et al., Robust Model Predictive Control for a Three-Phase PMSM Motor With Improved Control Precision, IEEE Trans. Ind. Electron. 11 (01) (2023) 862–873.
[119] S. Xiaodong, L. Xuwei, et al., An Improved Continuous-Control-Set Model Predictive Control for Dual Three-Phase PMSMs Based on Extended State Observer, IEEE Trans. Ind. Electron. 72 (04) (2024) 3449–3459.
[120] H.M. A, J. Imed, et al., Model-Free Predictive Current Control of Synchronous Reluctance Motors Based on a Recurrent Neural Network, IEEE Trans. Ind. Electron. 69 (11) (2022) 10984–10992.
[121] A. Xingke, L. Guohai, et al., Adjustable Model Predictive Control for IPMSM Drives Based on Online Stator Inductance Identification, IEEE Trans. Ind. Electron. 69 (04) (2022) 3368–3381.
[122] S.G. P, E. Kusuma, et al., A Modified Model Predictive Current Control of Permanent Magnet Synchronous Motor Drive, IEEE Trans. Ind. Electron. 68 (02) (2021) 1025–1034.
[123] W. Yao, Y. Hector, et al., Model-Free Predictive Current Control Using Extended Affine Ultralocal for PMSM Drives, IEEE Trans. Ind. Electron. 71 (07) (2024) 6719–6729.

CHAPTER 2

Research status and challenges of model predictive control in motor systems

2.1 Introduction

In the 1970s, model predictive control (MPC) was first applied to slow process control in the chemical industry due to its excellent dynamic control performance and multiobjective optimization ability [1,2]. The MPC method originated from the requirements of industrial control processes, with model algorithm control and dynamic matrix control as typical representatives. It can directly design predictive controllers using readily available nonparametric models such as impulse response or step response. The MPC method is an optimization control method based on the controlled object, which predicts the range of output variables for the next time step based on the current input variables and constraints of the controlled object. With the rapid improvement of microprocessor computing power, the MPC method has gradually attracted the attention of scholars in the field of motor control. Based on the discrete switching characteristics of the inverter in the motor drive system, Professor J. Rodriguez et al. applied the MPC method to the permanent magnet synchronous motor drive system, where the finite switching state of the inverter is used as the input to predict the variables that will control the motor system at the next moment [3,4]. The single-loop PID controller commonly used in motor systems is difficult to achieve good control performance due to the lack of coupling information, while the MPC method belongs to rolling optimization and can effectively solve the optimal control problem of motor systems under multivariable coupling and multiple constraint conditions [5]. Compared with other existing optimal control methods for motor systems, the MPC method replaces one-time global optimization with online finite time domain optimization, and each optimization step is calibrated with real-time feedback information, which can effectively avoid the difficulty

Robust Model Predictive Control of Permanent Magnet Synchronous Motor Under Parameter Mismatch
https://doi.org/10.1016/B978-0-443-44209-4.00007-5

of parameter model identification [6]. The MPC method has become one of the mainstream control methods for motor systems after field-oriented vector control and direct torque control.

The motor drive system is an electromechanical energy conversion system composed of a motor, an inverter, and a controller, where the controller is the key and core component of the entire drive system. As the structure of the motor body continues to be optimized and innovated for different application scenarios, high-performance motor control technology is also rapidly developing. To meet the quality requirements of dynamic response speed, speed range, and operating efficiency of motor systems, more and more advanced control technologies are being studied and researched in the field of motor control, such as relatively complex and computationally intensive adaptive control, fuzzy control, neural network control, and predictive control. With the continuous improvement of microprocessor computing speed and performance, more complex control algorithms can be implemented within one control cycle, and the application of these advanced control technologies can be realized in motor control systems. However, model predictive control has received widespread attention and research in the field of motor control due to its simple structure, fast dynamic response, and high control accuracy. It has become a representative example of the successful application of advanced motor control technology in the industrial field. This chapter will focus on the application and advantages of MPC technology in the field of motor control, discuss in detail the causes of parameter mismatch and its impact on predictive controllers, and provide the research status of robust predictive control technology for motor systems under parameter mismatch.

2.2 Application and advantages of MPC technology

The main feature of MPC technology is its use of predefined optimization algorithms to predict the future output variables of the controller based on the mathematical model of the controlled object. With the development of power electronics technology and microprocessor control technology, the application of predictive control technology in the field of motor control has also developed rapidly. The most widely used technology in the field of motor control is model predictive control. The MPC technology involves achieving optimal control of the system in the time domain based on the current state variables of the motor system and the discrete mathematical model of the motor. The MPC technology has the advantages of fast dynamic response speed and easy implementation of multiobjective collaborative control, and it is widely used in the field of motor control. The application of model predictive control technology in the field of motor control now encompasses various mainstream motor systems, such as permanent magnet synchronous motors, induction motors, linear motors, brushless DC motors, flux-switching motors, switched reluctance motors, and multiphase motors. It is regarded by researchers in the field of motors as the third high-performance motor control strategy, following the direct torque control strategy and the field-oriented vector control strategy [7–11]. This section will focus on the application and advantages of model predictive control technology in the field of motor control from the perspectives of different control objectives, different voltage vector control sets, and different control structures.

(1) According to the different control objectives of the motor, MPC technology can be divided into model predictive speed control, model predictive current control, and model predictive torque control, as shown in Fig. 2.1. Model predictive current control and model predictive speed control are based on the discrete mathematical model of the motor to calculate and obtain the ideal voltage vector that minimizes current error or speed error. Model predictive torque control is based on a discrete motor mathematical model to predict the electromagnetic torque and stator flux of future control cycles. By evaluating the cost function that includes stator flux error and torque error, the optimal voltage vector for controlling the motor is selected. In addition to the predictive control strategies for speed, current, and torque, researchers can also design different predictive controllers based on various application scenarios to achieve different predictive control objectives for the motor system. Regarding research on model predictive speed control, Gao et al. [12] proposed a novel direct predictive speed control method for permanent magnet synchronous motor

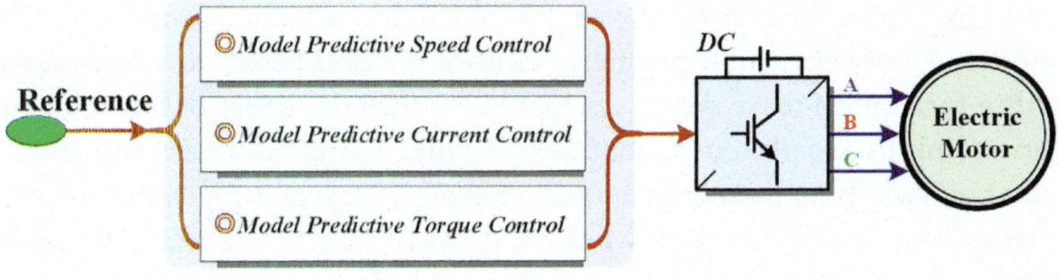

Fig. 2.1 Control block diagram of the motor system under different predictive control objectives.

(PMSM) drive systems, which can overcome the limitations of cascaded linear PI controllers and improve the dynamic and static performance of PMSM drive systems. Preindl et al. [13] proposed a model predictive direct speed control with a finite control set for PMSM drive systems, which can overcome the limitations of cascaded linear controllers. Kawai et al. [14] presented a direct speed control strategy based on finite control set model predictive control with a voltage smoother, which can effectively reduce the current ripple while achieving a high dynamic drive compared with the conventional model predictive speed control scheme. Garcia et al. [15] proposed a full predictive cascaded speed and current control of an induction machine, which has performance comparable to classical control strategies but without overshoot and provides better time response. Tarczewski et al. [16] proposed a constrained state feedback predictive speed control method for PMSM drive systems, which can solve the problem of low dynamic characteristics in traditional cascaded linear control systems and achieve fast response of PMSM drive systems in a short period of time.

Regarding the research on model predictive current control, Gao et al. [17] proposed a novel model predictive current compensation control strategy for PMSM drive systems, which can select the optimal switch state for optimal control in each control cycle. Turker et al. [18] presented a new discrete-time robust predictive current controller for PMSM drive systems, which can obtain smooth output torque and strong robustness in PMSM drive systems. Yuan et al. [19] proposed a novel current update mechanism based on the two most recent current variations to reconstruct the PMSM model, which can effectively reduce torque ripple and enhance current performance in contrast to conventional MPC. Zhang et al. [20] proposed an improved deadbeat predictive current control algorithm for PMSM drive systems, which can optimize the current-control performance of the PMSM with model parameter mismatch and one-step control delay. Jiang et al. [21] presented an improved deadbeat-based predictive current control scheme based on sliding mode, which can increase the speed robustness and current tracking accuracy of PMSM drive systems. Wang et al. [22] proposed a novel current predictive control method based on a fuzzy algorithm for PMSM drive systems, which can adjust the effect of the compensation link by weight coefficient in real time. Xiong et al. [23] proposed a constant switching frequency multivector model current prediction control method for multiphase PMSM drive systems, which can solve the problems of high computational complexity, variable switching frequency, and harmonic current in conventional predictive current control. Compared with conventional finite-set model predictive current control methods, this method maintains the advantages of a simple and intuitive control structure, fast dynamic response speed, and stronger parameter mismatch robustness.

Regarding the research on model predictive torque control, Ban et al. [24] present a novel predictive torque control strategy based on the finite control set for PMSM, which can overcome the large flux and torque ripples of conventional direct torque control. Liu et al. [25] proposed a Lyapunov-based finite control set model predictive direct torque control for PMSM drive systems, which can obtain the optimum voltage vector and minimize the torque ripple. Mora et al. [26] present a new and simple finite-control set model predictive control strategy based on minimizing a cost function, which can achieve good cogging torque estimation and accurate disturbance torque mitigation at low rotational speeds. Lv et al. [27] proposed a finite control set model predictive torque control scheme for switched reluctance motors based on a three-phase four-leg inverter, which can suppress torque ripple and reduce the sharp current drop at the initial stages of the commutation regions. Niu and Hernandez et al. [4,28] present a comprehensive evaluation of several direct torque control strategies for permanent magnet synchronous machines, evaluating them from several aspects such as torque ripple, stator flux ripple, and steady-state control performance. Ban et al. [24] proposed a novel predictive torque control strategy based on the finite control set for PMSM drive systems, which can overcome the large flux and torque ripples of conventional direct torque control.Yang et al. [29] proposed a novel model-based predictive current control algorithm with decoupling implementation, which can achieve low torque ripple and switching losses in PMSM drive systems. Ge et al. [30] proposed a novel model predictive torque control with a torque-sharing function and pulse width modulation (PWM) control signal based on linear polynomial fitting, which can achieve good reference torque tracking ability and torque ripple suppression performance. Cho et al. [31] proposed a simple control method for induction motors using a model-based predictive control algorithm, which can achieve very low steady-state torque ripple and fast transient dynamics. Mora et al. [26] proposed a new and simple finite-control set model predictive control strategy for PMSM drive systems, which can suppress the torque ripple of the motor system through torque ripple compensation. Li et al. [32] proposed an improved model predictive direct torque control strategy consisting of a quadratic evaluation method and a harmonic voltage elimination method, which can enable the motor system to maintain a fast dynamic response while maintaining good steady-state performance. Through comparative research, it can be concluded that the model predictive torque control method includes two different control objectives: magnetic flux and electromagnetic torque. Therefore it is necessary to introduce weight coefficients. Improper selection of weight coefficients will result in the prediction controller outputting an inappropriate amount of voltage that is not ideal, thereby deteriorating the predictive control performance of the motor system. Compared with model predictive torque control, model predictive speed control and model predictive current control do not require weight coefficients, making the control algorithm simple and easy to implement. However, they rely more on the mathematical model of the motor, and parameter mismatches can affect the performance of predictive control.

(2) According to different voltage vector control sets, model predictive control technology can be divided into continuous set model predictive control and finite set model predictive control. The logic diagram of the control system is shown in Fig. 2.2. Continuous set model predictive control constructs an optimization function based on the discrete mathematical model of the electric motor, which combines control objectives and constraints. This strategy obtains the optimal control voltage vector by solving the optimization function, which is consistent with the magnetic field-oriented vector control strategy. It requires the use of PWM modulation modules to synthesize the required optimal voltage vector. The continuous set model predictive control algorithm has the advantages of fast dynamic response and the ability to achieve multiple objective controls. According to the different prediction models and optimization methods, continuous set model predictive control can be divided into explicit model predictive control and generalized predictive control [33]. Generalized predictive control is a control algorithm for solving optimization problems online, which has the disadvantages of large computational complexity and difficulty in handling constraint conditions, making it difficult to

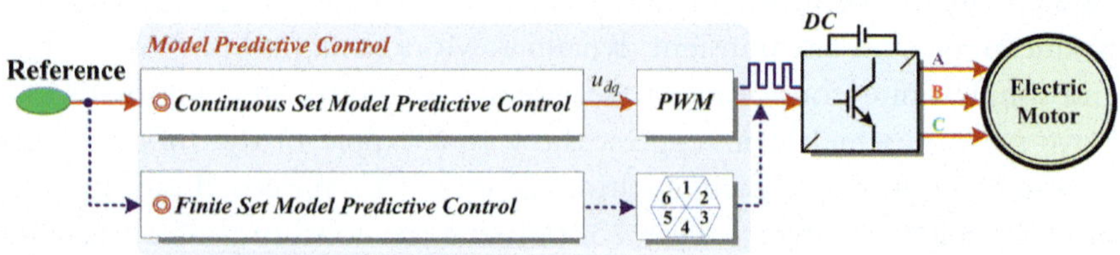

Fig. 2.2 Control block diagram of continuous-set model predictive control and finite-set model predictive control.

implement in practical motor control systems. Explicit model predictive control can significantly reduce computational complexity by solving the optimal voltage vector offline, but it has the problem of large data storage capacity. Belda et al. [34] proposed a specific explicit design of generalized predictive control for the speed and position of three-phase PMSM drive systems, which can solve the required magnetic field weakening and current limitation problems through specific programs. Li et al. [35] proposed a generalized observer-based robust predictive current control strategy for PMSM drive systems, which can effectively solve problems such as parameter mismatch, digital delay, and external interference. Cao et al. [36] developed a novel safety-critical generalized predictive control approach for PMSM drives, which can effectively solve the interaction problem between safety constraints and control performance in speed regulation. Wei et al. [37] proposed a model-free predictive control strategy using a sinusoidal generalized universal model for PMSM drive systems, which can achieve better model accuracy, current quality, and robustness compared to conventional control strategies. Wang et al. [38] presented a generalized predictive current control for dual-three-phase PMSM, which can enhance the output torque capability by accurately tracking the harmonic current reference in both phase and magnitude. Wei et al. [37] also proposed a generalized multiple-vector-based model predictive control for PMSM drives, which can effectively reduce complexity and computational burden by eliminating the enumeration-based prediction and complex calculations in conventional MPC methods. Liu et al. [39] proposed an improved predictive functional control method for PMSM servo systems by introducing an extended state observer, which can achieve a satisfying effect in the presence of strong disturbances.

Finite set model predictive control is based on the mathematical model of the control object. By fully utilizing the discrete characteristics of the inverter switching devices, it directly predicts the voltage vectors corresponding to all possible switching states of the inverter and calculates the optimal voltage vector required for the motor system in the next control cycle. Finite set model predictive control has received widespread attention from researchers in the field of modern motor control due to its advantages of simple control principles and convenient implementation [40,41]. Han et al. [42] presented a new multiobjective finite control set model predictive control using a novel delay compensation technique for PMSM, which can reduce the current and torque ripples during the control process. Zhao et al. [43] proposed a novel finite-control-set model predictive current control for a five-phase PM motor with continued modulation, which can effectively reduce computational burden and enhance engineering practicality. Gong et al. [44] developed a novel predicting plant model in the continuous-time domain based on the numerical solutions of the PMSM state-space model, which can completely eliminate the influence of the low control frequency. Zhang et al. [45] proposed a direct voltage-vector selection-based model-predictive direct-speed control method for PMSM drives, which can improve the robustness of the system and reduce the influence of measurement noise. Ahmed et al. [46] presented a comparative study of two predictive speed control schemes for induction machines in terms of their design and performance, which can achieve smaller current ripple and lower computational complexity. Gao et al. [17] proposed a new direct compensation method by predicting the current variation within the delay time, which can single out the best switching state of PMSM drive systems in each period for optimal control. Sun et al. [47] proposed a finite control set model-free predictive current control for dual voltage vector PMSM based on a super local model, which can improve the current tracking performance and eliminate the influence of motor parameter mismatches. Martín et al. [48] designed a five-phase induction motor rotor current observer for finite control set model predictive control, which can achieve sensorless observation of motor rotor position by adding a rotor current observer. Compared to continuous set model predictive control, finite set model predictive control does not require the use of modulation modules while maintaining the advantages of the continuous set model predictive control strategy. It has the advantages of a simple control principle, easy software and hardware execution, and better dynamic and static performance.

(3) According to different model predictive control schemes, model predictive control technology can be divided into cascaded structure model predictive control schemes and single-stage model predictive control schemes, as shown in Fig. 2.3. Taking the PMSM drive system as an example, the control system of the PMSM generally adopts a dual closed-loop control structure composed of a current inner loop and a speed outer loop, and the dual closed-loop control structure typically adopts the conventional proportional-integral controller. Model predictive control has attracted significant attention from researchers in the field of motor control due to its advantages, such as fast dynamic response speed and simple control structure. By retaining the commonly used cascade control structure, the model predictive controller is applied to the current inner loop control, while the

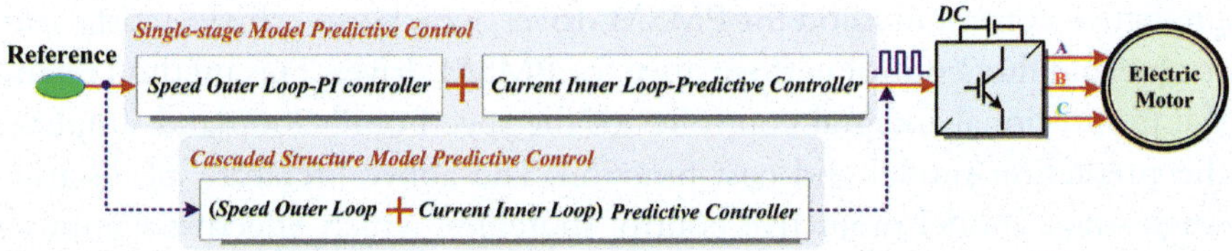

Fig. 2.3 Block diagram of the cascade structure MPC scheme and the single-stage MPC scheme.

speed outer loop still uses a proportional-integral controller or other types of controllers. This type of control scheme is summarized as a single-stage model predictive control scheme. Sun et al. [49] proposed a single-stage model predictive control structure based on an improved model predictive current control strategy. This strategy uses a traditional proportional controller for the outer loop of speed and an improved model predictive control for the inner loop of current, which can reduce the torque and current ripple of PMSM drives. Zhang et al. [50] presented an improved model-free predictive current control for PMSM drives based on an extended state observer, while still using traditional proportional controllers for the outer loop of speed. The proposed single-stage model predictive control scheme can eliminate the influence of parameter mismatch on the model predictive controller and improve the robustness of PMSM drives. Liu et al. [51] presented a novel finite-set sliding-mode model-free predictive current control for PMSM drives, which has better steady-state performance, good dynamic response, and very low computational burden. Wei et al. [52] presented a model-free predictive current control using adaptive ultra-localized time-series on PMSM drives, which can achieve better model accuracy and current quality with suitable robustness compared to conventional MPC. Li et al. [53] presented a model-free predictive current control algorithm based on online current gradient updates, which can reduce current fluctuations and enhance current tracking performance under parameter mismatches. Zhang et al. [54] proposed an improved model predictive current control method based on the incremental model for PMSM drives, which can effectively eliminate the influence of parameter mismatches and reduce parameter sensitivity. Luo et al. [55] proposed a low-complexity model predictive flux control for six-phase PMSM, which can reduce harmonic current and torque ripple while maintaining the fast dynamic response of predictive control. This strategy still uses traditional proportional controllers for the outer loop of speed and model predictive flux control for the inner loop of current. Luo et al. [56] proposed a simplified cost-function model predictive control method for PMSM drives, which can achieve satisfactory torque and flux control performance for PMSM drives. This method retains the proportional-integral controller of the speed outer loop and simplifies the prediction model and cost function. The above literature all discusses single-stage model predictive control strategies, which effectively improve the current inner loop control performance of the motor system, while the speed outer loop control still adopts the traditional proportional-integral controller.

Compared with the single-stage model predictive control scheme, to fully utilize the advantages of multiconstraint processing and multiobjective coordinated control of model predictive control, the cascade structure model predictive control scheme adopts the model predictive control algorithm to unify the speed outer loop and current inner loop. Fu et al. [57] developed a cascaded structure model-free predictive control algorithm that combines the current inner loop and the speed outer loop, which can improve the dynamic response of the speed and stator current by using an exponential high-gain extended state observer. Li et al. [58] proposed a robust continuous model predictive speed and current control for PMSM with an adaptive integral sliding-mode approach, which can achieve better performance and faster response speed under load torque disturbances. Yang et al. [59] proposed a new model predictive direct speed control with a novel cost function for PMSM drives, which can obtain high-speed dynamics and low steady current ripples by using the cascade structure model predictive control scheme. Yu et al. [60] presented a robust cascaded deadbeat predictive speed control method for the dual three-phase variable-flux PMSM drives, which can achieve high-performance control in terms of speed drop and recovery time during parameter variation and large magnetization current pulses. Gao et al. [61] proposed a model-free hybrid parallel predictive speed control based on the ultra-local model of PMSM drives, which can eliminate the influence of machine parameters and weighting factors on the model predictive controller. Zhao et al. [62] proposed an expanded limit boundary explicit model predictive-direct speed control method for PMSM drives, which can realize cascade-free speed control with a wider speed range and higher load capacity. Tu et al. [63] proposed a predictive speed and current control with multiple timescale optimization in a cascade architecture for PMSM drive systems, which can improve the accuracy of the slow-sampling model for fast prediction instants. Fuentes et al. [64] designed a torque inner loop predictive controller and a speed outer loop predictive controller based on a cascade structure model predictive control framework, which can achieve fast optimal control and improve the steady-state performance of PMSM drives. Garcia et al. [15] proposed a novel cascade structure model predictive control strategy for electric transmission systems by using the concept of model predictive control and cascade structure. The strategy includes an external loop and an internal loop, with the internal loop using finite control set model predictive control and the external loop using deadbeat predictive control. Darba et al. [65] proposed a model predictive speed and current control method based on a cascade structure to address the problem that traditional proportional-integral controllers cannot achieve high-performance control and wide-speed range operation of low-inertia motor systems. Hu et al. [66] proposed a novel predictive position control with current and speed limits for PMSM drives, which can obtain fewer static position errors and improve the practicality of the predictive position control method. Compared with the single-stage model predictive control scheme, the cascaded structure model predictive control scheme fully embodies the advantages of multiconstraint condition processing and multiobjective coordinated control in model predictive control, which can improve the bandwidth of the control system. However, there are problems such as complex cost function design and difficulty in determining weight factors.

Based on the above literature, it can be seen that traditional proportional-integral controllers, designed based on steady-state operating points, are no longer able to meet the wide speed range and high dynamic quality operation requirements of PMSM drive systems in different application fields. Model predictive control technology has been widely and deeply studied in the field of motor control due to its advantages of a simple and intuitive control structure, fast dynamic response speed, and high control accuracy.

2.3 Reasons and effects of parameter mismatch

Model predictive control technology is a high-performance control algorithm based on discrete mathematical models and is strictly dependent on the actual parameters of PMSM drive systems. The model predictive control algorithm uses discrete mathematical models to predict the operating state and dynamic characteristics of PMSM drive systems in the future. It has the advantages of fast dynamic response speed, simple algorithm implementation, high decoupling degree, zero steady-state error, and no overshoot. However, PMSM drive systems are high-order nonlinear systems with uncertain factors such as time-varying parameters and external disturbances, which can lead to inconsistencies between the actual parameters of the motor system and the motor parameters applied in the model predictive controller. This will cause the voltage vector output by the model predictive controller to deviate from the ideal voltage vector value, which will inevitably reduce the control accuracy and robustness of the system. It may lead to instability or even oscillation of PMSM drive systems under adverse conditions. In the field of industrial applications, the main reasons for the mismatch between model predictive controller parameters and motor drive system parameters are external disturbances, detection errors, mechanical failures, and other external uncertainties. For model predictive control technology that strictly relies on the actual parameters of the motor drive system, parameter mismatch can lead to inaccurate motor operation, thereby reducing the robustness and control accuracy of the control system. On the other hand, motor parameters will also change with different operating conditions or the degradation of their own material during the actual operation of the motor drive system, which will result in a mismatch between the model predictive controller parameters and the actual motor parameters, thereby reducing the operating quality and affecting the safe and stable operation of the motor drive systems. This section will elaborate on the causes of parameter mismatch and its impact on the model predictive controller and analyze the main reasons for the mismatch between the model predictive controller parameters and actual motor parameters, as shown in Fig. 2.4.

(1) Mechanical parameter perturbations result in a mismatch between the model predictive controller parameters and the actual motor parameters. The mechanical parameter size of the PMSM drive systems is directly related to the frequency domain width of the model predictive speed controller. In the position detection and speed control circuit of PMSM drive systems, mechanical parameters such as the friction coefficient, moment of inertia, and shaft stiffness are usually considered constants. However, in special application fields such as winding machines and high-precision robots, the inertia variation of the motor drive systems is several times that of the original system due to the time-varying inertia of the motor system. This results in a mismatch between the mechanical parameters applied in the model predictive speed controller and the actual motor parameters, leading to significant overshoot and a longer time required to reach stability in the response of the model predictive speed controller [67].

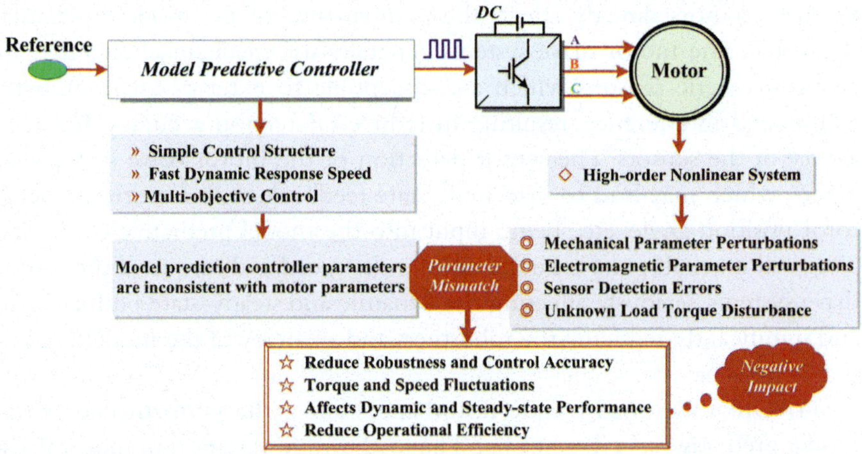

Fig. 2.4 The causes of parameter mismatch and their impact on model predictive controllers.

(2) Electromagnetic parameter perturbations result in a mismatch between the model predictive controller parameters and the actual motor parameters. The electromagnetic parameters of the PMSM drive systems include magnetic flux, resistance, inductance, and other parameters, which are also the main factors that determine the performance of the model predictive controller. The electromagnetic parameters of the motor drive systems can be considered constants within a certain temperature range. However, these parameters exhibit a nonlinear relationship with temperature changes under actual operating conditions. Temperature changes can cause a 20% to 40% perturbation in the resistance parameters, a 20% demagnetization phenomenon in the permanent magnet magnetic flux, and even permanent demagnetization faults [68,69]. In addition, the inductance parameter is a time-varying parameter that is difficult to obtain accurately. It is closely related to the current function of the motor drive systems and is easily affected by magnetic circuit saturation [70]. The variation in electromagnetic parameters can lead to a mismatch between the parameters used by the model predictive controller and the actual motor parameters, resulting in a decrease in the operating efficiency of the motor drive systems and an inability to output the rated torque.

(3) Sensor detection errors result in inaccurate motor system state feedback variables being received by the model predictive controller. Current sensors and angle sensors provide state feedback to the vector control of the motor drive systems, and the detection accuracy of system state variables

by these sensors directly affects the performance of the model predictive controller. The motor drive systems generally use mechanical encoders or electromagnetic sensors, which are susceptible to external environmental factors and interference, resulting in reduced detection accuracy and reliability of the sensors. The sensor detection of the motor drive system has errors, which may lead to inaccurate state feedback, such as current, speed, rotor position angle, etc., being input into the model predictive controller. This will inevitably cause fluctuations in the speed and torque of the motor drive systems, seriously affecting the dynamic and steady-state performance and significantly reducing the robustness and accuracy of the model predictive controller.

(4) Unknown load torque disturbances affect the performance of the model predictive speed controller. The load torque variation in motor drive systems is unknown and cannot be accurately measured directly, which significantly impacts the speed fluctuation of motor drive systems. Therefore online estimation of unknown load torque disturbances is crucial for predicting the performance of speed controllers. The design of speed controllers in motor drive systems relies on the premise that the load torque is a fixed value. However, the load torque in motor drive systems changes under actual operating conditions, and conventional model predictive controllers cannot effectively suppress the impact of load disturbances. If the load torque is not well estimated online or feedback-compensated, the model predictive controller will experience speed fluctuations, decreased control accuracy, and reduced robustness [71].

Based on the above analysis, it can be concluded that model predictive control technology has been widely applied in motor drive systems due to its advantages of fast dynamic response speed and simple control structure. However, model predictive control technology is limited by the model accuracy and the actual parameter accuracy of the motor drive systems. In practical engineering applications, mechanical parameter changes, electromagnetic parameter changes, sensor detection errors, and unknown load torque disturbances are inevitable. These factors result in a mismatch between the mathematical model in the model predictive controller and the actual motor, thereby affecting the accuracy of predictive control and deteriorating its performance. Therefore it is necessary to conduct research on robust model predictive control methods for motor drive systems under model parameter mismatch.

2.4 Research status of robust MPC technology

Model predictive control technology has the advantages of zero steady-state error and good dynamic response performance and has become a research hotspot in high-performance control technology in the field of motors. Although model predictive control technology has many advantages in the field of motor control, it is limited by the model accuracy and actual parameter accuracy of the motor drive system, especially relying on the parameters of permanent magnet flux and stator inductance. When the permanent magnet flux parameter in the model predictive controller is mismatched with the actual parameter of the PMSM drive system, it will cause static current differences and weaken the load capacity of the motor drive system. When the inductance parameters in the model predictive controller are mismatched with the actual parameters of the motor drive system, it may cause current oscillation and result in instability or loss of the model predictive controller. Therefore it is crucial to improve the robustness of model predictive control technology to the parameters of permanent magnet flux and stator inductance. To fully utilize the advantages of model predictive control technology and overcome its severe dependence on motor parameters, many robust model predictive control methods have been proposed for motor drive systems to enhance the robustness to parameter mismatches. At present, there are three main methods to improve the robustness of model predictive control technology for motor drive systems to parameter mismatches: the first type is improved model predictive control; the second type is modified predictive control based on parameter identification; the third type is compensation predictive control based on observer disturbance estimation.

(1) Improved model predictive control method. The implementation of the model predictive control algorithm on a hardware platform requires the discretization of continuous motor state equations, which not only considers the accuracy of discretization to improve predictive control performance but also takes into account the computational complexity of actual hardware facilities. The core idea of the improved model predictive control method is to weaken the influence of the model predictive controller on parameter mismatch by improving the predictive model of the motor drive system. To enhance the robustness of model predictive control technology to motor mathematical models and parameters, improved

model predictive control methods based on model-free predictive control technology have received significant attention in recent years. The idea of model-free predictive control technology is to treat the current as a linear change within a control cycle and replace the predictive control algorithm based on the motor mathematical model and parameters with the current difference within a control cycle to achieve model-free or nonparametric model predictive control of the motor drive system. Zhang et al. [50] proposed an improved model-free predictive current control method for motor drive systems based on extended state observers, which can solve the problem of parameter mismatch between model predictive control and actual motor parameters. This method only requires the input and output of the motor drive system without using any parameters, which can improve the robustness of the predictive control system to parameter mismatches. Tinazzi et al. [72] proposed a finite-set model predictive current control method for a motor drive system that does not require any motor parameters. This method retains the advantages of model predictive control technology while avoiding complex parameter estimation and self-debugging processes. Xu et al. [73] proposed a super-local model-free predictive current control method based on nonlinear disturbance compensation, which does not involve any motor parameters and has the advantages of strong antiinterference ability, small output ripple, and good dynamic characteristics. Carlet et al. [74] proposed a lightweight and computationally fast finite-set model-free current prediction control method that can bypass all the effects of parameter changes or mismatches, as well as model nonlinearity and uncertainty. Zhang et al. [75] proposed a novel model-free predictive control method that combines a super-local model with predictive current control. This method replaces the mathematical model of a doubly-fed generator with a super-local model and has strong parameter robustness. Lin et al. [76] proposed an improved model-free predictive control method for motor drive systems, which does not require specific motor mathematical models and parameters. It can effectively eliminate the effects of instantaneous switching and hysteresis current changes. Lin et al. [77] proposed a model-free predictive current control method for motor drive systems based on current deviation detection technology, which can reduce the need for excessive prior knowledge of the motor drive systems. This method only requires the use of stator current and the current difference under different switching states of the inverter to implement high-performance control of the motor drive systems. Wei et al. [37] proposed a model-free predictive control strategy using a sinusoidal generalized universal model, which has advantages in terms of model accuracy, current quality, and enhanced robustness compared to conventional predictive control strategies. Yin et al. [78] designed a switched ultra-local model-free predictive controller for motor drive systems, which exhibited favorable performance in acceleration, deceleration, and loading processes. The improved model predictive control method based on model-free predictive control requires two samplings of the current within one control cycle to obtain the required current difference, which undoubtedly increases the sampling frequency of the hardware and also places higher requirements on the sampling time and sampling accuracy of the software. In addition, the proposed model-free predictive control technology is based on a three-phase stationary coordinate system, which requires a large amount of space for storing current differential data.

(2) Modified predictive control based on parameter identification. The motor drive system primarily includes parameters such as stator resistance, stator inductance, and rotor permanent magnet flux linkage. These parameters vary to different degrees under various operating conditions due to external influences. To eliminate the negative impact of parameter mismatch on the performance of model predictive control, a model predictive control strategy based on the online real-time identification of motor parameters is one of the key approaches to enhancing the robustness against parameter mismatch. To accurately identify motor parameters in real-time online, the most mainstream solution is to inject external disturbance signals to achieve parameter identification; however, this approach causes output torque ripple and increases additional losses. Currently, parameter identification methods combined with model predictive control technology include recursive least squares, minimum mean square algorithms, model reference adaptive systems, extended Kalman filters, conformal mapping algorithms, neural networks, and other identification methods. Pulvirenti et al. [79] proposed a technique for the online identification of stator resistance and permanent magnet flux linkage in PMSM drive systems, which can achieve online identification of motor parameters over a wide operating range and avoid adverse effects caused by parameter mismatch on the output torque of the motor system. Dang et al. [80] proposed an online estimation method for motor parameters based on discrete dynamic models. This method utilizes the differences in the dynamic characteristics of motor parameters to form two affine projections, which are used to estimate stator inductance and resistance, rotor flux, and load torque, respectively. Liu et al. [81] proposed a neural network method for the online identification

of motor parameters to achieve multiparameter online identification. This method does not require any additional parameters; only the current and voltage need to be collected to simultaneously identify the inductance, resistance, and magnetic flux parameters online. Zhang et al. [82] proposed an improved model predictive current control method based on an incremental model, which can improve the anti-parameter interference ability of the model predictive control by identifying inductance parameters online and introducing an incremental predictive model. Yuan et al. [83] proposed an improved model predictive control method based on a current change update mechanism to suppress the interference of parameter mismatch. By identifying motor parameters online to calibrate the parameters used in the model predictive controller, the impact of parameter mismatch deterioration can be effectively suppressed. Yuan et al. [84] proposed an improved finite set model predictive current control method for motor drive systems to overcome the sensitivity to parameter mismatch. By introducing stationary and rotating correction coordinate systems to obtain real-time prediction parameters, the method can effectively suppress interference caused by parameter mismatch. Wang et al. [22] proposed a current prediction control method based on a fuzzy algorithm for motor drive systems, which can eliminate the influence of parameter mismatch on the prediction controller in real time by adjusting the weighting coefficient. Sawma et al. [85] proposed a model predictive current control algorithm based on minimum mean square identification, which can achieve the organic combination of the minimum mean square identification algorithm and the model predictive control algorithm. Anl et al. [86] proposed an adjustable model predictive control strategy for an interior PMSM drive system, which can effectively reduce torque ripple and current distortion through online parameter identification. An et al. [87] presented a new robust predictive current control method for a five-phase PMSM based on online d-q inductance identification, which can improve output performance and enhance parameter robustness under open circuit faults. Wang et al. [88] presented a novel online parameter identification algorithm for deadbeat current control of the PMSM drive system, which can obtain satisfactory performance under different working states. Xie et al. [89] presented a parameter identification method using online particle swarm optimization to obtain important motor parameters. This method combines the deadbeat prediction current control strategy to accurately predict the current and calculate the voltage vector of the PMSM drive system. Wang et al. [90] proposed an improved deadbeat predictive current control method based on parameter identification for PMSM drive systems, which can effectively solve the problem of underdetermined equations in multiparameter identification and improve the robustness of motor parameter mismatch. He et al. [91] proposed a cascade robust dual-loop predictive tolerant control for motor drive systems with multiparameter mismatch fault diagnosis, which can improve prediction accuracy and suppress current tracking errors by using a neural network multiparameter identification strategy. Zhang et al. [92] presented an automatic parameter identification method for deadbeat predictive current control based on the ordinary least square algorithm, which can obtain high steady-state and dynamic performance in PMSM drive systems. Chen et al. [93] proposed an adaptive predictive control method for motor drive systems with varying inductance. The method uses an adaptive observer and recursive algorithm to identify inductance online and combines it with finite set model predictive control. Although the parameter identification-based modified model predictive control method can eliminate the impact of parameter mismatch on motor performance, it relies on information such as the current, voltage, and rotor speed of the motor system.

(3) Compensation predictive control based on observer disturbance estimation. The core idea of the compensatory predictive control method is to consider all factors that cause parameter mismatch, such as external disturbances, motor parameter disturbances, measurement errors, etc., as disturbances, and construct an observer to estimate the disturbances in real time. Then, the online estimated interference is input as a feedback value into the predictive controller to achieve real-time compensation for parameter mismatch. The widely used observers in the field of motor system predictive control include Luenberger observer, extended Kalman filter, sliding mode observer, and adaptive observer. Zhang et al. [94] proposed a robust model predictive direct speed control method suitable for PMSM drive systems and designed a full-parameter sliding mode disturbance observer that simultaneously considers electrical parameter disturbances, mechanical parameter disturbances, and load torque changes. This method uses observer estimates to compensate for reference voltage errors in real time, which improves the robustness of the predictive controller to parameter mismatches and load torque disturbances. Wang et al. [95] proposed a model predictive control method based on a generalized proportional integral observer, which can eliminate the influence of uncertainties such as load torque disturbance and parameter time-variation on the motor system. Abdelrahem et al. [96] designed an extended Kalman filter to estimate

parameter changes and disturbances caused by inaccurate modeling of the motor system, which can achieve low torque ripple operation of the motor system in the case of parameter mismatch. Yin et al. [97] proposed a robust current prediction control method based on an immune optimization disturbance observer, which can achieve both stable transient response and high-precision steady-state performance under parameter mismatch. Wang et al. [98] studied a disturbance observer-based predictive torque control method for motor systems, which can improve anti-interference ability and robustness through feedforward compensation of the disturbance observer. Mwasilu et al. [99] designed a load torque disturbance estimator based on a two-stage extended Kalman filter for PMSM drive systems, which can greatly improve the prediction process of the model and enhance the robustness of the predictive control algorithm to changes in mechanical system parameters. Yang et al. [100] proposed a robust current prediction control algorithm for PMSM drive systems based on an incremental system model, which can improve robustness to inductance parameters by using an extended state observer. Yang et al. [101] proposed a high-precision continuous unbiased model predictive control method based on a disturbance observer, which can greatly reduce computational complexity and exhibit significant anti-interference performance. Xu et al. [102] proposed a direct predictive control method for deadbeat torque based on a disturbance input decoupled flux observer, which can achieve zero overshoot and steady-state error operation of the motor system in the case of parameter changes or inaccurate identification. Yang et al. [103] studied the robust control problem of nonlinear systems with disturbances and uncertainties. A robust control method based on a nonlinear disturbance observer was proposed to eliminate the effects of system mismatch disturbance and parameter mismatch. Wu et al. [104] proposed a new model-free predictive current control method based on an enhanced extended state observer, which can effectively improve the robustness of motor drive systems to inductance parameter mismatch. Yang et al. [105] proposed an improved model-free deadbeat predictive current control with a Luenberger disturbance observer method, which can improve the control performance degradation caused by parameter mismatch in motor drive systems. Zhou et al. [106] proposed an incremental deadbeat predictive current control strategy with a sliding mode disturbance observer, which can suppress the negative impact of parameter mismatch on the control performance of PMSM drive systems. Ke et al. [107] proposed a new predictive current control method that uses an extended sliding mode observer based on the Hurwitz power convergence law, which can achieve high robustness and excellent steady-state and dynamic performance of PMSM drive systems. Wu et al. [108] proposed a robust speed-stator flux-tolerant predictive control for PMSM drives with a composite discrete extended state observer, which can eliminate the influence of parameter disturbance and obtain excellent stator flux/torque control performance. Wang et al. [109] proposed a fast-terminal cost index-based continuous control set model predictive speed control for PMSM systems, which can minimize the lag error between the actual speed and reference speed. Luo et al. [110] proposed a Luenberger observer-based model predictive control for six-phase PMSM with localization error compensation, which can improve tracking capability and reduce harmonic currents under parameter mismatch. Compared with the parameter identification predictive control method, the compensation predictive control method based on observer disturbance estimation can simultaneously eliminate more external factors that affect the performance of the predictive controller. The parameter identification predictive control method is mainly used to eliminate the influence of parameter mismatch on the model predictive controller.

2.5 Conclusion

This section provides a detailed introduction to the definition and sources of model predictive control technology. The application and advantages of model predictive control technology in the field of motor control are described from the perspectives of different voltage vector control sets, different control objectives, and different control structures. The reasons for parameter mismatch and its impact on the model predictive controller are elaborated in detail, and the main reasons for the parameter mismatch between the predicted controller parameters and the actual parameters of the motor system are provided. Finally, the research status of robust predictive control technology under parameter mismatch is presented, and three robust predictive control methods are described in detail: improved model predictive control, modified predictive control based on parameter identification, and compensated predictive control based on observer disturbance estimation. In summary, the model predictive control technology has improved the performance of motor systems and expanded their application areas in manufacturing. However, the model predictive control technology heavily relies on motor system parameters and mathematical models, and further exploration and research are needed on robust model predictive control technology under parameter mismatch. On the other hand, by fully utilizing the advantages of model predictive control technology, it is necessary to combine robust model predictive control strategies with the structure and characteristics of motor systems to improve the control performance of different motors under parameter mismatches.

References

[1] M. Tomlinson, T. Mouton, et al., A fixed switching frequency scheme for finite-control-set model predictive control—Concept and Algorithm, IEEE Trans. Ind. Electron. 63 (12) (2016) 7662–7670.
[2] S. Vazquez, J. Rodriguez, Model predictive control for power converters and drives: advances and Trends, IEEE Trans. Ind. Electron. 64 (02) (2017) 935–947.
[3] J. Rodriguez, Predictive current control of a voltage source inverter, IEEE Trans. Ind. Electron. 54 (12) (2007) 495–503.
[4] F. Niu, B. Wang, et al., Comparative evaluation of direct torque control strategies for permanent magnet synchronous machines, IEEE Trans. Power Electron. 31 (02) (2016) 1408–1424.
[5] H. Nguyen, J. Jung, Finite control set model predictive control to guarantee stability and robustness for surface-mounted PM synchronous motors, IEEE Trans. Ind. Electron. 65 (11) (2018) 8510–8519.
[6] T. Geyer, G. Papafotiou, et al., Model predictive direct torque control—part I: concept, algorithm, and analysis, IEEE Trans. Ind. Electron. 56 (06) (2009) 1894–1905.
[7] J. Guzinski, H. Abu-Rub, Speed sensorless induction motor drive with predictive current controller, IEEE Trans. Ind. Electron. 60 (02) (2013) 699–709.
[8] F. Morel, X. Lin-Shi, et al., A comparative study of predictive current control schemes for a permanent-magnet synchronous machine drive, IEEE Trans. Ind. Electron. 56 (07) (2009) 2715–2728.
[9] A. Darba, F.D. Belie, et al., Improved dynamic behaviour in BLDC drives using model predictive speed and current control, IEEE Trans. Ind. Electron. 63 (02) (2016) 728–740.
[10] C.K. Lin, J.T. Yu, et al., Improved model-free predictive current control for synchronous reluctance motor drives, IEEE Trans. Ind. Electron. 63 (06) (2016) 3942–3953.
[11] M. Cheng, F. Yu, et al., Dynamic performance evaluation of a nine-phase flux switching permanent-magnet motor drive with model predictive control, IEEE Trans. Ind. Electron. 63 (07) (2016) 4539–4549.
[12] X. Gao, M. Abdelrahem, et al., Direct predictive speed control with a sliding manifold term for PMSM drives, IEEE J. Emerg. Sel. Top. Power Electron. 8 (02) (2020) 1258–1267.
[13] M. Preindl, S. Bolognani, Model predictive direct speed control with finite control set of PMSM drive systems, IEEE Trans. Power Electron. 28 (02) (2013) 1007–1015.
[14] H. Kawai, Z. Zhang, et al., Direct speed control based on finite control set model predictive control with voltage smoother, IEEE Trans. Ind. Electron. 70 (03) (2023) 2363–2372.
[15] C. Garcia, J. Rodriguez, et al., Full Predictive Cascaded Speed and Current Control of an Induction Machine, IEEE Trans. Energy Convers. 31 (03) (2016) 1059–1067.
[16] T. Tarczewski, L.M. Grzesiak, et al., Constrained state feedback speed control of PMSM based on model predictive approach, IEEE Trans. Ind. Electron. 63 (06) (2016) 3867–3875.

[17] J. Gao, C. Gong, et al., Novel compensation strategy for calculation delay of finite control set model predictive current control in PMSM, IEEE Trans. Ind. Electron. 67 (07) (2020) 5816–5819.
[18] T. Turker, U. Buyukkeles, et al., A robust predictive current controller for PMSM drives, IEEE Trans. Ind. Electron. 63 (06) (2016) 3906–3914.
[19] X. Yuan, S. Zhang, et al., Nonparametric predictive current control for PMSM, IEEE Trans. Power Electron. 35 (09) (2020) 9332–9341.
[20] X. Zhang, B. Hou, et al., Deadbeat predictive current control of permanent magnet synchronous motors with stator current and disturbance observer, IEEE Trans. Power Electron. 32 (05) (2016) 3818–3834.
[21] Y. Jiang, W. Xu, et al., Improved deadbeat predictive current control combined sliding mode strategy for PMSM drive system, IEEE Trans. Veh. Technol. 67 (01) (2017) 51–63.
[22] Z. Wang, A. Yu, et al., A novel current predictive control based on fuzzy algorithm for PMSM, IEEE J. Emerg. Sel. Top. Power Electron. 7 (02) (2019) 990–1001.
[23] C. Xiong, H. Xu, et al., A constant switching frequency multiple vector based model predictive current control of five-phase PMSM with non-sinusoidal back-EMF, IEEE Trans. Ind. Electron. 67 (03) (2019) 31695–31707.
[24] F. Ban, G. Lian, et al., Study on a novel predictive torque control strategy based on the finite control set for PMSM, IEEE Trans. Appl. Supercond. 29 (02) (2019) 1–6.
[25] Q. Liu, K. Hameyer, Torque ripple minimization for direct torque control of PMSM with modified FCSMPC, IEEE Trans. Ind. Appl. 52 (06) (2016) 4855–4864.
[26] A. Mora, A. Orellana, et al., Model predictive torque control for torque ripple compensation in variable-speed PMSMs, IEEE Trans. Ind. Electron. 63 (07) (2016) 4584–4592.
[27] D. Lv, W. Ding, et al., Finite control set model predictive torque control of switched reluctance motor based on three-phase four-leg inverter, IEEE Trans. Ind. Electron. 72 (10) (2025) 9931–9941.
[28] O. Sandre-Hernandez, J. Rangel-Magdaleno, et al., A comparison on finite set model predictive torque control schemes for PMSMs, IEEE Trans. Power Electron. 33 (10) (2018) 8838–8847.
[29] W. Yang, W. Xu, et al., Novel decoupling model-based predictive current control strategy for flux-switching permanent-magnet synchronous machines with low torque ripple and switching loss, IEEE Int. Conf. Appl. Supercond. Electromagn. Devices 31 (02) (2014) 25–27.
[30] L. Ge, J. Guo, et al., Model predictive torque control of switched reluctance machines with torque sharing function and PWM Control Signals Based on Linear Polynomial Fitting, IEEE Trans. Power Electron. 40 (01) (2024) 40–45.
[31] Y. Cho, Y. Bak, et al., Torque-Ripple reduction and fast torque response strategy for predictive torque control of induction motors, IEEE Trans. Appl. Supercond. 33 (03) (2017) 2458–2470.
[32] G. Li, J. Hu, Model predictive torque control for torque ripple compensation in variable-speed PMSMs, IEEE Trans. Ind. Electron. 66 (08) (2018) 5820–5829.
[33] A.W. Ordys, D.W. Clarke, A state-space description for GPC controllers, Int. J. Syst. Sci. 24 (09) (1993) 1727–1744.
[34] K. Belda, D. Vomik, Explicit generalized predictive control of speed and position of PMSM drives, IEEE Trans. Ind. Electron. 63 (06) (2016) 3889–3896.
[35] X. Li, W. Tian, A generalized observer-based robust predictive current control strategy for PMSM drive system, IEEE Trans. Ind. Electron. 69 (02) (2021) 1322–1332.
[36] Z. Cao, J. Mao, et al., Safety-Critical generalized predictive control for speed regulation of PMSM drives based on dynamic robust control barrier function, IEEE Trans. Appl. Supercond. 33 (03) (2017) 2458–2470.

[37] Y. Wei, H. Young, et al., Model-Free predictive control using sinusoidal generalized universal model for PMSM drives, IEEE Trans. Ind. Electron. 71 (11) (2024) 13720–13731.

[38] T. Wang, L. Guo, et al., Generalized predictive current control for dual-three-phase PMSM to achieve torque enhancement through harmonic injection, IEEE Trans. Power Electron. 38 (05) (2023) 6422–6433.

[39] H. Liu, S. Li, Speed control for PMSM servo System using predictive functional control and extended state observer, IEEE Trans. Ind. Electron. 17 (12) (2016) 3364–3379.

[40] P. Cortes, M.P. Kazmierkowski, et al., Predictive control in power electronics and drives, IEEE Trans. Ind. Electron. 55 (12) (2008) 4312–4324.

[41] J. Rodriguez, P. Cortes, et al., Model predictive control-a simple and powerful method to control power converters, IEEE Trans. Ind. Electron. 56 (06) (2009) 1826–1838.

[42] Y. Han, C. Gong, et al., Multiobjective finite control set model predictive control using novel delay compensation technique for PMSM, IEEE Trans. Power Electron. 35 (10) (2020) 11193–11204.

[43] W. Zhao, T. Tao, et al., A novel finite control set model predictive current control for five-phase PM motor with continued modulation, IEEE Trans. Power Electron. 35 (07) (2019) 7261–7270.

[44] C. Gong, Y. Hu, et al., Accurate FCS model predictive current control technique for surface-mounted PMSMs at low control frequency, IEEE Trans. Power Electron. 35 (06) (2019) 5567–5572.

[45] X. Zhang, Y. He, Direct voltage-selection based model predictive direct speed control for PMSM drives without weighting factor, IEEE Trans. Power Electron. 34 (08) (2019) 7838–7851.

[46] A. Ahmed, B. Koh, et al., A Comparison of Finite Control Set and Continuous Control Set Model Predictive Control Schemes for Speed Control of Induction Motors, IEEE Trans. Ind. Inform. 14 (04) (2017) 1334–1346.

[47] Z. Sun, Y. Deng, J Wang, Finite control set model-free predictive current control of PMSM With two voltage vectors based on ultralocal model, IEEE Trans. Power Electron. 38 (01) (2023) 776–788.

[48] C. Martín, M. Arahal, Five-Phase induction motor rotor current observer for finite control set model predictive control of stator current, IEEE Trans. Ind. Electron. 63 (07) (2020) 4527–4538.

[49] X. Sun, M. Wu, et al., An improved model predictive current control for PMSM drives based on current track circle, IEEE Trans. Ind. Electron. 68 (05) (2020) 3782–37933.

[50] Y. Zhang, J. Jin, et al., Model-free predictive current control of PMSM drives based on extended state observer using ultralocal model, IEEE Trans. Ind. Electron. 68 (02) (2021) 993–1003.

[51] X. Liu, H. Yang, et al., A Novel Finite-Set Sliding-Mode Model-Free Predictive Current Control for PMSM Drives Without DC-Link Voltage Sensor, IEEE Trans. Power Electron. 39 (01) (2024) 320–331.

[52] Y. Wei, H. Young, Adaptive ultralocalized time-series for improved model-free predictive current control on PMSM drives, IEEE Trans. Energy Convers. 39 (15) (2024) 5155–5165.

[53] H. Li, D. Lan, et al., Model-Free predictive current control of PMSM with online current gradient updating using ultra-local model, IEEE Trans. Energy Convers. 40 (03) (2025) 1898–1908.

[54] X. Zhang, L. Zhang, et al., Model predictive current control for PMSM drives with parameter robustness improvement, IEEE Trans. Power Electron. 34 (02) (2018) 1645–1657.

[55] Y. Luo, L. Chunhua, A flux constrained predictive control for a six-phase PMSM motor with lower complexity, IEEE Trans. Ind. Electron. 66 (07) (2019) 5081–5093.

[56] Y. Luo, C. Liu, et al., Model predictive control for a six-phase PMSM motor with a reduced dimension cost function, IEEE Trans. Ind. Electron. 17 (12) (2016) 3364–3379.
[57] R. Fu, Model-Free Predictive Current and Speed Control With Ultralocal Model for PMSM Drives Using Exponential High-Gain Extended State Observer, IEEE J. Emerg. Sel. Top. Power Electron. 67 (02) (2025) 6099–6108.
[58] Z. Li, F. Wang, et al., Robust Continuous Model Predictive Speed and Current Control for PMSM With Adaptive Integral Sliding-Mode Approach, IEEE Trans. Power Electron. 36 (12) (2021) 14398–14408.
[59] L. Yang, H. Li, et al., Model predictive direct speed control with novel cost function for SMPMSM drives, IEEE Trans. Power Electron. 37 (08) (2022) 9586–95959.
[60] K. Yu, Z. Wang, et al., Robust cascaded deadbeat predictive control for dual three-phase variable-flux PMSM considering intrinsic delay in speed loop, IEEE Trans. Ind. Electron. 69 (12) (2022) 12107–12118.
[61] S. Gao, Y. Wei, et al., Robust cascaded deadbeat predictive control for dual three-phase variable-flux PMSM considering intrinsic delay in speed loop, IEEE Trans. Ind. Electron. 69 (10) (2022) 9739–9748.
[62] M. Zhao, Y. Cao, et al., Expanded limit boundary explicit model predictive direct speed control for PMSMs, IEEE Trans. Power Electron. 39 (5) (2024) 6089–6101.
[63] W. Tu, G. Luo, et al., Predictive Cascaded Speed and Current Control for PMSM Drives With Multi-Timescale Optimization, IEEE Trans. Power Electron. 34 (11) (2019) 11046–11061.
[64] E. Fuentes, A. Silva, et al., MPC implementation of a quasi-time-optimal speed control for a PMSM drive with inner modulated-FS-MPC torque control, IEEE Trans. Ind. Electron. 63 (06) (2016) 3897–3905.
[65] A. Darba, P.D.H. Belie, et al., Improved dynamic behavior in BLDC drives using model predictive speed and current control, IEEE Trans. Ind. Electron. 63 (02) (2016) 728–740.
[66] J. Hu, C. He, et al., A Novel Predictive Position Control With Current and Speed Limits for PMSM Drives Based on Weighting Factors Elimination, IEEE Trans. Ind. Electron. 69 (12) (2022) 12458–12468.
[67] S. Li, H. Gu, et al., Fuzzy adaptive internal model control schemes for PMSM speed-regulation system, IEEE Trans. Ind. Inform. 8 (04) (2012) 767–779.
[68] M. Rashed, A.F. Stronach, et al., A stable back-EMF MRAS-based sensorless low-speed induction motor drive insensitive to stator resistance variation, IEE Proc. Electr. Power Appl. 15 (06) (2004) 685–693.
[69] S.J. Underwood, I. Husain, et al., Online parameter estimation and adaptive control of permanent-magnet synchronous machines, IEEE Trans. Ind. Electron. 57 (07) (2010) 2435–2443.
[70] K.D. Lee, H.W. Lee, et al., Analysis of motor performance according to the inductance design of IPMSM, IEEE Trans. Magn. 51 (03) (2015) 1–4.
[71] L. He, F. Wang, et al., Zynq implemented luenberger disturbance observer based predictive control scheme for PMSM drives, IEEE Trans. Power Electron. 35 (2) (2019) 1770–1778.
[72] F. Tinazzi, P.G. Carlet, et al., Motor parameter-free predictive current control of synchronous motors by recursive least-square self-commissioning model, IEEE Trans. Ind. Electron. 67 (11) (2020) 9093–9100.
[73] L. Xu, G. Chen, et al., Ultra-local model-free predictive current control based on nonlinear disturbance compensation for permanent magnet synchronous motor, IEEE Access 2 (08) (2020) 127690–127699.
[74] P.G. Carlet, F. Tinazzi, et al., An effective model-free predictive current control for synchronous reluctance motor drives, IEEE Trans. Ind. Electron. 55 (04) (2019) 3781–3790.

[75] Y. Zhang, T. Jiang, et al., Model-free predictive current control of a DFIG using an ultra local model for grid synchronization and power regulation, IEEE Trans. Ind. Electron. 35 (04) (2020) 2269–2280.
[76] C.K. Lin, J.T. Yu, et al., Improved model-free predictive current control for synchronous reluctance motor drives, IEEE Trans. Ind. Electron. 63 (04) (2016) 2269–2280.
[77] C.K. Lin, T.H. Liu, et al., Model-free predictive current control for interior permanent-magnet synchronous motor drives based on current difference detection technique, IEEE Trans. Ind. Electron. 61 (02) (2014) 667–681.
[78] Z. Yin, X. Wang, et al., A switched ultra-local model-free predictive controller for PMSMs, IEEE Trans. Power Electron. 39 (09) (2024) 10665–10669.
[79] M. Pulvirenti, G. Scarcella, et al., On-line stator resistance and permanent magnet flux linkage identification on open-end winding PMSM drives, IEEE Trans. Ind. Appl. 55 (01) (2019) 504–515.
[80] D.Q. Dang, M.S. Rafaq, et al., Online parameter estimation technique for adaptive control applications of interior PM synchronous motor drives, IEEE Trans. Ind. Electron. 63 (03) (2016) 1438–1449.
[81] K. Liu, Q. Zhang, et al., Online multiparameter estimation of nonsalient-pole PM synchronous machines with temperature variation tracking, IEEE Trans. Ind. Electron. 58 (05) (2011) 1776–1788.
[82] X. Zhang, L. Zhang, et al., Model predictive current control for PMSM drives with parameter robustness improvement, IEEE Trans. Power Electron. 34 (02) (2019) 1645–1657.
[83] X. Yuan, S. Zhang, et al., Improved model predictive current control for SPMSM drives with parameter mismatch, IEEE Trans. Ind. Electron. 67 (02) (2020) 852–862.
[84] X. Yuan, S. Zhang, et al., Improved model predictive current control for SPMSM drives using current update mechanism, IEEE Trans. Ind. Electron. 68 (03) (2020) 1938–1948.
[85] J. Sawma, F. Khatounian, et al., Analysis of the impact of online identification on model predictive current control applied to permanent magnet synchronous motors, IET Electr. Power Appl. 11 (05) (2017) 864–873.
[86] X. Anl, G. Liu, et al., Adjustable model predictive control for IPMSM drives based on online stator inductance identification, IEEE Trans. Ind. Electron. 69 (04) (2021) 3368–3381.
[87] X. An, G. Liu, et al., Robust predictive current control for fault-tolerant operation of five-phase PM motors based on online stator inductance identification, IEEE Trans. Power Electron. 36 (11) (2021) 13162–13175.
[88] Z. Wang, J. Chai, et al., A novel online parameter identification algorithm designed for deadbeat current control of the permanent-magnet synchronous motor, IEEE Trans. Ind. Appl. 58 (02) (2021) 2029–2041.
[89] C. Xie, S. Zhang, et al., Parameter identification for SPMSM with deadbeat predictive current control using online PSO, IEEE Trans. Transp. Electr. 10 (02) (2024) 4055–4064.
[90] L. Wang, S. Zhang, et al., Y Zhou, An improved deadbeat predictive current control based on parameter identification for PMSM, IEEE Trans. Transp. Electr. 10 (02) (2023) 2740–2753.
[91] J. He, R. Tang, et al., Robust dual-loop cascade predictive tolerant control for NPMSM drives with multiparameter mismatch fault diagnosis, IEEE Sens. J. 25 (20) (2025) 8405–38415.
[92] H. Zhang, T. Fan, et al., Polynomial estimation of flux linkage for predictive current control in PMSM, IEEE J. Emerg. Sel. Top. Power Electron. 10 (12) (2022) 6112–6122.
[93] Z. Chen, J. Qiu, et al., Adaptive finite-control-set model predictive current control for IPMSM drives with inductance variation, IET Electr. Power Appl. 11 (05) (2017) 874–884.

[94] X. Zhang, Y. Cheng, et al., He Y, Robust model predictive direct speed control for SPMSM drives based on full parameter disturbances and load observer, IEEE Trans. Power Electron. 35 (08) (2020) 8361–8373.
[95] J. Wang, F. Wang, et al., Generalized proportional integral observer based robust finite control set predictive current control for induction motor systems with time varying disturbances, IEEE Trans. Ind. Inform. 14 (09) (2018) 4159–41683.
[96] M. Abdelrahem, C. Hackl, et al., Robust predictive control for direct-driven surface-mounted permanent-magnet synchronous generators without mechanical sensors, IEEE Trans. Energy Convers. 33 (01) (2018) 179–189.
[97] Z. Yin, X. Han, et al., Research on model predictive current control for induction machine based on immune-optimized disturbance observer, IEEE J. Emerg. Sel. Top. Power Electron. 6 (04) (2018) 1699–1710.
[98] J. Wang, F. Wang, et al., Design and implementation of disturbance compensation-based enhanced robust finite control set predictive torque control for induction motor systems, IEEE Trans. Ind. Inform. 13 (05) (2017) 2645–2656.
[99] F. Mwasilu, H. Nguyen, et al., Finite set model predictive control of interior PM synchronous motor drives with an external disturbance rejection technique, IEEE/ASME Trans. Mechatron. 22 (02) (2017) 762–773.
[100] M. Yang, X. Lang, et al., Flux immunity robust predictive current control with incremental model and extended state observer for PMSM drive, IEEE Trans. Power Electron. 32 (12) (2017) 267–9279.
[101] J. Yang, W.X. Zheng, et al., Design of a prediction-accuracy-enhanced continuous time MPC for disturbed systems via a disturbance observer, IEEE Trans. Ind. Electron. 62 (09) (2015) 5807–5816.
[102] W. Xu, R.D. Lorenz, et al., Reduced parameter sensitivity stator flux linkage observer in deadbeat-direct torque and flux control for IPMSMs, IEEE Trans. Ind. Appl. 50 (04) (2014) 2626–26364.
[103] J. Yang, S. Li, et al., Sliding-mode control for systems with mismatched uncertainties via a disturbance observer, IEEE Trans. Ind. Electron. 60 (01) (2013) 60–169.
[104] X. Wu, P. Zou, Sliding-mode control for systems with mismatched uncertainties via a disturbance observer, IEEE Trans. Ind. Electron. 71 (04) (2023) 3461–3471.
[105] N. Yang, S. Zhang, et al., X Li, A new model-free deadbeat predictive current control for PMSM using parameter-free luenberger disturbance observer, IEEE J. Emerg. Sel. Top. Power Electron. 11 (01) (2022) 407–417.
[106] Y. Zhou, L. Yan, et al., M Hamad, Incremental deadbeat predictive current control strategy with disturbance observer for PMSM drive systems using four-leg inverters, IEEE Trans. Power Electron. 40 (11) (2025) 16483–16497.
[107] D. Ke, F. Wang, et al., Predictive Current Control for PMSM Systems Using Extended Sliding Mode Observer With Hurwitz-Based Power Reaching Law, IEEE Trans. Power Electron. 36 (06) (2020) 7223–7232.
[108] G. Wu, P. Guan, et al., Robust speed-stator flux tolerant predictive control for PMSM drives with parameters disturbance faults diagnosis, IEEE Sens. J. 40 (11) (2025) 34840–34849.
[109] F. Wang, L. He, et al., FPGA-Based predictive speed control for PMSM system using integral sliding-mode disturbance observer, IEEE Trans. Ind. Electron. 68 (02) (2020) 72–98.
[110] Y. Luo, K. Yang, et al., Luenberger observer-based model predictive control for six-phase PMSM motor with localization error compensation, IEEE Trans. Ind. Electron. 70 (11) (2022) 10800–108101.

CHAPTER 3

Novel predictive stator flux control techniques for three-phase PMSM considering flux linkage parameter mismatch

3.1 Introduction

Permanent magnet synchronous motor (PMSM) drives have been extensively employed in industrial applications, such as robots, electric vehicles, and wind power generators [1,2], due to their advantages of compact structure, excellent dynamic properties, and high-performance operations [3,4]. To achieve the desired servo control performance, the field-oriented control (FOC) strategy has been used in most PMSM drive systems. In an FOC-based PMSM drive, the double-loop controller is usually adopted. The outer loop controller is designed to control rotor speed, while the inner loop controller is responsible for controlling stator current. The predictive current control (PCC) has received extensive attention for the inner loop controller as it offers the advantages of excellent transient response characteristics and accurate reference current tracking [5,6].

The performance of the PCC method crucially depends on the accuracy of the stator inductance, flux linkage parameter, as well as rotor position. However, the permanent magnet (PM) flux linkage parameter may change due to temperature rise and magnet saturation, especially under high-temperature operating conditions [7]. In engineering, the temperature change will inevitably lead to permanent magnet demagnetization; a 20% flux reduction for a ferrite-based magnet occurs per 100°C increase in ambient temperature [8]. Although the permanent magnet has a positive coefficient of coercivity, the flux linkage parameter of the motor still changes with temperature variation, especially in motors used in tanks and high-speed trains. On the other hand, the rotor position measured by the resolver or the encoder attached to the PMSM shaft may deviate from the

Robust Model Predictive Control of Permanent Magnet Synchronous Motor Under Parameter Mismatch
https://doi.org/10.1016/B978-0-443-44209-4.10003-X

true position due to the misalignment of the mechanical sensor components mounted on the PMSM [9–11], which causes the Park coordinate transformation to provide a mistaken feedback current to the PCC algorithm. As a result, the control performance of the PMSM deteriorates.

The main approach to improving flux linkage parameter robustness for the PCC is based on identification. In [12], an improved deadbeat PCC algorithm based on stator current and a disturbance observer is proposed for PMSM drive systems, which can optimize the PCC performance of the PMSM system with flux linkage parameter mismatch. In [13], a Kalman filter and a Luenberger observer are designed to estimate rotor state variables. In [14,15], the robust PCC algorithm based on a disturbance observer is adopted for PMSM drives, which can overcome the effect of flux linkage parameter mismatch on the PCC and achieve high dynamic performance. In [16], a robust fault-tolerant PCC algorithm based on a composite sliding mode observer (SMO) is proposed for PMSM, which can guarantee the performance of the system regardless of parameter perturbation, permanent magnet demagnetization, and one-step delay. In [17], a model-free PCC method for interior PMSM drives based on a current difference detection technique is proposed, which does not require any motor parameters. However, the higher current sampling frequency tends to limit its industrial application. In [18,19], a flux immunity robust PCC algorithm for PMSM drives is proposed to eliminate the use of the PM flux linkage parameter in the prediction model, but the method ignores the effect caused by rotor position measurement error.

The accuracy of rotor position is crucial for FOC when minimum torque ripple is demanded over the entire operating range of the motor. In [9–11], experimental results show that the torque oscillation caused by other factors is almost negligible compared to the rotor position error. In [20], a PMSM torque predictive control scheme is proposed, which can eliminate high current harmonics and torque ripples caused by rotor position error. In [21], a new torque predictive control method for PMSM drives is proposed, which can control the torque by using the magnitude of the stator voltage vector. In [22], a sensorless predictive torque control scheme that relies on the anisotropy of the inductance is proposed, which can achieve low torque ripple during the full speed range. The control methods mentioned above provide low torque ripple under rotor position error. However, the flux linkage parameter is considered constant in these control methods, which cannot be guaranteed in engineering.

In this chapter, a novel PSFC technique is developed to guarantee the performance of PMSM drives regardless of flux linkage parameter perturbation, rotor position error, and one-step delay in digital control. In contrast to the conventional PCC method, the proposed PSFC method can achieve lower torque ripple and stator current harmonics by applying a novel composite discrete-time SMO based on the stator flux state. The performance of the proposed PSFC method is validated through simulation and experimental results. This chapter is organized as follows: The influence of flux linkage parameter perturbation and rotor position error on the conventional PCC method is analyzed in Section 3.2. The PSFC strategy is proposed in Section 3.3. The composite discrete-time SMO based on the stator flux state is designed in Section 3.4. The simulations are set up in Section 3.5. Section 3.6 concludes this chapter.

3.2 The influence of flux linkage parameter and rotor position

The voltage state-space equations of the three-phase PMSM are described in Eq. (3.1) [18].

$$\begin{cases} \dfrac{di_d}{dt} = -\dfrac{R}{L_d}i_d + \dfrac{L_q}{L_d}\omega i_q + \dfrac{1}{L_d}u_d \\ \dfrac{di_q}{dt} = -\dfrac{R}{L_q}i_q - \dfrac{L_d}{L_q}\omega i_d - \dfrac{\psi_{ro}\omega}{L_q} + \dfrac{1}{L_q}u_q \end{cases} \tag{3.1}$$

where, $\dfrac{L_q}{L_d}\omega i_q$ and $-\dfrac{L_d}{L_q}\omega i_d$ are d- and q-axis cross-coupling terms, respectively. According to Eq. (3.1), the voltage state-space equations of the SPMSM can be given as

$$\begin{cases} \dfrac{di_d}{dt} = -\dfrac{R}{L}i_d + \omega i_q + \dfrac{1}{L}u_d \\ \dfrac{di_q}{dt} = -\dfrac{R}{L}i_q - \omega i_d - \dfrac{\psi_{ro}\omega}{L} + \dfrac{1}{L}u_q \end{cases} \tag{3.2}$$

For surface-mounted PMSM, $L_d = L_q = L$. The d- and q-axis cross-coupling terms are ωi_q and $-\omega i_d$, respectively. According to Eq. (3.2), the d- and q-axis voltage equations of the SPMSM can be obtained as

$$\begin{cases} u_d = Ri_d + \dfrac{d\psi_d}{dt} - \omega\psi_q \\ u_q = Ri_q + \dfrac{d\psi_q}{dt} + \omega\psi_d \end{cases} \tag{3.3}$$

where,

$$\begin{cases} \psi_d = Li_d + \psi_{ro} \\ \psi_q = Li_q \end{cases}$$

The electromagnetic torque produced by the SPMSM can be expressed as

$$T_e = \frac{3}{2} n_p \psi_{ro} i_q = \frac{3}{2} n_p \psi_{ro} \frac{\psi_q}{L} \tag{3.4}$$

The mechanical dynamic model of the SPMSM can be described as

$$\frac{J}{n_p} \frac{d\omega}{dt} = T_e - T_L \tag{3.5}$$

where, u_d and u_q represent the *d*- and *q*-axis stator voltages, respectively; i_d and i_q are the *d*- and *q*-axis currents, respectively; ψ_d and ψ_q are the *d*- and *q*-axis stator flux components, respectively; R, L, and J are the stator resistance, stator inductance, and the moment of inertia, respectively; ω is the electrical rotor speed, ψ_{ro} is the flux linkage, n_p is the number of pole pairs, T_e is the electromagnetic torque, and T_L is the load torque.

The variation of flux linkage is illustrated in Fig. 3.1. Where, "θ_o" and "θ" are the actual value and measurement value of the rotor position, respectively. There exists an error $\Delta\theta$ between θ and θ_o when the rotor position measurement is inaccurate. "ψ_r" and "ψ_{ro}" are the actual and

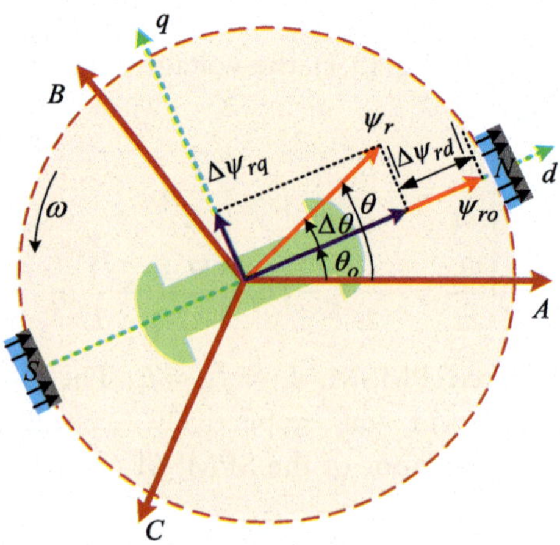

Fig. 3.1 Changes in flux linkage of a three-phase PMSM.

nominal values of the flux linkage parameters, respectively. There also exists an error between "ψ_r" and "ψ_{ro}" due to temperature rise and saturation. The flux linkage parameter perturbation rate is defined as

$$\lambda = \frac{\psi_{ro} - \psi_r}{\psi_{ro}} \tag{3.6}$$

The d- and q-axis flux linkage equations can be expressed as follows when there exist flux linkage parameter perturbations and rotor position errors.

$$\begin{cases} \psi_d = \psi_{ro} + Li_d + \Delta\psi_{rd} \\ \psi_q = Li_q + \Delta\psi_{rq} \end{cases} \tag{3.7}$$

with

$$\begin{cases} \Delta\psi_{rd} = \psi_r \cos(\Delta\theta) - \psi_{ro} = \psi_{ro}[(1-\lambda)\cos(\Delta\theta) - 1] \\ \Delta\psi_{rq} = \psi_r \sin(\Delta\theta) = \psi_{ro}(1-\lambda)\sin(\Delta\theta) \end{cases} \tag{3.8}$$

where, $\Delta\psi_{rd}$ and $\Delta\psi_{rq}$ are the d- and q-axis flux linkage components under flux linkage parameter perturbation and rotor position error, respectively.

In engineering, the time constant of the electrical system is much less than that of the mechanical system. Therefore $\Delta\psi_{rd}$ and $\Delta\psi_{rq}$ can be looked on as

$$\begin{cases} \dfrac{d\Delta\psi_{rd}}{dt} \approx 0 \\ \dfrac{d\Delta\psi_{rq}}{dt} \approx 0 \end{cases}$$

Substituting Eq. (3.7) into Eq. (3.3), we obtain:

$$\begin{bmatrix} \dfrac{di_d}{dt} \\ \dfrac{di_q}{dt} \end{bmatrix} = \begin{bmatrix} -\dfrac{R}{L} & \omega \\ -\omega & -\dfrac{R}{L} \end{bmatrix} \begin{bmatrix} i_d \\ i_q \end{bmatrix} + \begin{bmatrix} \dfrac{1}{L} & 0 \\ 0 & \dfrac{1}{L} \end{bmatrix} \begin{bmatrix} u_d \\ u_q \end{bmatrix} + \begin{bmatrix} 0 & \dfrac{\omega}{L} \\ -\dfrac{\omega}{L} & 0 \end{bmatrix} \begin{bmatrix} \psi_{ro} + \Delta\psi_{rd} \\ \Delta\psi_{rq} \end{bmatrix} \tag{3.9}$$

According to Eq. (3.7), Eq. (3.4) changes to:

$$T_e = \frac{3}{2} n_p (\psi_{ro} + \Delta\psi_{rd}) \frac{\psi_q + \Delta\psi_{rq}}{L} \tag{3.10}$$

Substituting Eq. (3.10) into Eq. (3.5), we obtain:

$$\frac{d\omega}{dt} = \frac{3n_p^2\psi_{ro}}{2JL}\psi_q - \frac{n_p}{J}T_L + \frac{n_p}{2JL}f_\omega \tag{3.11}$$

with

$$f_\omega = 3n_p\left[\psi_{ro}\Delta\psi_{rq} + \Delta\psi_{rd}\psi_q + \Delta\psi_{rd}\Delta\psi_{rq}\right] \tag{3.12}$$

where, f_ω is the equivalent disturbance caused by flux linkage parameter perturbation and rotor position error.

To evaluate the sensitivity of conventional PCC to the flux linkage parameter and rotor position, the influence of flux linkage parameter mismatch and rotor position error is analyzed in this section. The block diagram of conventional PCC is illustrated in Fig. 3.2.

The output voltage vectors of the PCC are expressed in [18].

$$\boldsymbol{u}(k) = \boldsymbol{F}^{-1}[\boldsymbol{i}^{ref}(k+1) - \mathbf{G}(k)i(k) - \boldsymbol{P}(k)] \tag{3.13}$$

where, $\boldsymbol{i}^{ref}(k+1) = \begin{bmatrix} i_d^{ref}(k+1) \\ i_q^{ref}(k+1) \end{bmatrix}$, $\mathbf{i}(k) = \begin{bmatrix} i_d(k) \\ i_q(k) \end{bmatrix}$, $\boldsymbol{u}(k) = \begin{bmatrix} u_d(k) \\ u_q(k) \end{bmatrix}$,

$$\boldsymbol{F} = \begin{bmatrix} \frac{T_s}{L} & 0 \\ 0 & \frac{T_s}{L} \end{bmatrix},\ \mathbf{G}(k) = \begin{bmatrix} 1-\frac{R}{L}T_s & T_s\omega(k) \\ -T_s\omega(k) & 1-\frac{R}{L}T_s \end{bmatrix},\ \boldsymbol{P}(k) = \begin{bmatrix} 0 \\ -\frac{\psi_{ro}}{L}T_s\omega(k) \end{bmatrix}.$$

According to Eq. (3.9), the discrete model of the PMSM under flux linkage parameter perturbation and rotor position error can be expressed as follows:

$$\boldsymbol{i}(k+1) = \mathbf{G}(k)\cdot\boldsymbol{i}(k) + \boldsymbol{F}\cdot\boldsymbol{u}(k) + \bar{\boldsymbol{P}}(k) \tag{3.14}$$

where, $\bar{\boldsymbol{P}}(k) = \begin{pmatrix} \frac{\Delta\psi_{rq}}{L}T_s\omega(k) \\ -\frac{\psi_{ro}}{L}T_s\omega(k) - \frac{\Delta\psi_{rd}}{L}T_s\omega(k) \end{pmatrix}$.

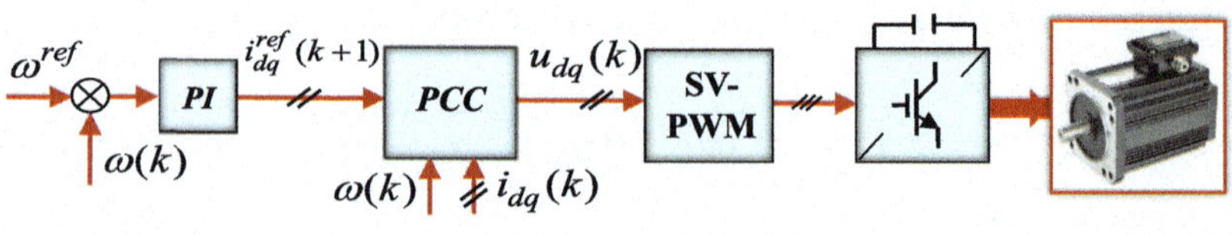

Fig. 3.2 Block diagram of the conventional PCC method.

For sampling delay, the conventional PCC method employed in the practical application belong to one beat delay control. At the $(k+1)T_s$ moment, the voltage vector $\mathbf{u}(k)$ is applied to the PMSM. Therefore substituting Eq. (3.13) into Eq. (3.14) yields

$$i_d(k+1) = i_d^{ref}(k+1) + \Delta i_d(k+1) \tag{3.15}$$

$$i_q(k+1) = i_q^{ref}(k+1) + \Delta i_q(k+1) \tag{3.16}$$

with

$$\begin{cases} \Delta i_d(k+1) = \dfrac{\Delta\psi_{rq}}{L}T_s\omega(k) = \dfrac{(1-\lambda)\sin(\Delta\theta)}{L}\psi_{ro}T_s\omega(k) \\ \Delta i_q(k+1) = -\dfrac{\Delta\psi_{rd}}{L}T_s\omega(k) = -\dfrac{(1-\lambda)\cos(\Delta\theta)-1}{L}\psi_{ro}T_s\omega(k) \end{cases} \tag{3.17}$$

From Eq. (3.15) and Eq. (3.16), it can be found that there exists an error between the current response value and the current reference value under flux linkage parameter perturbation and rotor position error. According to Eq. (3.17), the current error is related to the speed, the flux linkage parameter perturbation rate, and the rotor position error. The PMSM is uncontrollable if the rotor position error range exceeds ± 5° [23]. Therefore, in the case of the rotor position allowable error range (i.e., ± 5°), the influence of rotor position error on PCC is analyzed in this chapter.

The motor parameters are shown in Table 3.1. The *d*- and *q*-axis current errors are plotted under flux linkage parameter perturbation and rotor position error, as shown in Figs. 3.3 and 3.4. Fig. 3.3 shows the *d*- and *q*-axis current errors under flux linkage parameter mismatch. The *q*-axis current response value is greater than its reference, but the flux linkage parameter mismatch does not affect the *d*-axis current. Fig. 3.4 shows the relationship between the current error and the speed under flux linkage parameter mismatch and rotor position error. When the rotor position error range is positive, the *d*- and *q*-axis response currents are greater than

Table 3.1 Main parameters of SPMSM.

Parameters	Value	Parameters	Value
Stator phase resistance (R)	0.02 Ω	Inductances (L)	1 mH
Number of pole pairs (n_p)	4	Flux linkage of PM (Ψ_{ro})	0.892 Wb

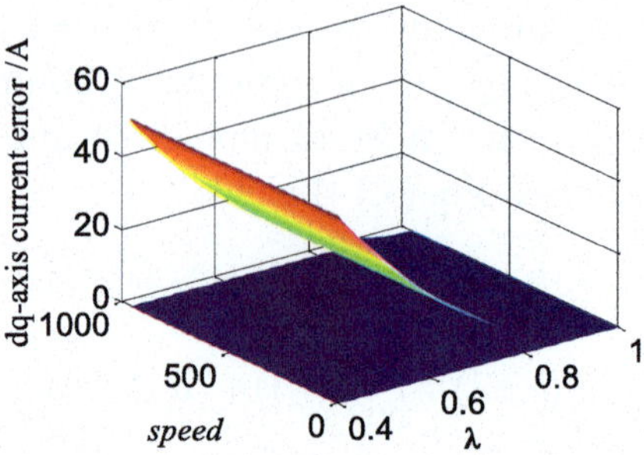

Fig. 3.3 The *d*- and *q*-axis current errors under flux linkage parameter mismatch.

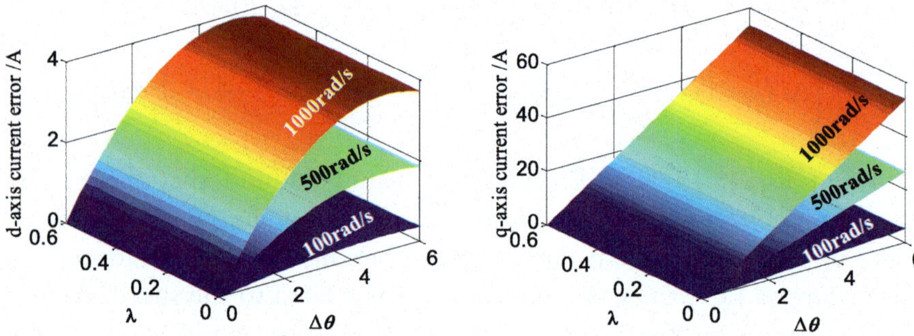

(a) d-axis current errors (b) q-axis current errors

Fig. 3.4 The *d*- and *q*-axis current errors under flux linkage parameter mismatch and rotor position error. (A) *d*-axis current errors. (B) *q*-axis current errors.

the current reference. When the rotor position error range is negative, the *q*-axis response current is greater than the current reference, and the *d*-axis response current is less than the current reference. The *q*-axis current error increases with the increase in speed under flux linkage parameter perturbation and rotor position error. Through the analysis above, it can be concluded that the *q*-axis current of conventional PCC is mostly sensitive to rotor position and flux linkage parameters, while the influence on the *d*-axis current can be neglected.

3.3 Predictive stator flux control strategy

According to Eqs. (3.9), (3.10) and (3.11), we can express the state equations of the three-phase PMSM as follows when flux linkage parameter mismatch and rotor position error occur.

$$\begin{cases} \dfrac{d\boldsymbol{x}}{dt} = \boldsymbol{Ax} + \boldsymbol{Bu} + \boldsymbol{C}\boldsymbol{f}_{\psi} + \boldsymbol{D}\boldsymbol{f}_{\boldsymbol{a}} \\ \boldsymbol{x} = \boldsymbol{E}\boldsymbol{x}_1 + \boldsymbol{f}_{\psi} \end{cases} \tag{3.18}$$

where, $\boldsymbol{x} = \begin{bmatrix} \psi_d \\ \psi_q \\ \omega \end{bmatrix}$, $\boldsymbol{x}_1 = \begin{bmatrix} i_d \\ i_q \\ \omega \end{bmatrix}$, $\boldsymbol{u} = \begin{bmatrix} u_d \\ u_q \\ \psi_q \end{bmatrix}$, $\boldsymbol{f_\psi} = \begin{bmatrix} \psi_{ro} \\ 0 \\ 0 \end{bmatrix}$, and $\boldsymbol{f_a} = \begin{bmatrix} \Delta\psi_{rd} \\ \Delta\psi_{rq} \\ T_L - \dfrac{f_\omega}{2L} \end{bmatrix}$

are state variables, system inputs, system outputs, PM flux linkage, and fault function, respectively. The coefficient matrixes of the state equations are $\boldsymbol{A} = \begin{bmatrix} -\dfrac{R}{L} & \omega & 0 \\ -\omega & -\dfrac{R}{L} & 0 \\ 0 & 0 & 0 \end{bmatrix}$, $\boldsymbol{B} = \begin{bmatrix} 1 & 0 & 0 \\ 0 & 1 & 0 \\ 0 & 0 & \dfrac{3n_p^2\psi_{ro}}{2JL} \end{bmatrix}$, $\boldsymbol{C} = \begin{bmatrix} \dfrac{R}{L} & 0 & 0 \\ 0 & 0 & 0 \\ 0 & 0 & 0 \end{bmatrix}$,

$\boldsymbol{D} = \begin{bmatrix} 0 & \omega & 0 \\ -\omega & 0 & 0 \\ 0 & 0 & -\dfrac{n_p}{J} \end{bmatrix}$, $\boldsymbol{E} = \begin{bmatrix} L & 0 & 0 \\ 0 & L & 0 \\ 0 & 0 & 1 \end{bmatrix}$, respectively.

The discrete expression of Eq. (3.18) is:

$$\begin{cases} \begin{bmatrix} u_d(k) \\ u_q(k) \end{bmatrix} = \begin{bmatrix} (\dfrac{R}{L} - \dfrac{1}{T_s}) & -\omega(k) \\ \omega(k) & (\dfrac{R}{L} - \dfrac{1}{T_s}) \end{bmatrix} \begin{bmatrix} \psi_d(k) \\ \psi_q(k) \end{bmatrix} \\ \qquad + \begin{bmatrix} \dfrac{1}{T_s}\psi_d(k+1) - \dfrac{R}{L}\psi_{ro} \\ \dfrac{1}{T_s}\psi_q(k+1) \end{bmatrix} + \begin{bmatrix} -\omega(k)\Delta\psi_{rq} \\ \omega(k)\Delta\psi_{rd} \end{bmatrix} \\ \begin{bmatrix} \psi_d(k) \\ \psi_q(k) \end{bmatrix} = \begin{bmatrix} L & 0 \\ 0 & L \end{bmatrix} \begin{bmatrix} i_d(k) \\ i_q(k) \end{bmatrix} + \begin{bmatrix} \psi_{ro} \\ 0 \end{bmatrix} \end{cases} \tag{3.19}$$

To reduce the influence of sampling delay on the proposed PSFC method, at the kT_s moment, we need to obtain the stator flux value $\psi_{dq}(k+1)$ of $(k+1)T_s$ moment, and then calculate the voltage vector $u_{dq}(k+1)$ by using the stator flux value $\psi_{dq}(k+1)$ of moment. Therefore the stator flux value $\psi_{dq}(k+1)$ of $(k+1)T_s$ moment is used to replace the $\psi_{dq}(k)$ in this chapter to compensate the sampling delay. These values for $\psi_d^{ref}(k+2)$ and $\psi_q^{ref}(k+2)$ are used as the stator flux reference in Eq. (3.19) by considering $\psi_d(k+2) = \psi_d^{ref}(k+2)$ and $\psi_q(k+2) = \psi_q^{ref}(k+2)$. The rotor speed can be considered as constant during one sampling period (i.e., $\omega(k) = \omega(k+1)$). Therefore, to eliminate the influence of flux linkage parameter mismatch, rotor position error and one step delay, the Eq. (3.19) is modified to

$$\begin{cases} u_d(k+1) = \dfrac{1}{T_s}\psi_d^{ref}(k+2) + (\dfrac{R}{L} - \dfrac{1}{T_s})\psi_d(k+1) - \omega(k)\psi_q(k+1) - \dfrac{R}{L}\psi_{ro} \\ \qquad - \omega(k)(1-\hat{\lambda})\psi_{ro}\sin(\Delta\hat{\theta}) \\ u_q(k+1) = \dfrac{1}{T_s}\psi_q^{ref}(k+2) + \omega(k)\psi_d(k+1) + (\dfrac{R}{L} - \dfrac{1}{T_s})\psi_q(k+1) \\ \qquad + \omega(k)\psi_{ro}\left[(1-\hat{\lambda})\cos(\Delta\hat{\theta}) - 1\right] \\ \psi_d(k+1) = \psi_{ro} + Li_d(k+1) \\ \psi_q(k+1) = Li_q(k+1) \end{cases} \tag{3.20}$$

Where $\hat{\lambda}$ is the observed value of the flux linkage parameter perturbation rate; $\Delta\hat{\theta}$ is the observed value of the rotor position error; $\psi_d^{ref}(k+2)$ and $\psi_q^{ref}(k+2)$ are the d- and q-axis stator flux references, respectively.

To achieve an accurate value of the d- and q-axis stator flux references (i.e., $\psi_d^{ref}(k+2)$, $\psi_q^{ref}(k+2)$), a second-order expansion for the rotor speed is proposed.

$$\omega(k+1) = \omega(k) + \frac{d\omega}{dt}\cdot T_{ds} + \frac{d^2\omega}{dt^2}\cdot\frac{T_{ds}^2}{2} \tag{3.21}$$

Where T_{ds} is the sampling time of the external speed loop, and $T_{ds} = 10T_s$.

According to Eq. (3.11), due to the load torque being invariant in the sampling time, and the change of flux linkage components being slow, therefore, $\frac{dT_L}{dt} = 0$ and $\frac{df_\omega}{dt} = 3n_p\Delta\psi_{rd}\frac{d\psi_q}{dt}$, one yields

$$\frac{d^2\omega}{dt^2} = \frac{3n_p^2\psi_{ro}}{2JL}\frac{d\psi_q}{dt} + 3n_p\Delta\psi_{rd}\frac{d\psi_q}{dt} \tag{3.22}$$

Substituting Eq. (3.11) and Eq. (3.22) into Eq. (3.21), we obtain:

$$\begin{aligned} \omega(k+1) = \omega(k) &+ \left[\frac{3n_p^2\psi_{ro}}{2JL}\psi_q(k) - \frac{n_p}{J}T_L + \frac{n_p}{2JL}f_\omega\right]\cdot T_{ds} \\ &+ \left[\frac{3n_p^2\psi_{ro}}{2JL} + 3n_p\Delta\psi_{rd}\right]\frac{\psi_q(k) - \psi_q(k-1)}{T_{ds}}\cdot\frac{T_{ds}^2}{2} \end{aligned} \tag{3.23}$$

Taking $\psi_q^{ref}(k)$ and ω^{ref} as the references of $\psi_q(k)$ and $\omega(k+1)$, respectively, the following equation can be obtained from Eq. (3.23).

$$\psi_q^{ref}(k) = \frac{\omega^{ref} - \omega(k) + \dfrac{3n_p^2\psi_{ro}T_{ds} + 6JLn_p\Delta\widehat{\psi}_{rd}T_{ds}}{4JL}\psi_q(k-1) + \dfrac{n_pT_{ds}}{J}\widehat{T}_L - \dfrac{n_pT_{ds}}{2JL}\widehat{f}_\omega}{\dfrac{9n_p^2\psi_{ro}T_{ds} + 6JLn_p\Delta\widehat{\psi}_{rd}T_{ds}}{4JL}} \tag{3.24}$$

where $\psi_q^{ref}(k)$ and ω^{ref} are the stator flux reference and speed reference, respectively; $\Delta\hat{\psi}_{rd} = \psi_{ro}\left[(1-\hat{\lambda})\cos(\Delta\hat{\theta}) - 1\right]$; $\hat{T}_L$ and $\hat{f}_\omega$ are the observed values of the load torque and unknown external disturbances, respectively.

Then, the stator flux reference is calculated using Lagrange extrapolation under $i_d^{ref}(k) = 0$, as expressed by

$$\begin{cases} \psi_d^{ref}(k+2) = \psi_d^{ref}(k+1) = \psi_d^{ref}(k) = \psi_{ro} + Li_d^{ref}(k) = \psi_{ro} \\ \psi_q^{ref}(k+2) = 3\psi_q^{ref}(k) - 3\psi_q^{ref}(k-1) + \psi_q^{ref}(k-2) \end{cases} \tag{3.25}$$

The block diagram of the PMSM drive system with the PSFC method is shown in Fig. 3.5. The proposed PSFC method is used to substitute the conventional PCC to overcome the influence of flux linkage parameter mismatch, rotor position error, and one-step delay.

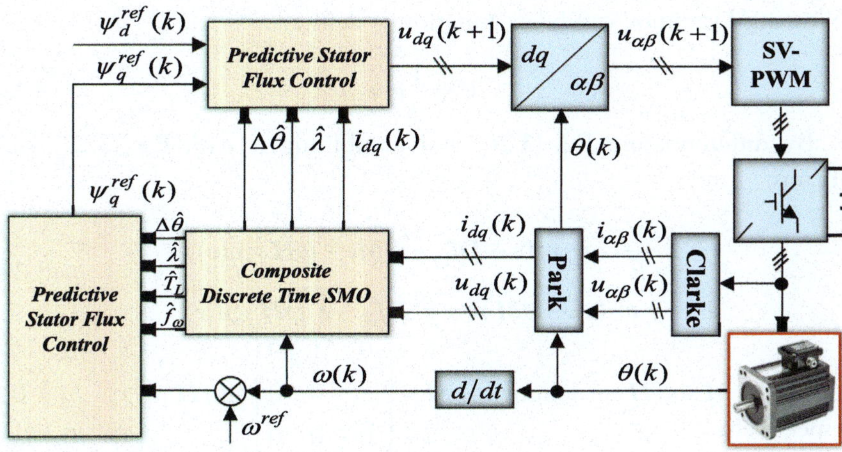

Fig. 3.5 Block diagram of the PMSM drive system with the PSFC method.

3.4 Design of composite discrete-time SMO

The key to realizing the PSFC method is to estimate the flux linkage parameter perturbation rate, rotor position error, and load torque. The composite discrete-time SMO can simultaneously estimate these unknown factors. Based on Eq. (3.18), the composite SMO is designed as:

$$\begin{cases} \frac{d\hat{\boldsymbol{x}}}{dt} = \boldsymbol{A}\hat{\boldsymbol{x}} + \boldsymbol{B}\boldsymbol{u} + \boldsymbol{C}\boldsymbol{f}_{\psi} + \omega\boldsymbol{M}\boldsymbol{e} + \omega\boldsymbol{H}\,\mathrm{sgn}(\boldsymbol{e}) \\ \hat{\boldsymbol{x}} = \boldsymbol{E}\hat{\boldsymbol{x}}_1 + \boldsymbol{f}_{\psi} \end{cases} \tag{3.26}$$

where, $\hat{\boldsymbol{x}}$ is the observed value of $\boldsymbol{x}$; sgn() is the sliding mode control function; $\boldsymbol{M}$ and $\boldsymbol{H}$ are the designed gain matrixes of composite SMO.

The time-varying sliding surface is defined as. $\boldsymbol{e} = \boldsymbol{x} - \hat{\boldsymbol{x}} = \boldsymbol{E}(\boldsymbol{x}_1 - \hat{\boldsymbol{x}}_1) = \begin{bmatrix} e_1 \\ e_2 \\ e_3 \end{bmatrix} = \begin{bmatrix} \psi_d - \hat{\psi}_d \\ \psi_q - \hat{\psi}_q \\ \omega - \hat{\omega} \end{bmatrix} = \boldsymbol{E}\begin{bmatrix} i_d - \hat{i}_d \\ i_q - \hat{i}_q \\ \omega - \hat{\omega} \end{bmatrix}$ The dynamic equation, considering the estimation error, is obtained by subtracting Eq. (3.18) from Eq. (3.26).

$$\frac{d\boldsymbol{e}}{dt} = \boldsymbol{A}\boldsymbol{e} + \boldsymbol{D}\boldsymbol{f}_{\boldsymbol{a}} - \omega\boldsymbol{M}\boldsymbol{e} - \omega\boldsymbol{H}\,\mathrm{sgn}(\boldsymbol{e}) \tag{3.27}$$

Stability analysis of the proposed composite SMO

To analyze the asymptotic stability of the proposed composite SMO, the following Lyapunov candidate function is defined as:

$$V = \frac{1}{2}\boldsymbol{e}^T\boldsymbol{e} \tag{3.28}$$

By differentiating Eq. (3.28) and combining it with Eq. (3.27), one obtains:

$$\begin{aligned} \frac{dV}{dt} &= \boldsymbol{e}^T\left(\boldsymbol{A}\boldsymbol{e} + \boldsymbol{D}\boldsymbol{f}_{\boldsymbol{a}} - \omega\boldsymbol{M}\boldsymbol{e} - \omega\boldsymbol{H}\,\mathrm{sgn}(\boldsymbol{e})\right) \\ &= \boldsymbol{e}^T\left(\boldsymbol{A} - \omega\boldsymbol{M}\right)\boldsymbol{e} + \boldsymbol{e}^T\boldsymbol{D}\boldsymbol{f}_{\boldsymbol{a}} - \boldsymbol{e}^T\omega\boldsymbol{H}\,\mathrm{sgn}(\boldsymbol{e}) \end{aligned} \tag{3.29}$$

where, $\boldsymbol{M}$ and $\boldsymbol{H}$ are designed as $\boldsymbol{M} = \begin{bmatrix} 0 & 1 & 0 \\ -1 & 0 & 0 \\ 0 & 0 & 0 \end{bmatrix}$ and $\boldsymbol{H} = \begin{bmatrix} h_1 & 0 & 0 \\ 0 & h_2 & 0 \\ 0 & 0 & h_3 \end{bmatrix}$, respectively.

Simplifying Eq. (3.29) yields

$$\begin{aligned}\frac{dV}{dt}&=-\frac{R}{L}e_1^2-\frac{R}{L}e_2^2+\boldsymbol{e}^T\boldsymbol{D}\boldsymbol{f_a}-\boldsymbol{e}^T\omega\boldsymbol{H}\operatorname{sgn}(\boldsymbol{e})\\&\le\boldsymbol{e}^T\boldsymbol{D}\boldsymbol{f_a}-\boldsymbol{e}^T\omega\boldsymbol{H}\operatorname{sgn}(\boldsymbol{e})\\&=\left(\omega\Delta\psi_{rq}e_1-\omega h_1|e_1|\right)+(-\omega\Delta\psi_{rd}e_2-\omega h_2|e_2|)\\&\quad+\left(-\frac{n_p}{J}(T_L-\frac{f_\omega}{2L})e_3-\omega h_3|e_3|\right)\end{aligned}\tag{3.30}$$

In engineering, the fault function $\boldsymbol{f_a}$ is bounded, that is, $|\Delta\psi_{rd}|\le N_1$, $|\Delta\psi_{rq}|\le N_2$, $\left|T_L-\frac{f_\omega}{2L}\right|\le N_3$, where, N_1, N_2, N_3 are the boundary values. If $\boldsymbol{H}$ satisfies $h_1\ge N_2\ge|\Delta\psi_{rq}|$, $h_2\ge N_1\ge|\Delta\psi_{rd}|$, $h_3\ge\frac{n_pN_3}{J\omega}\ge\left|\frac{n_p}{J\omega}(T_L-\frac{f_\omega}{2L})\right|$, then ${dV}/{dt}\le 0$. Therefore the stability and convergence of the composite SMO is guaranteed. When the composite SMO is stable, the observed value will converge to the actual value.

Expression of composite discrete-time SMO

Since the proposed PSFC is computed at discrete instants, it is necessary to deduce the discrete expression of the composite SMO. If the sampling period is short enough, the composite discrete-time SMO can be expressed by Eq. (3.31) according to Eq. (3.26).

$$\begin{cases}\widehat{\psi}_d(k+1)=\left(1-\frac{R}{L}T_s\right)\widehat{\psi}_d(k)+\omega(k)T_s\widehat{\psi}_q(k)+T_su_d(k)\\\qquad+\frac{R}{L}T_s\psi_{ro}+\omega(k)T_se_2(k)+h_1\omega(k)T_s\operatorname{sgn}(e_1(k))\\\widehat{\psi}_q(k+1)=\left(1-\frac{R}{L}T_s\right)\widehat{\psi}_q(k)-\omega(k)T_s\widehat{\psi}_d(k)+T_su_q(k)-\omega(k)T_se_1(k)\\\qquad+h_2\omega(k)T_s\operatorname{sgn}(e_2(k))\\\widehat{\omega}(k+1)=\widehat{\omega}(k)+\frac{3n_p^2\psi_{ro}T_s}{2JL}\psi_q(k)+h_3\omega(k)T_s\operatorname{sgn}(e_3(k))\end{cases}\tag{3.31}$$

Estimation of flux linkage parameter perturbation rate, rotor position error, equivalent disturbance, and load torque

According to Eq. (3.27), when the system reaches the sliding mode surface, that is $de/dt = \boldsymbol{e} = 0$, the discrete form of fault function is obtained as follows

$$\boldsymbol{D}\boldsymbol{f_a} = \omega(k)\boldsymbol{H}\,\mathrm{sgn}(\boldsymbol{e}(k)) \tag{3.32}$$

From Eq. (3.32), the estimated load torque and new flux-linkage components are obtained.

$$\begin{cases} \Delta\widehat{\psi}_{rd} = -h_2\,\mathrm{sgn}(e_2(k)) \\ \Delta\widehat{\psi}_{rq} = h_1\,\mathrm{sgn}(e_1(k)) \\ \widehat{T}_L = -\dfrac{J\omega(k)}{n_p}h_3\,\mathrm{sgn}(e_3(k)) + \dfrac{\widehat{f}_\omega}{2L} \end{cases} \tag{3.33}$$

From Eq. (3.12), the equivalent disturbance is expressed as:

$$\hat{f}_\omega = 3n_p\left[\psi_{ro}\Delta\hat{\psi}_{rq} + \Delta\hat{\psi}_{rd}\psi_q(k) + \Delta\hat{\psi}_{rd}\Delta\hat{\psi}_{rq}\right] \tag{3.34}$$

To eliminate the chattering of the sign function, the sign function is replaced with a hyperbolic tangent function with smooth continuity. The hyperbolic tangent function is defined as:

$$\mathrm{sgn}(v) = \tanh(v) = \frac{\mathrm{e}^{\rho v} - 1}{\mathrm{e}^{\rho v} + 1} \tag{3.35}$$

where ρ is a coefficient of regulation, and must be greater than zero.

According to Eq. (3.33) and Fig. 3.1, we can deduce the following equations for the rotor position error and the PM flux linkage parameter.

$$\Delta\hat{\theta} = \tan^{-1}\left(\frac{\Delta\hat{\psi}_{rq}}{\psi_{ro} - |\Delta\hat{\psi}_{rd}|}\right) \tag{3.36}$$

$$\hat{\psi}_r = \frac{\Delta\hat{\psi}_{rq}}{\sin(\Delta\hat{\theta})} \tag{3.37}$$

According to Eq. (3.6), the flux linkage parameter perturbation rate is expressed by

$$\hat{\lambda} = \frac{\psi_{ro} - \hat{\psi}_r}{\psi_{ro}} \tag{3.38}$$

3.5 Simulation results

To verify the effectiveness of the predictive stator flux control (PSFC) algorithm, the PMSM drive system was built in Simulink Mat-lab. The parameters of PMSM used in the simulation are shown in Table 3.1. The proposed composite discrete-time SMO parameters are $h_1 = 2.5$, $h_2 = 1.5$, $h_3 = 50$ and $\rho = 2$. At 0 s, the speed reference steps from 0 to 1000 rad/s. And at 0.5 and 1.1 s, the load torque is increased suddenly from no-load to rated load and decreased from rated load to no-load, respectively.

Simulation results for PMSM under flux linkage parameter mismatch

In this simulation, the control performance comparison between the conventional PCC method and the proposed PSFC method under flux linkage parameter mismatch is investigated. The simulation results of the conventional PCC method and the proposed PSFC method under flux linkage parameter mismatch are shown in Figs. 3.6 and 3.7, respectively.

The conventional PCC method demonstrates the same superior speed control performance as the proposed PSFC method under flux linkage parameter mismatch, as clearly shown in Fig. 3.6A. From Fig. 3.6B, it can be observed that the q-axis current response exceeds the current reference during the transient process under flux linkage parameter mismatch. In addition, an apparent static error between the q-axis current response and the current reference is evident under flux linkage parameter mismatch. Based on the simulation results, it is clear that flux linkage parameter

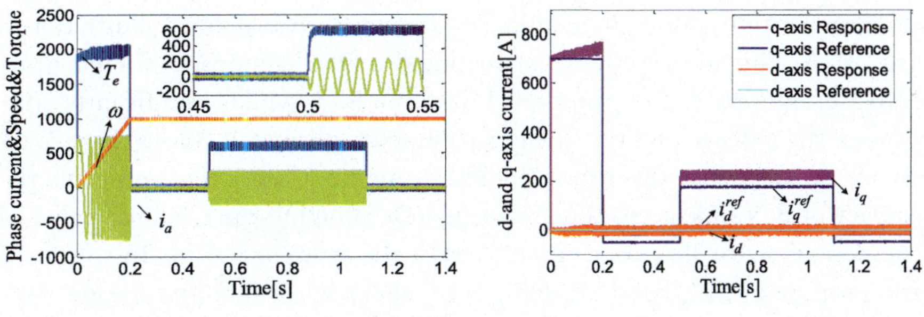

(a) The phase current, speed and torque

(b) The d-and q-axis current

Fig. 3.6 Simulation results of the conventional PCC method under flux linkage parameter mismatch. (A) The phase current, speed, and torque. (B) The d- and q-axis currents.

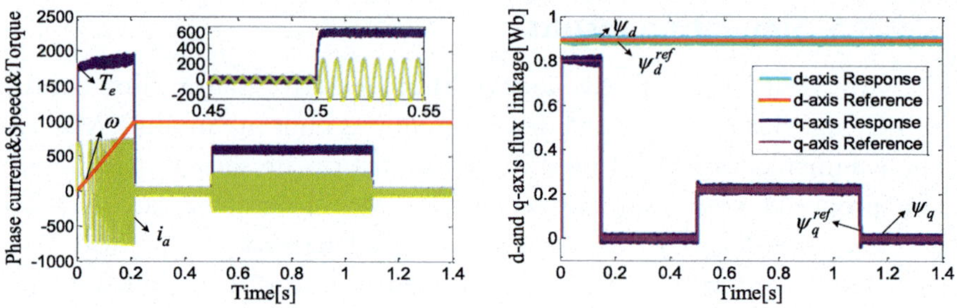

(a) The phase current, speed and torque (b) The d-and q-axis stator flux linkage

Fig. 3.7 Simulation results of the proposed PSFC method under flux linkage parameter mismatch. (A) The phase current, speed, and torque. (B) The *d*- and *q*-axis stator flux linkage.

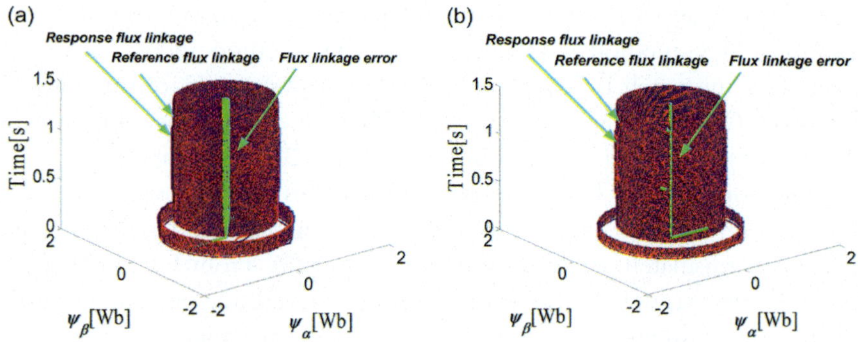

Fig. 3.8 The three-dimensional rotor flux trajectories under flux linkage parameter mismatch: (A) Conventional PCC method, (B) Proposed PSFC method.

mismatch affects the *q*-axis current response in the conventional PCC method. In comparison, the proposed PSFC method achieves a lower level of torque ripple under flux linkage parameter mismatch, as illustrated in Fig. 3.7A. This improvement is attributed to the composite discrete-time SMO employed in the proposed PSFC method, which significantly suppresses the effects of flux linkage parameter mismatch. From Fig. 3.7B, it can be seen that the proposed PSFC method effectively mitigates the impact of flux linkage parameter mismatch, allowing the response value of the *dq*-axis stator flux to accurately track the reference value. In summary, the proposed PSFC method delivers excellent stator flux and torque control performance, as the designed composite discrete-time SMO compensates for flux linkage parameter mismatch.

Fig. 3.8 depicts the three-dimensional rotor flux trajectories of the conventional PCC method and the proposed PSFC method under flux linkage parameter mismatch. The stator flux linkage errors of the conventional PCC method and the proposed PSFC method are 3.4% and 1.5%, respectively, under flux linkage parameter mismatch. The static error of the q-axis current in the conventional PCC method, shown in Fig. 3.6B, is caused by the large flux linkage error in Fig. 3.8A. In contrast, in the proposed PSFC method, the flux linkage error is drastically reduced because it can track the stator flux reference accurately, as shown in Figs. 3.7B and 3.8B. Note that there is a large overshoot in the three-dimensional rotor flux trajectories at 0.5 s and 1.1 s; this is because the mismatching moment of the flux linkage parameter affects the observation accuracy of the designed composite discrete-time SMO. The simulation results of the reference and estimated values of the conventional PCC method and the proposed PSFC method are illustrated in Fig. 3.9. Fig. 3.9A shows the unknown disturbance observation value of the proposed PSFC method using the designed composite discrete-time SMO. Fig. 3.9B shows the load torque observation value of the proposed PSFC method using the designed composite discrete-time SMO. According to Eq. (3.27), the error equation of the composite discrete-time SMO is based on the d- and q-axis currents. During the motor start-up process, the d- and q-axis currents are not stable, and the observed values of the d- and q-axis currents deviate greatly from the actual values. Therefore there is a large estimation error in the initial load torque observation. The results reflect that the proposed composite discrete-time SMO can accurately estimate the unknown disturbance and load torque when the motor reaches a steady state.

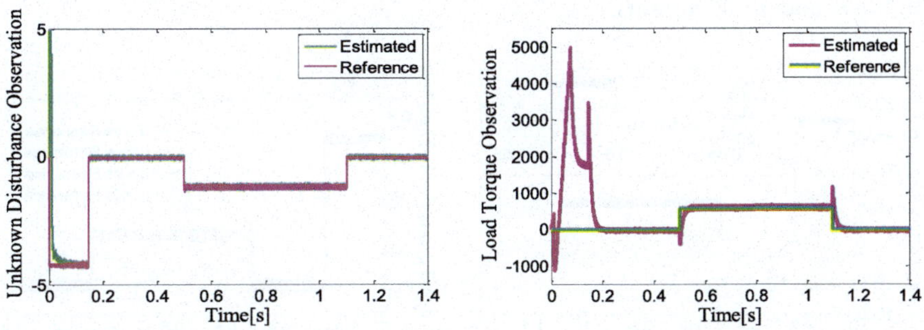

(a) Unknown disturbance observation value (b) Load torque observation value

Fig. 3.9 Simulation results of the reference and estimated values under flux linkage parameter mismatch: (A) Unknown disturbance observation value, (B) Load torque observation value.

Simulation for PMSM under flux linkage parameter mismatch and rotor position errors

Figs. 3.10 and 3.11 show the simulated results of the conventional PCC method and the proposed PSFC method under flux linkage parameter mismatch and rotor position error. A steady-state error exists in the *q*-axis current when using the conventional PCC method under flux linkage parameter mismatch and rotor position error, as shown in Fig. 3.10B. The reason for the steady-state error is that the conventional PCC method relies on the precise mathematical model of the PMSM. The proposed PSFC method demonstrates the same superior speed control performance as the conventional PCC method under flux linkage parameter mismatch and rotor position error, as clearly shown in Fig. 3.10A. From Fig. 3.10, it can be observed that the current and torque control performance of the

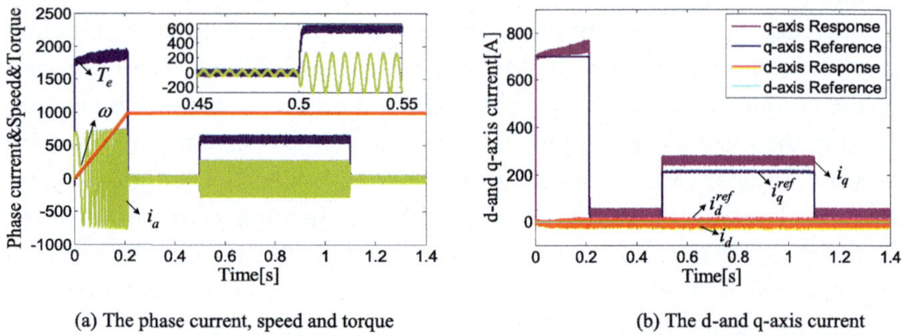

Fig. 3.10 Simulation results of the conventional PCC method under flux linkage parameter mismatch and rotor position error. (A) The phase current, speed, and torque. (B) The *d*- and *q*-axis currents.

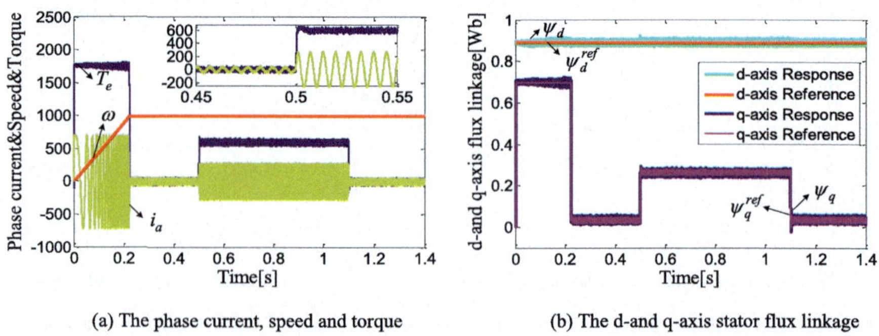

Fig. 3.11 Simulation results of the proposed PSFC method under flux linkage parameter mismatch and rotor position error. (A) The phase current, speed, and torque. (B) The *d*- and *q*-axis stator flux linkage.

conventional PCC method is severely affected by flux linkage parameter mismatch and rotor position error. In comparison, the proposed PSFC method achieves excellent torque control performance in the presence of flux linkage parameter mismatch and rotor position error, as shown in Fig. 3.11A. From Fig. 3.11B, it can be seen that the proposed PSFC method accurately tracks stator flux linkage references under flux linkage parameter mismatch and rotor position error. In addition, it is evident that the composite discrete-time SMO designed can compensate for flux linkage parameter mismatch and rotor position error, thereby effectively suppressing torque ripple and stator flux errors.

Fig. 3.12 depicts the three-dimensional rotor flux trajectories of the conventional PCC method and the proposed PSFC method under flux linkage parameter mismatch and rotor position error. The stator flux linkage errors of the conventional PCC method and the proposed PSFC method are 3.2% and 1.8%, respectively, under flux linkage parameter mismatch and rotor position error, as shown in Fig. 3.12. It can be observed that the proposed PSFC method exhibits satisfying steady-state performance (smaller α-β stator flux linkage error) compared to the conventional PCC method. The simulation results of the reference and estimated values for the conventional PCC method and the proposed PSFC method are illustrated under flux linkage parameter mismatch and rotor position error, as shown in Fig. 3.13. Fig. 3.13A demonstrates that the designed composite discrete-time SMO can accurately observe unknown disturbances in the proposed PSFC method. Fig. 3.13B shows that the load torque observation value of the proposed PSFC method can be accurately obtained using the designed composite discrete-time SMO. From Fig. 3.13, it can be seen that

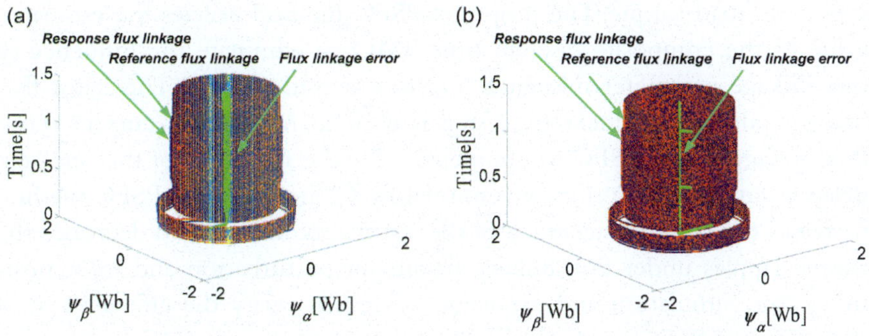

Fig. 3.12 The three-dimensional rotor flux trajectories under flux linkage parameter mismatch and rotor position error. (A) Conventional PCC method. (B) Proposed PSFC method.

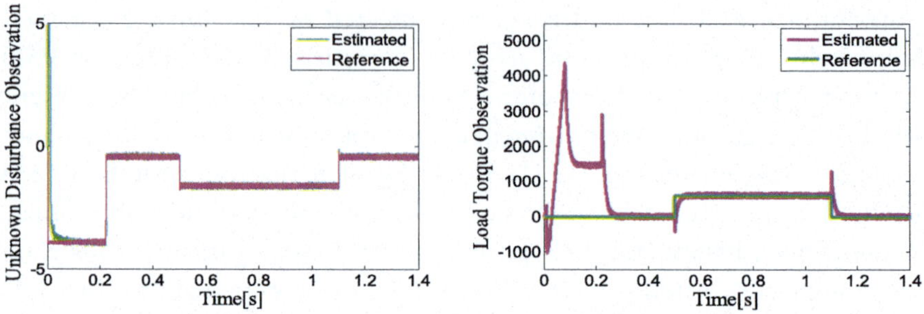

(a) Unknown disturbance observation value (d) Load torque observation value

Fig. 3.13 Simulation results of the reference and estimated values under flux linkage parameter mismatch and rotor position error: (A) Unknown disturbance observation value, (B) Load torque observation value.

the composite discrete-time SMO, based on stator flux state, can simultaneously estimate the unknown disturbance and load torque in real time. In summary, the flux linkage static error and torque ripple of the proposed PSFC method can be effectively suppressed by using the designed composite discrete-time SMO in the case of flux linkage parameter mismatch and rotor position error.

3.6 Conclusion

In this chapter, a novel PSFC method has been proposed for the PMSM drive system, which can effectively enhance robustness against flux linkage parameter mismatch and rotor position error. Furthermore, a composite discrete-time SMO based on the stator flux state is designed to estimate the flux linkage parameter perturbation rate, rotor position error, and load torque. The proposed PSFC method utilizes the estimated value of the composite discrete-time SMO to eliminate the influence of flux linkage parameter mismatch and rotor position error. The stator flux linkage value of the next time step is used to replace the current stator flux linkage in the PSFC to compensate for the influence of the one-step delay. Compared with the conventional PCC, the proposed PSFC method increases the tracking accuracy of the stator flux linkage and decreases the torque ripples under flux linkage parameter perturbation and rotor position error. Simulation and experimental results verify the effectiveness of the proposed PSFC method, which demonstrates its superiority in control precision and disturbance rejection under flux linkage parameter mismatch and rotor position error, while achieving the merits of low torque ripple and excellent steady-state performance.

References

[1] X. Liu, H. Chen, et al., Research on the performances and parameters of interior PMSM used for electric vehicles, IEEE Trans. Ind. Electron. 63 (06) (2016) 3533–3545.

[2] S.K. Kommuri, M. Defoort, et al., A robust observer-based sensor fault-tolerant control for PMSM in electric vehicles, IEEE Trans. Ind. Electron. 63 (12) (2016) 7671–7681.

[3] G.S. Buja, M.P. Kazmierkowski, et al., Direct torque control of PWM inverter-fed ac motors—A survey, IEEE Trans. Ind. Electron. 51 (04) (2004) 744–757.

[4] T. Ohtani, N. Takada, et al., Vector control of induction motor without shaft encoder, IEEE Trans. Ind. Appl. 28 (01) (1992) 157–165.

[5] T. Türker, U. Buyukkeles, et al., A robust predictive current controller for PMSM drives, IEEE Trans. Ind. Electron. 63 (06) (2016) 3906–3914.

[6] C. Shan, L. Wang, et al., A cascade MPC control structure for a PMSM with speed ripple minimization, IEEE Trans. Ind. Electron. 60 (06) (2013) 2978–2987.

[7] A.M. Aljehaimi, P. Pillay, et al., Novel flux linkage estimation algorithm for a variable flux PMSM, IEEE Trans. Ind. Appl. 54 (03) (2018) 2319–2335.

[8] H.J. Ahn, D.M. Lee, et al., A new bump less rotor-flux position estimation scheme for vector-controlled washing machine, IEEE Trans. Ind. Electron. 12 (02) (2016) 466–473.

[9] Y. Xu, N. Parspour, et al., Torque ripple minimization using online estimation of the stator resistances with consideration of magnetic saturation, IEEE Trans. Ind. Electron. 61 (09) (2014) 5105–5114.

[10] A. Gebregergis, M.H. Chowdhury, et al., Modeling of permanent-magnet synchronous machine including torque ripple effects, IEEE Trans. Ind. Appl. 51 (01) (2015) 232–239.

[11] S. Chen, C. Namuduri, et al., Controller-induced parasitic torque ripples in a PM synchronous motor, IEEE Trans. Ind. Appl. 38 (05) (2002) 1273–1281.

[12] X. Zhang, B. Hou, et al., Deadbeat Predictive Current control of permanent-magnet synchronous motors with stator current and disturbance observer, IEEE Trans. Power Electron. 32 (05) (2017) 3818–3834.

[13] J. Rodas, F. Barrero, et al., Online estimation of rotor variables in predictive current controllers: a case study using five-phase induction machines, IEEE Trans. Ind. Electron. 63 (09) (2016) 5348–5356.

[14] J. Yang, W.X. Zheng, et al., Design of a prediction accuracy-enhanced continuous-time MPC for disturbed systems via a disturbance observer, IEEE Trans. Ind. Electron. 62 (09) (2016) 5807–5816.

[15] J. Yang, J. Su, et al., High-order mismatched disturbance compensation for motion control systems via a continuous dynamic sliding mode approach, IEEE Trans. Ind. Electron. 10 (01) (2014) 604–614.

[16] C. Zhang, G. Wu, et al., Robust fault-tolerant predictive current control for permanent magnet synchronous motors considering demagnetization fault, IEEE Trans. Ind. Electron. 65 (07) (2017) 5324–5534.

[17] C.K. Lin, T.H. Liu, et al., Model-free predictive current control for interior permanent-magnet synchronous motor drives based on current difference detection technique, IEEE Trans. Ind. Electron. 61 (02) (2014) 667–681.

[18] M. Yang, X. Lang, et al., A flux immunity robust predictive current control with incremental model and extended state observer for PMSM drive, IEEE Trans. Power Electron. 32 (12) (2017) 9267–9279.

[19] X. Zhang, L. Zhang, et al., Model predictive current control for PMSM drives with parameter robustness improvement, IEEE Trans. Power Electron. 63 (12) (2018) 1–6.

[20] H. Zhu, X. Xiao, et al., Torque ripple reduction of the torque predictive control scheme for permanent-magnet synchronous motors, IEEE Trans. Ind. Electron. 59 (02) (2012) 871–877.

[21] Y. Cho, K.B. Lee, et al., Torque-ripple minimization and fast dynamic scheme for torque predictive control of permanent magnet synchronous motors, IEEE Trans. Power Electron. 30 (04) (2012) 2182–2190.
[22] P. Landsmann, R. Kennel, Saliency based sensorless predictive torque control with reduced torque ripple, IEEE Trans. Power Electron. 27 (10) (2012) 4311–4320.
[23] J. Lara, J. Xu, A. Chandra, Effects of rotor position error in the performance of field oriented controlled PMSM drives for electric vehicle traction applications, IEEE Trans. Ind. Electron. 63 (08) (2016) 4738–4751.

CHAPTER 4

Robust fault-tolerant predictive current control for three-phase PMSM considering demagnetization fault

4.1 Introduction

Permanent magnet synchronous motors (PMSMs) have been widely adopted in industrial applications due to their definite advantages of high power density and high efficiency. PMSM is a nonlinear and strongly coupled multivariable system, and field orientation control is commonly used to obtain the direct relationship between current and electromagnetic torque [1,2]. However, fast and accurate torque control and low torque ripple of electrical drives are important indexes for measuring the performance of servo systems [3]. The dynamic characteristics of a current control loop are directly related to electromagnetic torque, which has been considered the core factor that determines the dynamic quality of servo systems.

The high-performance control of PMSM drives requires the rapid response of current control loops; thus many advanced current control strategies, such as predictive current control (PCC) [4,5], hysteresis control [6,7], and fuzzy PI control [8,9], have been proposed. By comparison, the PCC method has been extensively investigated because of its ability to track the current reference rapidly and accurately, and it takes into account both stability and small current harmonic components [10,11].

PCC is a control algorithm based on the discrete mathematical model of PMSM, which theoretically improves the bandwidth of current control loops and the dynamic performance of motors. However, the inductance, flux linkage, and other parameters of motor models should be accurately utilized when PCC is applied, and the current operating state of motors should be accurately determined [12]. In practice, PMSM is a time-varying

Robust Model Predictive Control of Permanent Magnet Synchronous Motor Under Parameter Mismatch
https://doi.org/10.1016/B978-0-443-44209-4.10004-1

control object. Thus the parameters of the motor will change due to the deterioration of materials or operating conditions. For these reasons, parameter perturbation of resistance and inductance, as well as demagnetization of the permanent magnet, will occur [13,14]. Parameter perturbation and demagnetization consequently result in model parameter mismatch, which affects the performance of control systems and leads to the divergence of these systems. In addition, the current sampling delay of a digital control system greatly reduces the performance of PCC [15,16]. These problems seriously limit the industrial applications of PCC.

Some methods have been presented to enhance control performance and compensate for the effects of one-step delay. In [17–19], the system delay caused by modulation and code calculation is considered in system modeling, which can thus effectively reduce the influence of delay on the performance of a control system. In [20], a robust PCC algorithm has been proposed to address prediction performance deterioration caused by inductance parameter mismatch, but the proposed algorithm disregards the effects of permanent magnet demagnetization and resistance mismatch. In [21], a model-free PCC method for interior PMSM drives is proposed, which only utilizes the stator currents and the current differences. This method does not require any motor parameters. However, it is difficult to apply in engineering applications because the method requires a higher sampling frequency of current. In [22], the nonlinear model of PMSM has been established on the basis of parameter perturbation, and a robust PCC has been designed to eliminate the effect of model parameter mismatch on PCC performance. In [23–25], the model PCC algorithm based on a disturbance observer is adopted to control the current loop of PMSM, which can achieve high dynamic performance and strong robustness by compensating for the influence of parameter perturbation. However, the effect of permanent magnet demagnetization on PCC is not considered in motor operations [22–25].

The demagnetization phenomenon is due to the armature reaction, especially under operating conditions requiring strong torque, e.g., high loads [26]. This phenomenon is also caused by high-temperature operations. In a PMSM, a 20% flux reduction for a ferrite-based magnet occurs per 100°C increase in ambient temperature [27]. However, heavy-haul locomotives often run under high loads and poor environments; in such cases, demagnetization occurs. Demagnetization means that the strength of the rotor flux inside the PMSM decreases. As a result of rotor flux reductions, the rotor flux of the controller fails to match the real flux, thereby leading to static errors or oscillatory currents.

In order to overcome the effect of permanent magnet demagnetization on the PPC performance, many methods have been investigated. In [28], a Kalman filter and Luenberger observer were used with PPC to reconstruct the permanent magnet demagnetization of a rotor, and satisfactory results were obtained under permanent magnet demagnetization. In [29], a flux immunity robust predictive current control algorithm for PMSM drives was proposed; this algorithm can operate without knowledge of the rotor flux. In [30], a novel algorithm was proposed to eliminate the *d-q* axis static current error when the rotor flux used in the controller presents a parameter mismatch with the real flux. However, these methods do not consider the coupling between perturbed inductance and demagnetization resulting from the use of conventional PPC.

In this chapter, an RFT-PCC scheme is developed to guarantee the performance of the system regardless of parameter perturbation, permanent magnet demagnetization, and one-step delay. A composite observer, based on a sliding mode observer and Luenberger observer, is designed to predict the stator current for the next time step and compensate for the output voltage of the conventional PCC. The proposed RFT-PCC method can reduce the steady-state errors of the stator current and enhance the robustness of current control loops effectively under parameter perturbation and permanent magnet demagnetization. This chapter is organized as follows: The PMSM mathematical model of parameter perturbation and permanent magnet demagnetization is introduced in Section 4.2. The influence of model parameter mismatch on the current vector is analyzed in Section 4.3. The RFT-PCC algorithm, based on the composite observer, is proposed in Section 4.4. The effectiveness of the proposed RFT-PCC algorithm is verified through simulations and experiments in Section 4.5. The conclusion is presented in Section 4.6.

4.2 Mathematical model of nonlinear PMSM under demagnetization fault

The voltage equations of the healthy PMSM in the synchronous rotating frame can be expressed as follows [31].

$$\begin{cases} u_d = Ri_d + \dfrac{d\psi_d}{dt} - \omega\psi_q \\ u_q = Ri_q + \dfrac{d\psi_q}{dt} + \omega\psi_d \end{cases} \tag{4.1}$$

where u_d and u_q, respectively, represent the d- and q-axis stator voltages; i_d and i_q are the d- and q-axis currents, respectively; ψ_d and ψ_q are the d- and q-axis flux linkages, respectively; R is the stator resistance, and ω is the electrical rotor speed.

In the fault operation of the PMSM, resistance and inductance perturbation and permanent magnet demagnetization occur because of the influence of temperature and other external factors. Therefore the control system performance based on Eq. (4.1) is affected. The flux linkage amplitude varies from initial ψ_{ro} to ψ_r when demagnetization fault occurs, and a deviation angle γ between the directions of the rotor flux and the d-axis of the synchronous rotating frame exists [12,32]. The variation of PMSM flux linkage is illustrated in Fig. 4.1.

The flux linkage equation is expressed as follows when parameter perturbation and permanent magnet demagnetization are considered as

$$\begin{cases} \psi_d = \psi_{ro} + \Delta\psi_{rd} + (L_{do} + \Delta L_d)i_d \\ \psi_q = \Delta\psi_{rq} + (L_{qo} + \Delta L_q)i_q \end{cases} \tag{4.2}$$

where L_d and L_q are d- and q-axis stator inductances, respectively; ψ_{ro} is the flux linkage of permanent magnets. Introducing $\Delta L_d = L_d - L_{do}$, $\Delta L_q = L_q - L_{qo}$, and $\Delta R = R - R_o$, where R_o, L_{do} and L_{qo} are nominal values and ΔL_d, ΔL_q, and ΔR are perturbation values of the corresponding model parameter, and defining $\Delta\psi_{rd}$ and $\Delta\psi_{rq}$ as the new flux linkage components of d- and q-axis when demagnetization fault occurs, the new flux linkage components can be expressed as

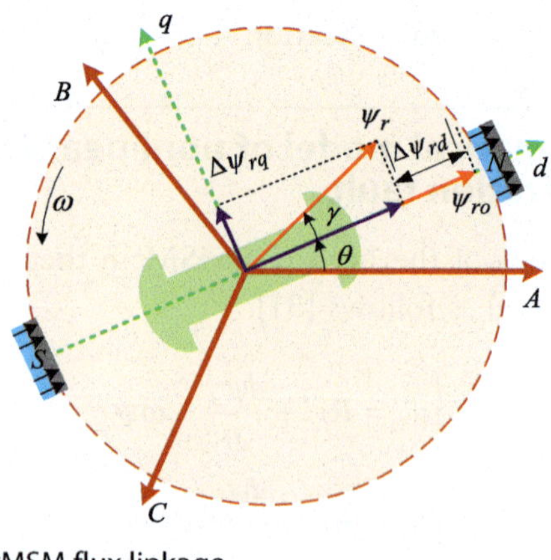

Fig. 4.1 Variation of PMSM flux linkage.

$$\begin{cases} \Delta\psi_{rd} = \psi_r \cos(\gamma) - \psi_{ro} \\ \Delta\psi_{rq} = \psi_r \sin(\gamma) \end{cases} \tag{4.3}$$

In an actual engineering system, the time constant of the electrical system is much less than the mechanical system time constant. Therefore the new flux linkage components can be considered as follows: $\frac{d\Delta\psi_{rd}}{dt} \approx 0$ and $\frac{d\Delta\psi_{rq}}{dt} \approx 0$.

Substituting Eq. (4.2) into Eq. (4.1), we can express the state equations of the PMSM as follows when parameter perturbation and permanent magnet demagnetization occur, as shown in Eqs. (4.4) and (4.5).

$$\frac{d\boldsymbol{x}}{dt} = \boldsymbol{A}\boldsymbol{x} + \boldsymbol{B}\boldsymbol{u} + \boldsymbol{D}\boldsymbol{f_a} - \boldsymbol{B}\boldsymbol{f} \tag{4.4}$$

$$\boldsymbol{y} = \boldsymbol{C}\boldsymbol{x} \tag{4.5}$$

where $\boldsymbol{x}, \boldsymbol{u}, \boldsymbol{y}, \boldsymbol{f_a}$ and $\boldsymbol{f}$ are state variables, system inputs, system outputs, permanent magnet flux linkage, and fault function, which are defined as, $\boldsymbol{x} = \begin{bmatrix} i_d \\ i_q \end{bmatrix}$, $\boldsymbol{u} = \begin{bmatrix} u_d \\ u_q \end{bmatrix}$, $\boldsymbol{y} = \begin{bmatrix} i_d \\ i_q \end{bmatrix}$, $\boldsymbol{f_a} = \begin{bmatrix} \psi_{ro} \\ 0 \end{bmatrix}$, $\boldsymbol{f} = \begin{bmatrix} f_d \\ f_q \end{bmatrix}$.

where f_d and f_q are the *d*- and *q*-axis fault functions.

$$\begin{aligned} f_d &= \frac{\Delta L_d}{(L_{do} + \Delta L_d)}(u_d - R_o i_d - \Delta R i_d + L_{qo}\omega i_q + \Delta L_q \omega i_q + \omega\Delta\psi_{rq}) \\ &\quad + (\Delta R i_d - \Delta L_q \omega i_q - \omega\Delta\psi_{rq}) \\ f_q &= \frac{\Delta L_q}{(L_{qo} + \Delta L_q)}(u_q - R_o i_q - \Delta R i_q - L_{do}\omega i_d - \Delta L_d \omega i_d - \omega\psi_{ro} - \omega\Delta\psi_{rd}) \\ &\quad + (\Delta R i_q + \Delta L_d \omega i_d + \omega\Delta\psi_{rd}) \end{aligned} \tag{4.6}$$

The coefficient matrices of the state equations are

$$\boldsymbol{A} = \begin{bmatrix} -\frac{R_o}{L_{do}} & \omega\frac{L_{qo}}{L_{do}} \\ -\omega\frac{L_{do}}{L_{qo}} & -\frac{R_o}{L_{qo}} \end{bmatrix}, \boldsymbol{B} = \begin{bmatrix} \frac{1}{L_{do}} & 0 \\ 0 & \frac{1}{L_{qo}} \end{bmatrix}, \boldsymbol{C} = \begin{bmatrix} 1 & 0 \\ 0 & 1 \end{bmatrix}, \boldsymbol{D} = \begin{bmatrix} 0 & \frac{\omega}{L_{do}} \\ -\frac{\omega}{L_{qo}} & 0 \end{bmatrix}.$$

4.3 Parameter sensitivity analysis of conventional PCC

The PCC is a discrete model control algorithm based on the precise model of a motor. In order to evaluate the sensitivity of conventional PCC to a parameter, the influence of model parameter mismatch is analyzed in this section.

Fig. 4.2 demonstrates the PWM duty update and the time sequence of current sampling. The PMSM under goes normal operations at the kT_s moment, and the model parameter mismatch occurs at the $(k+1)T_s$ moment. PCC methods employed in the practical application belong to one-beat delay control because of current sampling delay. At the kT_s moment, the PCC calculates the voltage vector $\boldsymbol{u}^*(k)$ according to the normal operation state of the PMSM. At the $(k+1)T_s$ moment, the voltage vector $\boldsymbol{u}^*(k)$ of the kT_s moment is applied to the PMSM mathematical model of parameter perturbation and permanent magnet demagnetization.

If the PMSM is controlled by conventional PCC method, through transformation of the discrete predictive model, the output voltage vector of the predictive controller at the kT_s moment can be expressed as [24]

$$\boldsymbol{u}^*(k) = \boldsymbol{F}_o^{-1}[\boldsymbol{i}^*(k+1) - \boldsymbol{E}_o(k)\boldsymbol{i}(k) - \boldsymbol{P}_o(k)] \tag{4.7}$$

where $\boldsymbol{i}^*(k+1)$ and $\boldsymbol{i}(k)$ are the current references and sampled current, respectively, $\boldsymbol{i}(k) = \begin{bmatrix} i_d(k) & i_q(k) \end{bmatrix}^{\mathrm{T}}$,

$$\boldsymbol{u}(k) = \begin{bmatrix} u_d(k) & u_q(k) \end{bmatrix}^{\mathrm{T}},\ \boldsymbol{E}_o(\boldsymbol{k}) = \begin{bmatrix} 1 - \dfrac{R_o}{L_{do}}T_s & \dfrac{L_{qo}}{L_{do}}T_s\omega(k) \\ -\dfrac{L_{do}}{L_{qo}}T_s\omega(k) & 1 - \dfrac{R_o}{L_{qo}}T_s \end{bmatrix},$$

$$\boldsymbol{F}_o = \begin{bmatrix} \dfrac{T_s}{L_{do}} & 0 \\ 0 & \dfrac{T_s}{L_{qo}} \end{bmatrix},\ \boldsymbol{P}_o(\boldsymbol{k}) = \begin{bmatrix} 0 \\ -\dfrac{\psi_{ro}}{L_{qo}}T_s\omega(k) \end{bmatrix}.$$

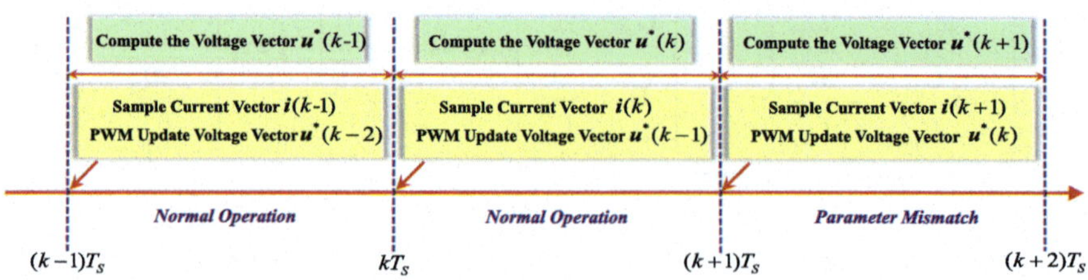

Fig. 4.2 Time sequence of duty cycle update.

From Eq. (4.7), it can be observed that the PCC models include three motor parameters (stator inductance, stator resistance, and flux linkage of permanent magnets), which means that the PCC is sensitive to the parameters of the motor model.

At the $(k+1)T_s$ moment, the discrete current predictive model of the PMSM under parameter perturbation and permanent magnet demagnetization can be expressed as follows according to Eq. (4.4).

$$\boldsymbol{i}(k+1) = \boldsymbol{E}_o(k)\cdot\boldsymbol{i}(k) + \boldsymbol{F}_o\cdot\boldsymbol{u}(k) + \boldsymbol{P}_o(k) - \boldsymbol{F}_o\boldsymbol{f}(k) \tag{4.8}$$

where the fault function is $\boldsymbol{f}(k) = \begin{bmatrix} f_d(k) \\ f_q(k) \end{bmatrix}$.

At the $(k+1)T_s$ moment, the voltage vector $\boldsymbol{u}^*(k)$ of the kT_s moment is applied to the PMSM because of one-step delay. Therefore substituting Eq. (4.7) into Eq. (4.8) yields

$$\boldsymbol{i}(k+1) = \boldsymbol{i}^*(k+1) - \boldsymbol{F}_o\boldsymbol{f}(k) \tag{4.9}$$

From Eq. (4.9), the current response vector $\boldsymbol{i}(k+1)$ can keep a good tracking of the current references vector $\boldsymbol{i}^*(k+1)$ when the PMSM is in normal operation (i.e., $\boldsymbol{f}(k)=0$). However, a current error exists between the current response vector $\boldsymbol{i}(k+1)$ and the current references vector $\boldsymbol{i}^*(k+1)$ under the parameters perturbation and permanent magnet demagnetization (i.e., $\boldsymbol{f}(k)\neq 0$). The current error can be expressed as

$$\Delta\boldsymbol{i}(k+1) = \boldsymbol{i}(k+1) - \boldsymbol{i}^*(k+1) = -\boldsymbol{F}_o\boldsymbol{f}(k) \tag{4.10}$$

With

$$-\boldsymbol{F}_o\boldsymbol{f}(k) = \begin{bmatrix} \Delta i_d \\ \Delta i_q \end{bmatrix} = -\begin{bmatrix} \dfrac{T_s}{L_{do}} f_d(k) \\ \dfrac{T_s}{L_{qo}} f_q(k) \end{bmatrix} \tag{4.11}$$

Where Δi_d and Δi_q are the d- and q-axis current error, respectively.

According to Eq. (4.1), the stationary equations can be given as follows because the PMSM is in normal operation at the kT_s moment [33]

$$\begin{cases} u_d(k) = R_o i_d(k) - L_{qo}\omega i_q(k) \\ u_q(k) = R_o i_q(k) + L_{do}\omega i_d(k) + \omega\psi_{ro} \end{cases} \tag{4.12}$$

The electromagnetic torque produced by the machine is

$$T_e(k) = \frac{3}{2} n_p \psi_{ro} i_q(k) \tag{4.13}$$

where n_p is the number of pole pairs.

The $i_d = 0$ control strategy is mainly studied in this chapter. Thus, compared with q-axis current, the function of d-axis current can be ignored basically. Combining with the Eq. (4.3), Eq. (4.6), Eq. (4.11), and Eq. (4.12), the d- and q-axis current errors can be simplified as

$$\begin{cases} \Delta i_d = \dfrac{\Delta L_q}{(L_{do} + \Delta L_d)} T_s \omega i_q(k) + \dfrac{\psi_r \sin(\gamma)}{(L_{do} + \Delta L_d)} T_s \omega \\ \Delta i_q = -\dfrac{\Delta R}{(L_{qo} + \Delta L_q)} T_s i_q(k) - \dfrac{\psi_r \cos(\gamma) - \psi_{ro}}{(L_{qo} + \Delta L_q)} T_s \omega \end{cases} \tag{4.14}$$

The motor parameters are shown in Table 4.1. When parameter perturbation and permanent magnet demagnetization occur, according to Eqs. (4.13) and (4.14), the d- and q-axis current errors can be plotted, as shown in Figs. 4.3–4.6. Fig. 4.3 shows the d- and q-axis current errors under resistance parameter mismatch. The resistance parameter mismatch does not affect the d-axis current, and the influence on the q-axis current can

Table 4.1 Main parameters of PMSM.

Parameters	Value	Parameters	Value
Stator phase resistance (R_o)	0.02 Ω	d-axis inductances (L_{do})	0.001 H
Flux linkage (Ψ_{ro})	0.892 Wb	q-axis inductances (L_{qo})	0.003572 H
Number of pole pairs (n_p)	4	Rotational inertia (J)	100 kg·m^2

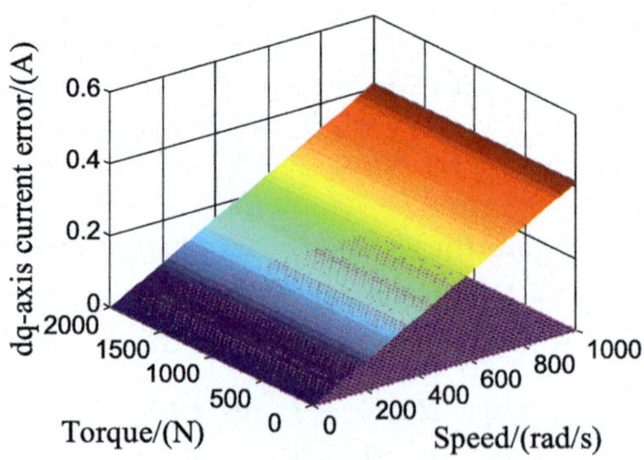

Fig. 4.3 The d- and q-axis current errors under resistance parameter mismatch ($\Delta R = -50\% R_o$).

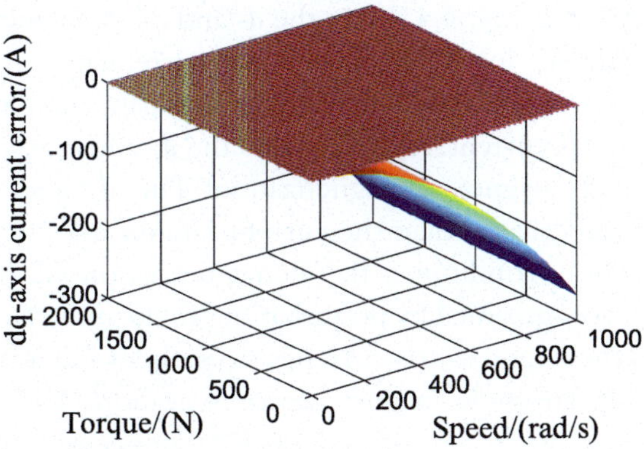

Fig. 4.4 The *d*- and *q*-axis current errors under inductance parameter mismatch ($\Delta L_d = -50\% L_{do}$, $\Delta L_q = -50\% L_{qo}$).

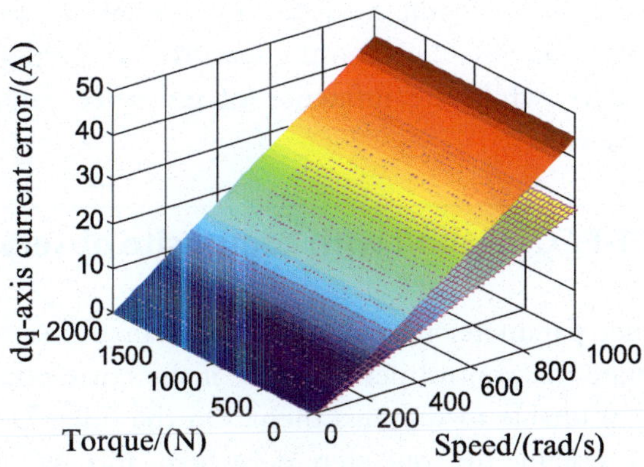

Fig. 4.5 The *d*- and *q*-axis current errors under permanent magnet demagnetization ($\psi_r = 50\% \psi_{ro}$, $\gamma = 30^0$).

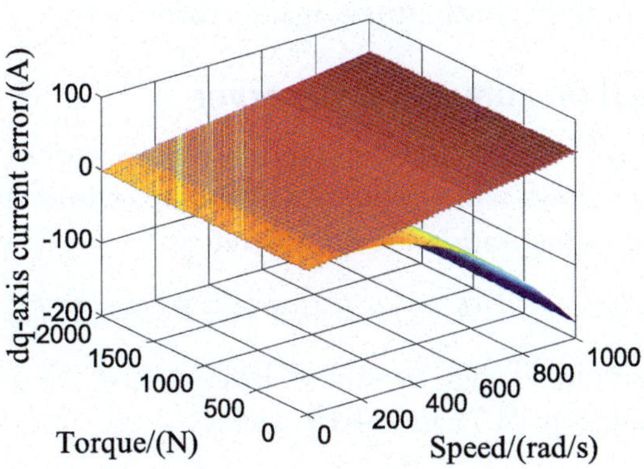

Fig. 4.6 The *d*- and *q*-axis current errors under inductance parameter mismatch and permanent magnet demagnetization ($\psi_r = 50\% \psi_{ro}$, $\gamma = 30^0$, $\Delta L_d = -50\% L_{do}$, $\Delta L_q = -50\% L_{qo}$).

basically be neglected. Fig. 4.4 shows the *d*- and *q*-axis current errors under inductance parameter mismatch. The *d*-axis current response value is less than the current reference, but the inductance parameter mismatch does not affect the *q*-axis current. Fig. 4.5 shows the *d*- and *q*-axis current errors under permanent magnet demagnetization. The *d*- and *q*-axis current response values are greater than the current references.

Under practical conditions, the demagnetization of the permanent magnet is often accompanied by perturbations in the inductance parameters. Fig. 4.6 shows the *d*- and *q*-axis current errors under inductance parameter mismatch and permanent magnet demagnetization. The *d*- and *q*-axis current errors are up to − 147A and 46A, respectively. At a speed of 400 rad/s, the *d*-axis current response value is less than the current reference, whereas the q-axis current response value is greater than the current reference. Through the analysis of the results, it can be observed that the conventional PCC is mostly sensitive to inductance and flux linkage parameters, while the influence of resistance parameters on the PCC can be neglected.

4.4 RFT-PCC method with composite observer

Inductance parameter mismatch and permanent magnet demagnetization deteriorate the control performance if the corresponding compensation method is unable to suppress them. On the other hand, in a motor digital control system, the one-step delay can also reduce the control performance of the PMSM if the delay problem is not considered in the design of the controller. Therefore this section presents the RFT-PCC method, which improves robustness against these factors.

RFT-PCC based on composite observer

According to Eq. (4.8), the expected input voltage vector of the PMSM can be expressed as follows when inductance parameter mismatch and permanent magnet demagnetization are considered as

$$\boldsymbol{u}(k)=\boldsymbol{F}_o^{-1}[\boldsymbol{i}(k+1)-\boldsymbol{E}_o(k)\boldsymbol{i}(k)-\boldsymbol{P}_o(k)+\boldsymbol{F}_o\boldsymbol{f}(k)] \tag{4.15}$$

The actual input voltage vector of PMSM is $\boldsymbol{u}^*(k)$ according to Eq. (4.7). Combining Eqs. (4.7) and (4.15) yields

$$\boldsymbol{u}(k)-\boldsymbol{u}^*(k)-\boldsymbol{f}(k)=\boldsymbol{F}_o^{-1}\Delta\boldsymbol{i}(k+1) \tag{4.16}$$

In order to make the current response values accurately track the current reference values under inductance parameter mismatch and permanent magnet demagnetization. The current error must satisfy $\Delta\boldsymbol{i}(k+1)=0$, and the expected input voltage vector of PMSM is calculated according to Eq. (4.16).

$$\boldsymbol{u}(k)=\boldsymbol{u}^{*}(k)+\boldsymbol{f}(k) \tag{4.17}$$

From Eq. (4.17), it can be found that the fault function $\boldsymbol{f}(k)$ is the compensation voltage. Thus the required compensation voltage can be accurately determined by observing the fault function $\boldsymbol{f}(k)$.

The PWM duty update and time sequence of current sampling are illustrated in Fig. 4.2. At the kT_s moment, combining the current sampling $\boldsymbol{i}(k)$ and current reference $\boldsymbol{i}^{*}(k+1)$, the voltage reference $\boldsymbol{u}^{*}(k)$ can be computed. As a result of the delays in the digital control system, the voltage reference $\boldsymbol{u}^{*}(k)$ should be updated by the inverter at the $(k+1)T_s$ moment, thereby leading to a one-beat delay of the control voltage. To reduce the influence of delays on the control performance of the system at the kT_s moment, we need to obtain the predictive current value $\hat{\boldsymbol{i}}(k+1)$ and then calculate the voltage vector $\boldsymbol{u}^{*}(k+1)$ by using the predictive current value $\hat{\boldsymbol{i}}(k+1)$. Therefore the predictive current value $\hat{\boldsymbol{i}}(k+1)$ is used to replace the sampled current $\boldsymbol{i}(k)$ in this work to compensate for the one-step delay.

The compensation voltage and current prediction value of the next time are simultaneously observed in Part B. The RFT-PCC adopts the observed compensation voltage $\hat{\boldsymbol{f}}(k)$ to replace the fault function $\boldsymbol{f}(k)$, and the observed current prediction value $\hat{\boldsymbol{i}}(k+1)$ of the next time is used for replacing the sampled current $\boldsymbol{i}(k)$. According to Eqs. (4.7) and (4.17), the expected input voltage vector of PMSM can be expressed as

$$\boldsymbol{u}(k)=\boldsymbol{F}_o^{-1}[\boldsymbol{i}^{*}(k+1)-\boldsymbol{E}_o(k)\hat{\boldsymbol{i}}(k+1)-\boldsymbol{P}_o(k)]+\hat{\boldsymbol{f}}(k) \tag{4.18}$$

The structural diagram of the PMSM drive system with RFT-PCC is shown in Fig. 4.7. The implementation of the proposed RFT-PCC method in this chapter is as simple as that of conventional PCC. The major difference lies in adding the composite observer to estimate the compensation voltage and the current prediction value for the next time step.

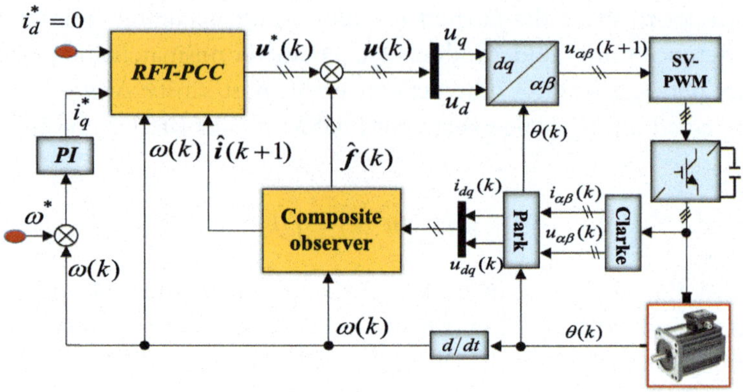

Fig. 4.7 Structural diagram of the PMSM drive system with RFT-PCC.

Design of a composite observer based on a sliding mode observer and a Luenberger observer

The sliding surface is selected as

$$s = \hat{x} - x = \begin{bmatrix} s_1 \\ s_2 \end{bmatrix} = \begin{bmatrix} \hat{i}_d - i_d \\ \hat{i}_q - i_q \end{bmatrix}$$

According to Eqs. (4.4) and (4.5), the composite observer, based on the sliding mode observer and the Luenberger observer, can be designed to observe the compensation voltage and the current prediction value for the next time step.

$$\frac{d\hat{x}}{dt} = A\hat{x} + Bu + Df_a + L\omega(\hat{y} - y) - Msgn(\hat{x} - \mathbf{x}) \tag{4.19}$$

where $\hat{x}$ is the observed value of x; $sgn(\cdot)$ is the sign function; $v = Msgn(x - \hat{x})$ is the sliding mode control function; M and L are the designed gain matrixes of Luenberger observer and sliding mode observer, respectively.

In practical applications, chattering suppression is important to designing observers because the chattering of the sign function can easily result in the chattering of the system. For this purpose, the sign function is replaced with a hyperbolic tangent function with smooth continuity. The hyperbolic tangent function can be expressed as $H(s) = [H_1(s_1) \quad H_2(s_2)]^T$, where $H_1(s_1)$ and $H_2(s_2)$ are given as

$$\begin{cases} H_1(s_1) = \tanh(s_1) = \dfrac{e^{2s_1} - 1}{e^{2s_1} + 1} \\ H_2(s_2) = \tanh(s_2) = \dfrac{e^{2s_2} - 1}{e^{2s_2} + 1} \end{cases}$$

The hyperbolic tangent function is used instead of the sign function, and the composite observer can be expressed according to Eq. (4.19).

$$\frac{d\hat{\boldsymbol{x}}}{dt} = \boldsymbol{A}\hat{\boldsymbol{x}} + \boldsymbol{Bu} + \boldsymbol{D}\boldsymbol{f}_{\boldsymbol{a}} + \boldsymbol{L}\omega(\hat{\boldsymbol{y}} - \boldsymbol{y}) - \boldsymbol{MH}(\boldsymbol{s}) \tag{4.20}$$

The error equation of the composite observer can be obtained by subtracting Eq. (4.4) from Eq. (4.20).

$$\frac{d\boldsymbol{s}}{dt} = \boldsymbol{A}_{1}\boldsymbol{s} + \boldsymbol{Bf} + (\boldsymbol{A}_2 + \boldsymbol{LC})\omega\boldsymbol{s} - \boldsymbol{MH}(\boldsymbol{s}) \tag{4.21}$$

Where $\boldsymbol{A}_1 = \begin{bmatrix} -\dfrac{R_o}{L_{do}} & 0 \\ 0 & -\dfrac{R_o}{L_{qo}} \end{bmatrix}$, $\boldsymbol{A}_2 = \begin{bmatrix} 0 & \dfrac{L_{qo}}{L_{do}} \\ -\dfrac{L_{do}}{L_{qo}} & 0 \end{bmatrix}$ and $\boldsymbol{A} = \boldsymbol{A}_1 + \boldsymbol{A}_2\omega$.

From Eq. (4.21), appropriate values must be selected for $\boldsymbol{M}$ and $\boldsymbol{L}$, to ensure the convergence of the composite observer error. From Eq. (4.21), it can be found that the change of the motor speed will affect the error equation of the composite observer. $\boldsymbol{L}$ should be set in accordance with $(\boldsymbol{A}_2 + \boldsymbol{LC}) = 0$ to isolate the influence of the motor speed. Therefore the gain matrixes of Luenberger observer can be designed as $\boldsymbol{L} = \begin{bmatrix} 0 & l_{12} \\ l_{21} & 0 \end{bmatrix}$, where $l_{12} = -\dfrac{L_{qo}}{L_{do}}$ and $l_{21} = \dfrac{L_{do}}{L_{qo}}$. After the gain matrix of Luenberger observer is designed, the error equation Eq. (4.21) of the composite observer can be simplified as

$$\frac{d\boldsymbol{s}}{dt} = \boldsymbol{A}_1\boldsymbol{s} + \boldsymbol{Bf} - \boldsymbol{MH}(\boldsymbol{s}) \tag{4.22}$$

From Eq. (4.22), the setting of $\boldsymbol{L}$ successfully isolates the influence of the motor speed on the error equations of the composite observer and enhances the robustness of the composite observer.

Consider the following Lyapunov function candidate.

$$V = \frac{1}{2}\boldsymbol{s}^{\mathrm{T}}\boldsymbol{s} \tag{4.23}$$

By differentiating the Lyapunov function in Eq. (4.23) and combining the obtained result with the result in Eq. (4.22), one obtains:

$$\begin{aligned} \frac{dV}{dt} &= \boldsymbol{s}^{\mathrm{T}}(\boldsymbol{A}_1\boldsymbol{s} + \boldsymbol{Bf} - \boldsymbol{MH}(\boldsymbol{s})) \\ &= \boldsymbol{s}^{\mathrm{T}}\boldsymbol{A}_1\boldsymbol{s} + \boldsymbol{s}^{\mathrm{T}}\boldsymbol{Bf} - \boldsymbol{s}^{\mathrm{T}}\boldsymbol{MH}(\mathbf{s}) \\ &= -\frac{R_o}{L_{do}}s_1^2 - \frac{R_o}{L_{qo}}s_2^2 + \boldsymbol{s}^{\mathrm{T}}\boldsymbol{Bf} - m_1 s_1 H_1(s_1) - m_2 s_2 H_2(s_2) \end{aligned} \tag{4.24}$$

Setting $\boldsymbol{M}$ satisfies $\boldsymbol{M} = \begin{bmatrix} m_1 & 0 \\ 0 & m_2 \end{bmatrix}$.

$$\begin{aligned} \frac{dV}{dt} &\le \boldsymbol{s}^{\mathrm{T}}\boldsymbol{Bf} - m_1 s_1 H_1(s_1) - m_2 s_2 H_2(s_2) \\ &\le \boldsymbol{s}^{\mathrm{T}}\boldsymbol{Bf} - m_1 \| s_1 \| - m_2 \| s_2 \| \\ &\le \| \boldsymbol{s} \| \| \boldsymbol{B} \| \| \boldsymbol{f} \| - m_3 \| \boldsymbol{s} \| = \| \boldsymbol{s} \| (\| \boldsymbol{B} \| \| \boldsymbol{f} \| - m_3) \end{aligned} \tag{4.25}$$

where $m_3 = \min\{m_1, m_2\}$, $m_1 > 0$, and $m_2 > 0$.

According to an engineering point of view, the fault function $\boldsymbol{f}$ should be bounded, that is, there exists a normal value N satisfies $\| \boldsymbol{f} \| \le N$. According to Eq. (4.25), if m_3 satisfies the condition of $m_3 \ge \| \boldsymbol{B} \| N + \eta$, where $\eta > 0$, then $\frac{dV}{dt} \le -\eta \| \boldsymbol{s} \|$. Therefore the stability and convergence of the composite observer is guaranteed based on the sliding-mode control theory. This completes the proof.

According to the sliding mode equivalent principle, when the system reaches the sliding-mode surface that is $\frac{d\boldsymbol{s}}{dt} = \boldsymbol{s} = 0$. From Eq. (4.21), discrete forms of fault function can be obtained as

$$\begin{cases} \hat{f}_d(k) = m_1 L_{do} H_1(s_1(k)) \\ \hat{f}_q(k) = m_2 L_{qo} H_2(s_2(k)) \end{cases} \tag{4.26}$$

where $\hat{f}_d(k)$, $\hat{f}_q(k)$ are the observed values of f_d, f_q.

Since the proposed composite observer will be only computed at discrete instants, the discrete expression of this observer is deduced. If the sampling period T_s is short enough, the discrete forms of Eq. (4.20) can be expressed as

$$\begin{cases} \hat{i}_d(k+1) = (1 - \frac{R_o}{L_{do}} T_s)\hat{i}_d(k) + \frac{T_s}{L_{do}} u_d(k) - l_{12} T_s \omega(k) i_q(k) - m_1 T_s H_1(s_1(k)) \\ \hat{i}_q(k+1) = (1 - \frac{R_o}{L_{qo}} T_s)\hat{i}_q(k) + \frac{T_s}{L_{qo}} u_q(k) - \frac{\psi_{ro}}{L_{qo}} T_s \omega(k) - l_{21} T_s \omega(k) i_d(k) - m_2 T_s H_2(s_2(k)) \end{cases} \tag{4.27}$$

where $\hat{i}_d(k+1)$, $\hat{i}_q(k+1)$ are the observed values of the prediction value of the next time.

4.5 Simulation results

Simulations are established in MATLAB/Simulink, the $i_d = 0$ control strategy is adopted, the sampling period T_s is set as 5 kHz, the DC voltage is 1500 V, and the amplitude value of the stator current is 350 A. The *PI* controller is adopted by the speed control loop, in which the proportion coefficient and integral coefficient are 800 and 0.3, respectively. In addition, the Luenberger observer gains are $l_{12} = -3.572$ and $l_{21} = 0.28$, which are calculated by $l_{12} = -\frac{L_{qo}}{L_{do}}$ and $l_{21} = \frac{L_{do}}{L_{qo}}$, respectively. In accordance with $m_3 \geq \| \boldsymbol{B} \| N + \eta \approx 1.7 \times 10^5$ and the actual simulation debugging, the sliding mode observer gains are set as $m_1 = 5 \times 10^5$ and $m_2 = 5 \times 10^5$. Consequently, the stability and convergence of the composite observer are guaranteed. The given speed is 400 rad/s, and the given load torque is 800 N. To realize the inductance parameter perturbation and permanent magnet demagnetization in MATLAB/Simulink, we reconstruct the PMSM module in accordance with Eq. (4.4) by using the step module to simulate the inductance parameter perturbation and permanent magnet demagnetization. The main parameters of PMSM used in the simulation are shown in Table 4.1.

Simulation results for PMSM under inductance parameter mismatch

Fig. 4.8 shows the simulation results of the PMSM drive system under inductance parameter mismatch using the conventional PCC method. From Fig. 4.8A, it can be seen that the inductance parameter mismatch leads to an increase in torque ripple in the PMSM drive system, indicating that the parameter mismatch seriously affects the control performance of

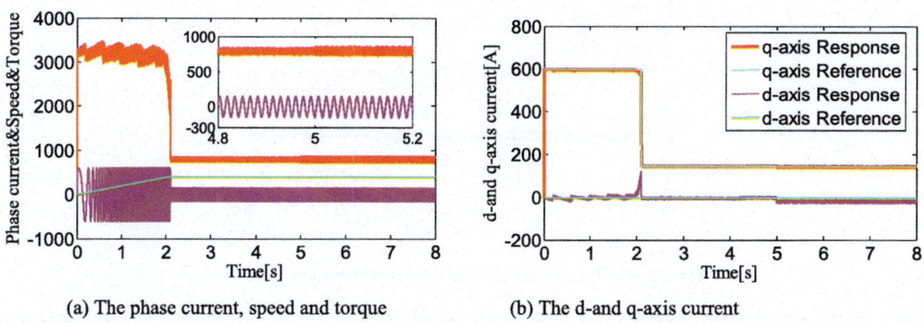

Fig. 4.8 Simulation results of the conventional PCC method under inductance parameter mismatch. (a) The phase current, speed, and torque. (b) The *d*- and *q*-axis currents.

the conventional PCC method. From Fig. 4.8B, it can be observed that the d-axis current response value of the PMSM drive system is significantly lower than the command value under inductance parameter mismatch, while the impact on the q-axis current is not significant. The above analysis demonstrates that the inductance parameter mismatch seriously affects the performance of the conventional PCC method on the d-axis current of the PMSM drive system, while the control performance on the q-axis current can be disregarded. This conclusion is consistent with the analysis results of current control performance under inductance parameter mismatch shown in Fig. 4.4. Fig. 4.9 presents the simulation results of the PMSM drive system using the proposed RFT-PCC method under inductance parameter mismatch. Comparing the simulation results of Fig. 4.8A and Fig. 4.9A, it can be observed that the proposed RFT-PCC method significantly improves the impact of inductance parameter mismatch on the torque control performance of the PMSM drive system, as the designed composite observer compensates for the inductance parameter mismatch. Compared with the conventional PCC method, the proposed RFT-PCC

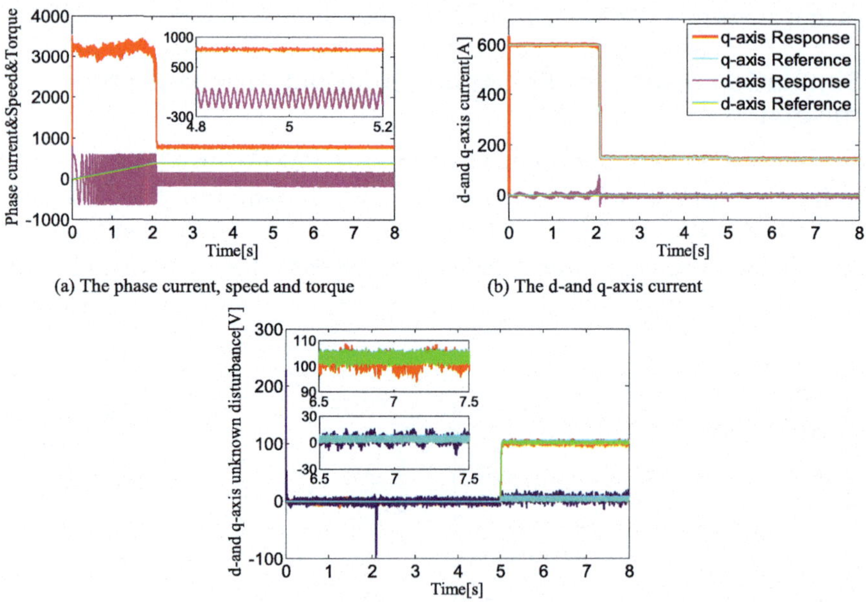

(a) The phase current, speed and torque

(b) The d-and q-axis current

(c) The d-axis and q-axis unknown disturbance compensation

Fig. 4.9 Simulation results of the proposed RFT-PCC method under inductance parameter mismatch. (a) The phase current, speed, and torque. (b) The d- and q-axis currents. (c) The d- and q-axis unknown disturbance compensation.

method effectively suppresses the torque ripple of the PMSM drive system under inductance parameter mismatch, and the speed control performance remains unaffected by the inductance parameter mismatch.

From Fig. 4.9B, it can be seen that the proposed RFT-PCC method, based on the designed composite observer, can significantly eliminate the d-axis current deviation caused by inductance parameter mismatch. By comparing Fig. 4.8B and Fig. 4.9B, it is evident that the proposed RFT-PCC method can significantly mitigate the impact of inductance parameter mismatch on the d-axis current of the PMSM drive system. From Fig. 4.9C, it can be observed that the designed composite observer can accurately detect the *d*- and *q*-axis unknown disturbances of the PMSM drive system caused by inductance parameter mismatch. The composite observer can accurately confirm that the d-axis and *q*-axis unknown disturbances are zero during normal operation. Furthermore, the composite observer can detect the d-axis unknown disturbance in the case of inductance parameter mismatch and provide feedback on the observed values to the proposed RFT-PCC method.

Simulation results for PMSM under permanent magnet demagnetization

Fig. 4.10 presents the simulation results of the PMSM drive system using the conventional PCC method under permanent magnet demagnetization. As shown in Fig. 4.10A, the control performance of the conventional PCC method deteriorates due to permanent magnet demagnetization, and the torque ripple of the PMSM drive system significantly increases when using the conventional PCC method. In addition, as shown in Fig. 4.10A, the stator current significantly increases under permanent magnet demagnetization. The reason for this phenomenon is that, to maintain the torque of the PMSM drive system unchanged, the magnetic flux is reduced, and the stator current is increased. Fig. 4.10B clearly shows that permanent magnet demagnetization causes the *d*- and *q*-axis response values of the PMSM drive system to be significantly greater than the reference values, indicating that permanent magnet demagnetization seriously affects the performance of the conventional PCC method. The reason for this phenomenon is that permanent magnet demagnetization leads to inaccurate voltage vector output in the conventional PCC method. The *d*- and *q*-axis current deviation of the conventional PCC method caused by permanent magnet demagnetization in the PMSM drive system is consistent with the results analyzed in Fig. 4.5.

Fig. 4.11 presents the simulation results of the PMSM drive system using the proposed RFT-PCC method under permanent magnet demagnetization. As shown in Fig. 4.11A, compared with the conventional PCC method, the proposed RFT-PCC method can effectively suppress the torque ripple of the PMSM drive system caused by permanent magnet demagnetization. By comparing the conventional PCC method with the

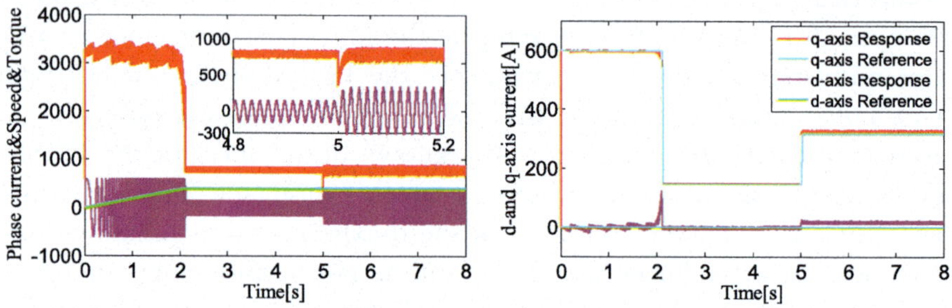

(a) The phase current, speed and torque (b) The d-and q-axis current

Fig. 4.10 Simulation results of the conventional PCC method under permanent magnet demagnetization. (a) The phase current, speed, and torque. (b) The *d*- and *q*-axis current.

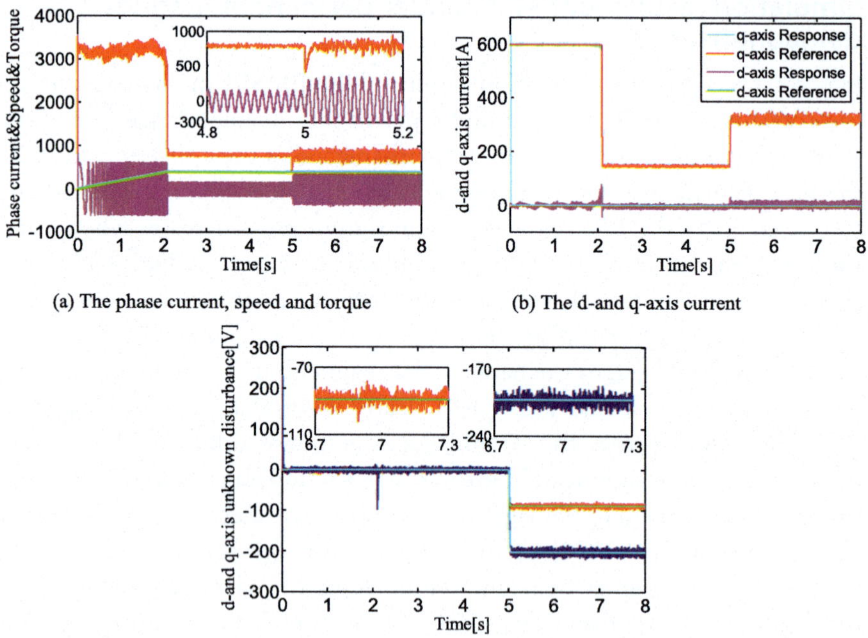

(c) The d-axis and q-axis unknown disturbance compensation

Fig. 4.11 Simulation results of the proposed RFT-PCC method under permanent magnet demagnetization. (a) The phase current, speed, and torque. (b) The *d*- and *q*-axis currents. (c) The *d*- and *q*-axis unknown disturbance compensation.

proposed RFT-PCC method, it can be concluded that permanent magnet demagnetization does not affect the speed control performance of the PMSM drive system, as shown in Fig. 4.10A and Fig. 4.11A. From Fig. 4.11B, it can be seen that the *d*- and *q*-axis current deviation of the PMSM drive system caused by permanent magnet demagnetization is eliminated after adopting the proposed RFT-PCC method, and the *d*- and *q*-axis response values of the PMSM drive system can accurately track the reference values. The reason for achieving this control effect is that the designed composite observer compensates for permanent magnet demagnetization. Fig. 4.11C clearly shows that the designed composite observer can accurately observe *d*- and *q*-axis unknown disturbances caused by permanent magnet demagnetization. Permanent magnet demagnetization can cause deviations in the *d*- and *q*-axis currents of the PMSM drive system, but the composite observer can compensate for both *d*- and *q*-axis currents simultaneously in the proposed RFT-PCC method.

Simulation results for PMSM under inductance parameter mismatch and permanent magnet demagnetization

Fig. 4.12 shows the simulation results of the PMSM drive system using the conventional PCC method under inductance parameter mismatch and permanent magnet demagnetization. From Fig. 4.12A, it can be seen that when the inductance parameter mismatch and permanent magnet demagnetization occur simultaneously, the torque ripple of the PMSM drive system using the conventional PCC method significantly increases. The reason for this phenomenon is that the conventional PCC method strictly relies on the mathematical parameters of the PMSM

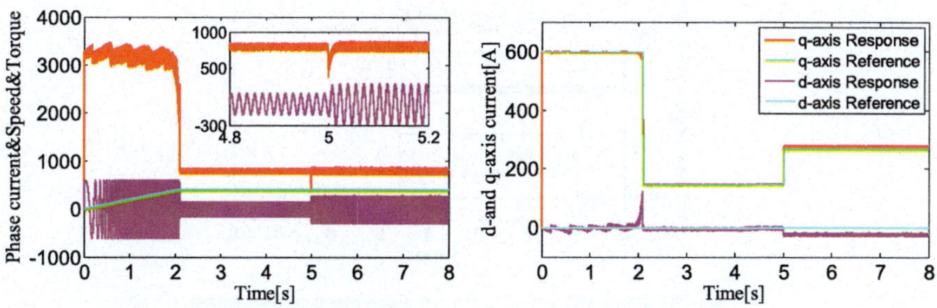

(a) The phase current, speed and torque (b) The d-and q-axis current

Fig. 4.12 Simulation results of the conventional PCC method under inductance parameter mismatch and permanent magnet demagnetization. (a) The phase current, speed, and torque. (b) The *d*- and *q*-axis currents.

drive system. It is worth noting that, due to the occurrence of inductance parameter mismatch and permanent magnet demagnetization at this time, there is a significant overshoot in torque ripple at 5 s. Fig. 4.12B clearly shows that the inductance parameter mismatch and permanent magnet demagnetization can cause the *d*-axis response current value to be less than the reference current value, and the *q*-axis response current value to be greater than the reference current value. The theoretical analysis that caused this phenomenon has been presented in Fig. 4.6, and the results of both analyses are consistent.

Fig. 4.13 presents the simulation results of the PMSM drive system using the proposed RFT-PCC method under inductance parameter mismatch and permanent magnet demagnetization. From Fig. 4.13A, it can be seen that, in the case of inductance parameter mismatch and permanent magnet demagnetization, the torque ripple of the PMSM drive system is significantly suppressed after adopting the proposed RFT-PCC

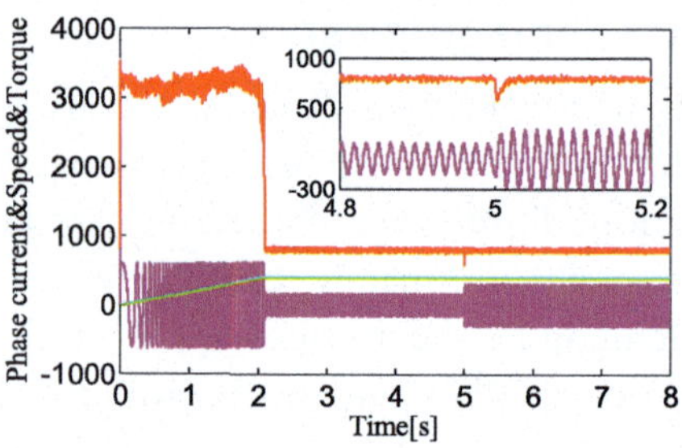

(a) The phase current, speed and torque

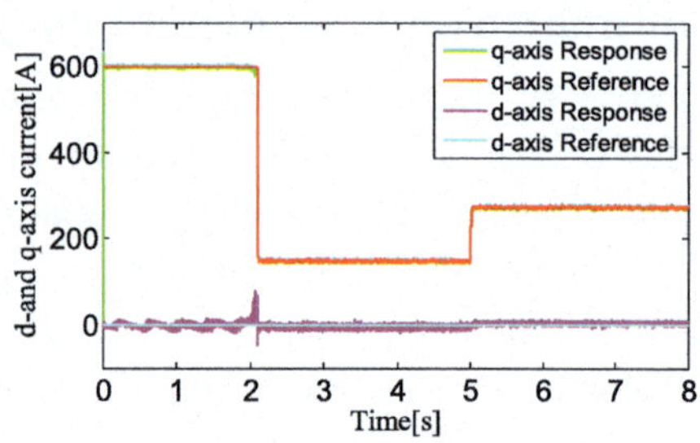

(b) The d-and q-axis current

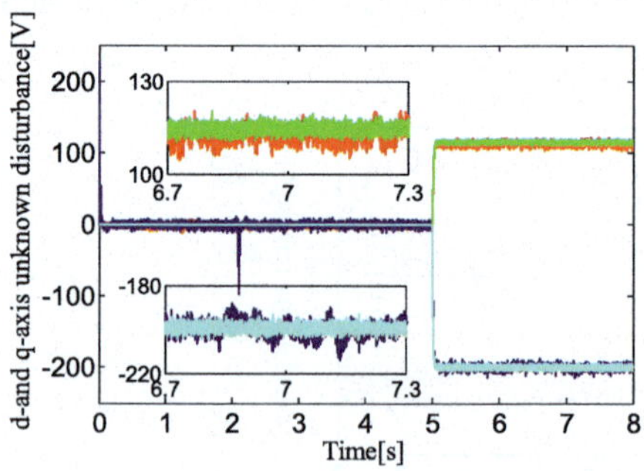

(c) The d-axis and q-axis unknown disturbance compensation

Fig. 4.13 Simulation results of the proposed RFT-PCC method under inductance parameter mismatch and permanent magnet demagnetization. (a) The phase current, speed, and torque. (b) The *d*- and *q*-axis currents. (c) The *d*- and *q*-axis unknown disturbance compensation.

method. This indicates that the proposed RFT-PCC method can effectively mitigate the impact of inductance parameter mismatch and permanent magnet demagnetization on the PMSM drive system. The *d*- and *q*- axis response currents of the PMSM drive system can accurately track the reference currents, and the deviation between the response currents and the reference currents is essentially zero, as clearly shown in Fig. 4.13B. This result indicates that the proposed RFT-PCC method can effectively achieve current tracking control under inductance parameter mismatch and permanent magnet demagnetization. The designed composite observer can accurately observe *d*- and *q*-axis unknown disturbances caused by inductance parameter mismatch and permanent magnet demagnetization, as clearly shown in Fig. 4.13C. According to the current deviation of the PMSM drive system, the proposed RFT-PCC method, based on a composite observer, requires a positive d-axis voltage compensation value and a negative *q*-axis voltage compensation value.

4.6 Conclusion

The novel RFT-PCC algorithm for PMSM drives is proposed in this chapter, which can improve the robustness against motor parameter perturbation and permanent magnet demagnetization. The coupling between the perturbed inductance and the demagnetization effect on conventional PCC is analyzed in detail. A composite observer, based on a sliding mode observer and a Luenberger observer, is designed to predict the stator current for the next time period and compensate for the output voltage of the conventional PCC. Compared with the conventional PCC, the proposed RFT-PCC method increases the tracking accuracy of the stator current and decreases the torque ripple under parameter perturbation and permanent magnet demagnetization. Simulation results indicate that the RFT-PCC controller operates well in steady-state processes while maintaining a quick dynamic response for the current control loops.

References

[1] C. Hung-Chi, C. Keng-Yuan, et al., High-efficiency current control methods based on multidimensional feedback quantization and its application to three-phase PMSM, IEEE Trans. Ind. Electron. 61 (11) (2014) 5820–5829.

[2] P. Kakosimos, et al., Predictive Speed Control with Short Prediction Horizon for Permanent Magnet Synchronous Motor Drives, IEEE Trans. Power Electron. 32 (10) (2017) 1–7.

[3] M. Habibullah, D. Xiao, et al., Finite-State Predictive Torque Control of Induction Motor Supplied from a Three-Level NPC Voltage Source Inverter, IEEE Trans. Power Electron. 21 (01) (2017) 79–489.
[4] A.Young Hector, M.A. P, et al., Analysis of finite-control-set model predictive current control with model parameter mismatch in a three-phase inverter, IEEE Trans. Ind. Electron. 63 (05) (2016) 3100–3107.
[5] P. Cortés, J. Rodríguez, Q.D. E, et al., Predictive current control strategy with imposed load current spectrum, IEEE Trans. Power Electron. 23 (02) (2008) 612–618.
[6] D. A, R.P. P, R. R, et al., A space-vector-based hysteresis current controller for a general n-level inverter-fed drive with nearly constant switching frequency control, IEEE Trans. Ind. Electron. 60 (05) (2013) 1989–1998.
[7] S.J. A, K. Ljokelsoy, T. Midtsund, et al., Synchronous reference frame hysteresis current control for grid converter applications, IEEE Trans. Ind. Electron. 47 (05) (2011) 2183–2194.
[8] J.J. W, C.Y. S, L.V. Q, et al., Fuzzy PI-type current controllers for permanent magnet synchronous motors, IEEE Trans. Ind. Appl. 5 (01) (2011) 143–152.
[9] J.J. W, L.V. Q, D.T. D, et al., Adaptive PID speed control design for permanent magnet synchronous motor drives, IEEE Trans. Power Electron. 32 (02) (2015) 900–908.
[10] S.C. Carpiuc, C. Lazar, Fast real-time constrained predictive current control in permanent magnet synchronous machine-based automotive traction drives, IEEE Trans. Ind. Electron. 1 (04) (2015) 326–335.
[11] M. F, L.S. X, R.J. M, et al., A comparative study of predictive current control schemes for a permanent-magnet synchronous machine drive, IEEE Trans. Ind. Electron. 56 (07) (2015) 2715–2728.
[12] T. L, Z. X, F. S, C.Y, K.Y, et al., An SRF-PLL-based sensorless vector control using the predictive deadbeat algorithm for the direct-driven permanent magnet synchronous generator, IEEE Trans. Power Electron. 29 (06) (2014) 2837–2849.
[13] M. S, L. J, J. H, et al., Demagnetization fault diagnosis of a PMSM based on structure analysis of motor inductance, IEEE Trans. Ind. Electron. 63 (06) (2016) 3795–3803.
[14] H.J. Ahn, D.M. Lee, A New Bumpless Rotor-Flux Position Estimation Scheme for Vector-Controlled Washing Machine, IEEE Trans. Ind. Electron. 12 (02) (2016) 466–473.
[15] T. L, Z. X, F.S. S, et al., An SRF-PLL-based sensorless vector control using the predictive deadbeat algorithm for the direct-driven permanent magnet synchronous generator, IEEE Trans. Power Electron. 29 (06) (2014) 2837–2849.
[16] C. P, R. J, S. C, et al., Delay compensation in model predictive current control of a three-phase inverter, IEEE Trans. Ind. Electron. 59 (02) (2012) 1323–1325.
[17] Y. Sozer, D.A. Torrey, E. Mese, Adaptive predictive current control technique for permanent magnet synchronous motors, IEEE Trans. Power Electron. 6 (01) (2013) 9–19.
[18] P. Karamanakos, T. Geyer, S. Manias, Direct model predictive current control strategy of DC–DC boost converters, IEEE J. Emerg. Sel. Top. Power Electron. 1 (04) (2013) 337–346.
[19] G. H, D.M. J, B. F, et al., Speed control of five-phase induction motors with integrated open-phase fault operation using model-based predictive current control techniques, IEEE Trans. Ind. Electron. 61 (09) (2013) 4474–4484.
[20] S. M, K.D. A, A. A, et al., Robustness improvement of predictive current control using prediction error correction for permanent-magnet synchronous machines, IEEE Trans. Ind. Electron. 63 (06) (2016) 3458–3466.
[21] L.C. K, L.T. H, F.L. C, et al., Model-free predictive current control for interior permanent-magnet synchronous motor drives based on current difference detection technique, IEEE Trans. Ind. Electron. 61 (02) (2014) 667–681.

[22] T. Türker, U. Buyukkeles, A.F. Bakan, A robust predictive current controller for PMSM drives, IEEE Trans. Ind. Electron. 63 (06) (2016) 3906–3914.
[23] Y. J, Z.W. X, L. S, et al., Design of a prediction-accuracy-enhanced continuous-time MPC for disturbed systems via a disturbance observer, IEEE Trans. Ind. Electron. 62 (09) (2016) 5807–5816.
[24] Z. Xiaoguang, B. Hou, Y. Mei, Deadbeat Predictive Current Control of Permanent-Magnet Synchronous Motors with Stator Current and Disturbance Observer, IEEE Trans. Power Electron. 32 (05) (2017) 3818–3834.
[25] Y. J, S. J, L. S, et al., High-order mismatched disturbance compensation for motion control systems via a continuous dynamic sliding-mode approach, IEEE Trans. Ind. Electron. 10 (01) (2014) 604–614.
[26] M. S, J. H, L. H, et al., Detection and Classification of Demagnetization and Interturn Short fault of IPMSMs, IEEE Trans. Ind. Electron. 10 (11) (2017) 207–216.
[27] H.J. Ahn, D.M. Lee, A New Bumpless Rotor-Flux Position Estimation Scheme for Vector-Controlled Washing Machine, IEEE Trans. Ind. Electron. 12 (02) (2017) 466–473.
[28] R. J, B. F, A.M. R, et al., Online estimation of rotor variables in predictive current controllers: a case study using five-phase induction machines, IEEE Trans. Ind. Electron. 63 (09) (2016) 5348–5356.
[29] Y. M, L. X, L. J, et al., A Flux Immunity Robust Predictive Current Control With Incremental Model and Extended State Observer for PMSM Drive, IEEE Trans. Power Electron. 3 (06) (2017) 255–267.
[30] W. G, Y. M, N. L, et al., Improved predictive current control with static current error elimination for permanent magnet synchronous machine, IEEE Trans. Ind. Electron. 13 (12) (2014) 661–667.
[31] N. F, W. B, B.A. S, et al., Comparative evaluation of direct torque control strategies for permanent magnet synchronous machines, IEEE Trans. Power Electron. 31 (02) (2016) 1408–1424.
[32] B.M. Ebrahimi, J. Faiz, Demagnetization fault diagnosis in surface mounted permanent magnet synchronous motors, IEEE Trans. Ind. Electron. 49 (03) (2013) 1185–1192.
[33] R. Jan, M. Doppelbauer, Predictive trajectory control of permanent-magnet synchronous machines with nonlinear magnetic, IEEE Trans. Ind. Electron. 63 (06) (2016) 3915–3924.

CHAPTER 5

Cascaded robust fault-tolerant predictive control for permanent magnet synchronous motor drives

5.1 Introduction

Permanent magnet synchronous motor (PMSM) drives based on field-oriented control (FOC) have been steadily adopted in industrial applications due to the fast and fully decoupled control of torque and flux [1–3]. In an FOC-based PMSM drive, the double-loop structure is usually adopted. The internal current loop regulates the stator current to track its reference, while the external speed loop adjusts the machine speed [4]. Thus the dynamic response and stability of the double-loop are the key factors that determine the dynamic quality of the whole drive system.

A classical approach is based on FOC realized by a cascade configuration of the well-known PI controllers [5]. This approach has been adopted in various application areas, but it also has its limitations. Usually, the parameter settings of the PI controllers only correspond to a specific working range. Thus this causes problems with the controllers when the working state of the motor changes [6]. In addition, the PMSM control system is a nonlinear system with parameter variations, as well as permanent magnet demagnetization [7]. In a PMSM, a 20% flux reduction for a ferrite-based magnet occurs per 100°C increase in ambient temperature [8]. This makes it difficult for PI control algorithms to achieve satisfying dynamic performance across the entire operating range of the PMSM [9].

To achieve high-performance control for PMSM drives, the predictive current control (PCC) approach has been adopted in a wide range of applications. PCC can calculate the required command voltage based on the discrete mathematical model of PMSM and make the feedback current optimally track its reference [10]. Compared with the well-known PI

Robust Model Predictive Control of Permanent Magnet Synchronous Motor Under Parameter Mismatch
https://doi.org/10.1016/B978-0-443-44209-4.00012-9

controllers, PCC theoretically improves the bandwidth of current control loops and the dynamic performance of motors [11,12]. However, PCC is absolutely dependent on an exact PMSM model, which means that parameter perturbation and permanent magnet demagnetization could deteriorate the performance of the PCC algorithm.

Some robust PCC methods have been presented to eliminate the influence of motor parameter perturbation and permanent magnet demagnetization. In [13], an extension of the PCC method is presented to improve the prediction accuracy for a PMSM. The proposed strategy not only reduces the current ripple but also improves the robustness of the PCC against parameter uncertainties. In [14], a flux-immunity robust PCC for PMSM drives is proposed, which can operate without requiring knowledge of the rotor flux. In [15], an improved deadbeat PCC algorithm for PMSM drive systems is proposed to optimize the current control performance of the PMSM with model parameter mismatch and one-step control delay. In [16], a discrete-time nonlinear robust predictive controller is proposed for the current loop of PMSM drives, which improves robustness against parametric uncertainties. In [13–16], the predictive control performance of the internal current loop has been greatly improved, but a classical PI controller is still used for the external speed loop.

A high-performance servo application must have a fast dynamic response, preferably without overshoot, and high steady-state accuracy. Recently, full predictive controls of speed and current for motor drives have been reported in the literature [17–21]. In [19], a new speed control strategy for the two-mass system based on the model predictive control scheme is presented. In [20], a control strategy based on finite-set model predictive control for the speed control of the PMSM is developed. In [21], a cascaded predictive speed and current control is proposed based on the explicit inversion of the mechanical model. In [19–21], these methods have some advantages, such as fast transient response and simple implementation. However, they are sensitive to system parameters, and permanent magnet demagnetization could lead to inaccurate prediction of motor behavior.

In this chapter, the CRFTPC with an integral terminal sliding mode observer is proposed for PMSM drives. The major contributions of this chapter are as follows: (1) This work proposes a CRFTPC for the speed and current of PMSM drives, making it possible to avoid the use of the conventional speed PI controller. (2) To strengthen the robustness of the proposed CRFTPC strategy, a novel integral terminal sliding mode observer is designed for estimating disturbances. (3) The optimal control law for high-performance speed and current loops is derived, eliminating the need to introduce a weighting factor. It is noted that the proposed speed loop can achieve a satisfying effect in cases of load perturbation and permanent magnet demagnetization. Furthermore, the proposed current loop is able to effectively enhance robustness against parameter perturbation and permanent magnet demagnetization. This chapter is organized as follows. The nonlinear mathematical model of PMSM drives is introduced in Section 5.2. The disturbances of PMSM drives under parameter mismatch are analyzed in Section 5.3. The cascaded robust fault-tolerant predictive control method algorithm is proposed in Section 5.4. The effectiveness of the proposed CRFTPC algorithm is verified through simulations in Section 5.5. The conclusion is presented in Section 5.6.

5.2 Nonlinear mathematical model of PMSM drives

Machine model description

The well-known electrical PMSMs are developed by using two-axis d–q current components. The $i_d = 0$ control strategy is mainly studied in this chapter. Thus the main equations used to model the PMSM in synchronous rotating frame can be described as [22]

$$\begin{cases} \dfrac{di_d}{dt} = -\dfrac{R}{L_d} i_d + \dfrac{1}{L_d} u_d + \omega \dfrac{L_q}{L_d} i_q \\ \dfrac{di_q}{dt} = -\dfrac{R}{L_q} i_q + \dfrac{1}{L_q} u_q - \omega \dfrac{L_d}{L_q} i_d - \omega \dfrac{\psi_r}{L_q} \\ \dfrac{J}{n_p} \dfrac{d\omega}{dt} = T_e - B\dfrac{\omega}{n_p} - T_L \\ T_e = 1.5 n_p \psi_r i_q \end{cases} \tag{5.1}$$

where u_d and u_q, respectively, represent the d- and q-axis stator voltages; i_d and i_q are the d- and q-axis currents, respectively; R, L_d, L_q, and J are the nominal value of stator resistance, the d-axis inductance, the q-axis inductance and the moment of inertia, respectively. B is the friction coefficient, and ω is the electrical rotor speed. ψ_r is the flux linkage established by the permanent magnets. n_p is the number of pole pairs, T_e is the electromagnetic torque, and T_L is the load torque.

PMSM model with parameter change

In the operation of the PMSM, resistance and inductance perturbations, as well as permanent magnet demagnetization, occur due to the influence of temperature and other external factors. Therefore, the control system's performance, based on Eq. (5.1), is affected.

Introducing $\Delta L_d = L_d - L_{do}$, $\Delta L_q = L_q - L_{qo}$, and $\Delta R = R - R_o$, where R_o, L_{do}, and L_{qo} are nominal values and ΔL_d, ΔL_q, and ΔR are perturbation values of the corresponding model parameter. The flux linkage amplitude varies from initial ψ_{ro} to ψ_r when demagnetization fault occurs, and defining $\Delta\psi_{rd} = \psi_r - \psi_{ro}$ as the new flux linkage components of *d*-axis. Therefore, the voltage equations are expressed as follows when parameter perturbation and permanent magnet demagnetization are considered as

$$\begin{bmatrix} \dfrac{di_d}{dt} \\ \dfrac{di_q}{dt} \end{bmatrix} = \begin{bmatrix} -\dfrac{R_o}{L_{do}} & \omega\dfrac{L_{qo}}{L_{do}} \\ -\omega\dfrac{L_{do}}{L_{qo}} & -\dfrac{R_o}{L_{qo}} \end{bmatrix} \begin{bmatrix} i_d \\ i_q \end{bmatrix} + \begin{bmatrix} \dfrac{1}{L_{do}} & 0 \\ 0 & \dfrac{1}{L_{qo}} \end{bmatrix} \begin{bmatrix} u_d \\ u_q \end{bmatrix} + \begin{bmatrix} 0 \\ -\dfrac{\psi_{ro}}{L_{qo}}\omega \end{bmatrix} - \begin{bmatrix} \dfrac{1}{L_{do}} & 0 \\ 0 & \dfrac{1}{L_{qo}} \end{bmatrix} \begin{bmatrix} \delta_d \\ \delta_q \end{bmatrix} \tag{5.2}$$

where δ_d and δ_q represent the unknown disturbances caused by parameter perturbation and permanent magnet demagnetization, they can be defined as

$$\begin{cases} \delta_d = \dfrac{\Delta L_d}{(L_{do} + \Delta L_d)}(u_d - R_o i_d - \Delta R i_d + L_{qo}\omega i_q + \Delta L_q \omega i_q) \\ \quad + (\Delta R i_d - \Delta L_q \omega i_q) \\ \delta_q = \dfrac{\Delta L_q}{(L_{qo} + \Delta L_q)}(u_q - R_o i_q - \Delta R i_q - L_{do}\omega i_d - \Delta L_d \omega i_d - \omega\psi_{ro} - \omega\Delta\psi_{rd}) \\ \quad + (\Delta R i_q + \Delta L_d \omega i_d + \omega\Delta\psi_{rd}) \end{cases}$$

When permanent magnet demagnetization occurs, the electromagnetic torque produced by the machine is

$$T_e = 1.5 n_p (\psi_{ro} + \Delta\psi_{rd}) i_q \tag{5.3}$$

The mechanical dynamics of the PMSM model are then

$$\frac{d\omega}{dt} = -\frac{B}{J}\omega + \frac{1.5n_p^2\psi_{ro}}{J}i_q - \frac{n_p}{J}\delta_\omega \tag{5.4}$$

where $\delta_\omega = -\Delta T_e + T_L$ represents unknown external disturbances. $\Delta T_e = 1.5n_p\Delta\psi_{rd}i_q$ is the unknown load disturbance caused by permanent magnet demagnetization.

According to Eqs. (5.2), (5.3), and (5.4), the nonlinear mathematical model of the PMSM can be established when parameter perturbation and permanent magnet demagnetization occur

$$\begin{cases} \dfrac{d\boldsymbol{x}(t)}{dt} = \boldsymbol{Ax}(t) + \boldsymbol{Bu}(t) + \boldsymbol{Df}_\psi - \boldsymbol{G}\delta(t) \\ \boldsymbol{y}(t) = \boldsymbol{Ex}(t) \end{cases} \tag{5.5}$$

where $\boldsymbol{x}(t) = \begin{bmatrix} i_d \\ i_q \\ \omega \end{bmatrix}$, $\boldsymbol{u}(t) = \begin{bmatrix} u_d \\ u_q \\ i_q \end{bmatrix}$, $\boldsymbol{y}(t) = \begin{bmatrix} i_d \\ i_q \\ \omega \end{bmatrix}$, and $\boldsymbol{\delta}(t) = \begin{bmatrix} \delta_d \\ \delta_q \\ \delta_\omega \end{bmatrix}$ are state variables, system inputs, system outputs, and unknown disturbances. The coefficient matrixes of the state equations are

$$\boldsymbol{A} = \begin{bmatrix} -\dfrac{R_o}{L_{do}} & \omega\dfrac{L_{qo}}{L_{do}} & 0 \\ -\omega\dfrac{L_{do}}{L_{qo}} & -\dfrac{R_o}{L_{qo}} & 0 \\ 0 & 0 & -\dfrac{B}{J} \end{bmatrix}, \quad \boldsymbol{B} = \begin{bmatrix} \dfrac{1}{L_{do}} & 0 & 0 \\ 0 & \dfrac{1}{L_{qo}} & 0 \\ 0 & 0 & \dfrac{1.5n_p^2\psi_{ro}}{J} \end{bmatrix},$$

$$\boldsymbol{f}_\psi = \begin{bmatrix} 0 \\ \psi_{ro} \\ 0 \end{bmatrix}, \quad \boldsymbol{D} = \begin{bmatrix} 0 & 0 & 0 \\ 0 & -\dfrac{\omega}{L_{qo}} & 0 \\ 0 & 0 & 0 \end{bmatrix}, \quad \boldsymbol{G} = \begin{bmatrix} \dfrac{1}{L_{do}} & 0 & 0 \\ 0 & \dfrac{1}{L_{qo}} & 0 \\ 0 & 0 & \dfrac{n_p}{J} \end{bmatrix}, \quad \boldsymbol{E} = \begin{bmatrix} 1 & 0 & 0 \\ 0 & 1 & 0 \\ 0 & 0 & 1 \end{bmatrix}.$$

5.3 Disturbance analysis of PMSM drives under parameter mismatch

According to Eq. (5.2), the stationary equations are given as:

$$\begin{cases} u_d = R_o i_d - L_{qo}\omega i_q \\ u_q = R_o i_q + L_{do}\omega i_d + \omega\psi_{ro} \end{cases} \tag{5.6}$$

The $i_d = 0$ control strategy is mainly studied in this chapter. Thus, compared with q-axis current, the function of d-axis current can be basically ignored. Combining with Eq. (5.5), the d- and q-axis disturbance can be simplified as

$$\begin{gathered} \delta_d = -\frac{L_{do}\Delta L_q}{(L_{do} + \Delta L_d)}\omega i_q \\ \delta_q = \frac{L_{qo}\Delta R}{(L_{qo} + \Delta L_q)} i_q + \frac{L_{qo}\Delta\psi_{rd}}{(L_{qo} + \Delta L_q)}\omega \end{gathered} \tag{5.7}$$

The motor parameters are shown in Table 5.1. According to Eq. (5.7), the d- and q-axis disturbances can be plotted, as shown in Figs.5.1–5.2. Fig. 5.1 shows the d- and q-axis disturbances under resistance parameter perturbation. The d-axis disturbance is zero, and the q-axis disturbance is negative. Fig. 5.1 illustrates that the resistance parameter perturbation does not affect the d-axis current, and the influence of resistance parameter perturbation can be neglected compared to the influence of inductance and flux linkage parameters on the q-axis current. Fig. 5.2 shows the d- and q-axis disturbances under inductance parameter perturbation. The q-axis disturbance is zero. The d-axis disturbance is positive and increases with the rise in speed and torque. Fig. 5.2 illustrates that the inductance parameter perturbation does not affect the q-axis current but seriously affects the d-axis current. Fig. 5.3 shows the d- and q-axis disturbances under permanent magnet demagnetization. The d-axis disturbance is zero. The q-axis disturbance is negative and increases with the growth in speed. Fig. 5.3 illustrates that permanent magnet demagnetization does not affect

Table 5.1 Main Parameters of PMSM.

Parameters	Value	Parameters	Value
Stator phase resistance (R_o)	0.02 Ω	d-axis inductances (L_{do})	0.001 H
Flux linkage (Ψ_{ro})	0.892 Wb	q-axis inductances (L_{qo})	0.003572 H

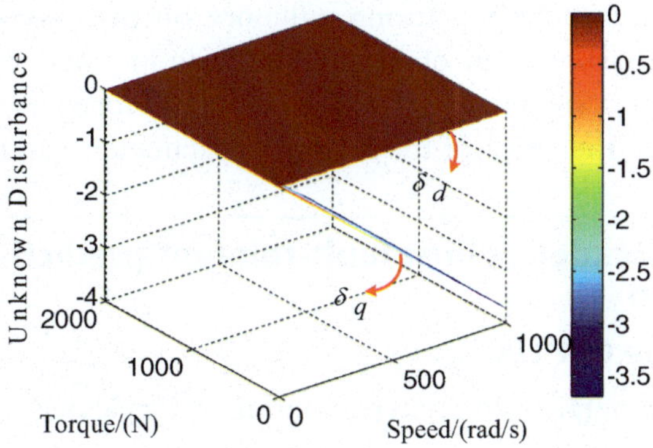

Fig. 5.1 The *d*- and *q*-axis unknown disturbance under resistance parameter perturbation. ($\Delta R = -50\% R_o$).

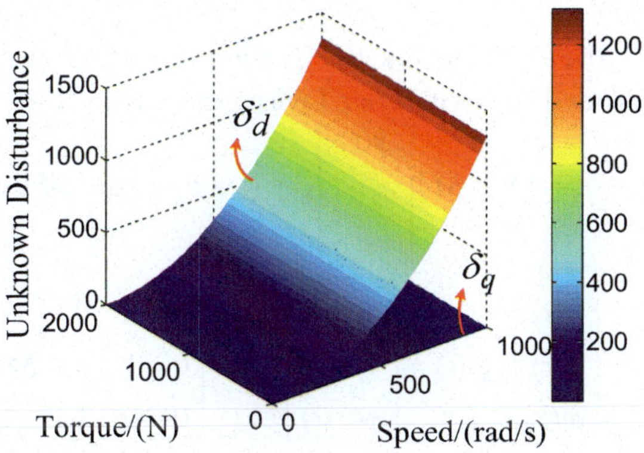

Fig. 5.2 The *d*- and *q*-axis unknown disturbance under inductance parameter perturbation. ($\Delta L_{dq} = -50\% L_{dqo}$).

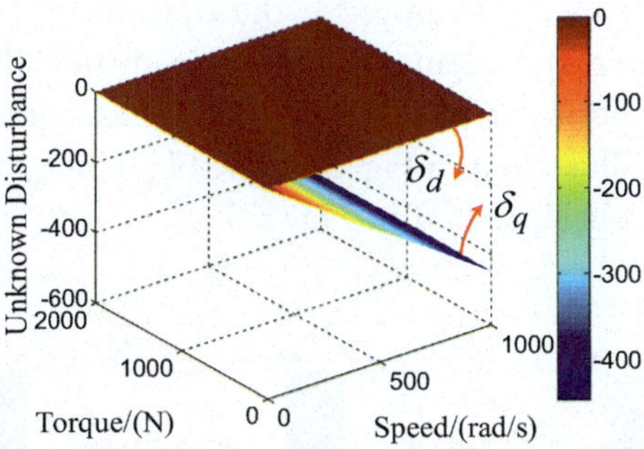

Fig. 5.3 The *d*- and *q*-axis unknown disturbance under permanent magnet demagnetization. ($\Delta \psi_{rd} = -50\% \psi_{ro}$).

the d-axis current but has a strong influence on the q-axis current. From the above analysis, we can observe that the inductance and flux linkage parameters have a significant influence on the performance of the control system, whereas the effect of the resistance parameters can be neglected.

5.4 Cascaded robust fault-tolerant predictive control method

Design of the CRFTPC

The predictive output and reference output are separately defined as

$$\boldsymbol{y}(t+\tau)=\begin{bmatrix} i_d(t+\tau) \\ i_q(t+\tau) \\ \omega(t+\tau) \end{bmatrix};\ \boldsymbol{y}_r(t+\tau)=\begin{bmatrix} i_d^{ref}(t+\tau) \\ i_q^{ref}(t+\tau) \\ \omega^{ref}(t+\tau) \end{bmatrix}$$

To simplify the calculations, the predicted output and the reference output are expanded into 1st-order Taylor series as

$$\boldsymbol{y}(t+\tau)=[\boldsymbol{E}\quad \boldsymbol{M}]\begin{bmatrix} \boldsymbol{y}(t) \\ \dfrac{d\boldsymbol{y}(t)}{dt} \end{bmatrix};\ \boldsymbol{y}_r(t+\tau)=[\boldsymbol{E}\quad \boldsymbol{M}]\begin{bmatrix} \boldsymbol{y}_r(t) \\ \dfrac{d\boldsymbol{y}_r(t)}{dt} \end{bmatrix}$$

where $\boldsymbol{y}(t)=\begin{bmatrix} i_d(t) \\ i_q(t) \\ \omega(t) \end{bmatrix}, \boldsymbol{y}_r(t)=\begin{bmatrix} i_d^{ref}(t) \\ i_q^{ref}(t) \\ \omega^{ref}(t) \end{bmatrix}, \boldsymbol{E}=\begin{bmatrix} 1 & 0 & 0 \\ 0 & 1 & 0 \\ 0 & 0 & 1 \end{bmatrix}, \boldsymbol{M}=\begin{bmatrix} \tau & 0 & 0 \\ 0 & \tau & 0 \\ 0 & 0 & \tau \end{bmatrix}.$

The selection of the cost function reflects the requirements of the control system's performance. Servo control belongs to the tracking control system, and it expects that the controlled output will track the reference input at the fastest speed. To improve the dynamic response of predictive control performance, the optimal control law in this chapter is calculated in a half-sampling period. The block diagram of the conventional predictive control method is illustrated in Fig. 5.4. The block diagram of the proposed CRFTPC method is illustrated in Fig. 5.5.

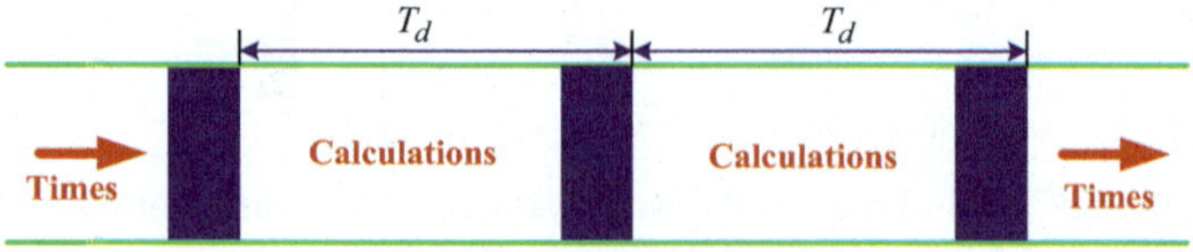

Fig. 5.4 Time diagram of the conventional predictive control method implementation.

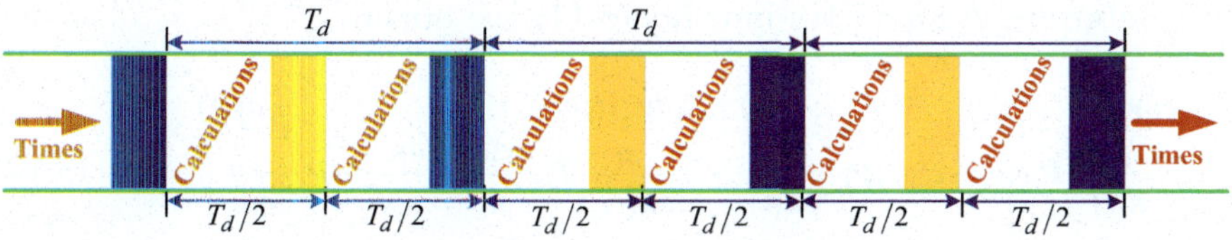

Fig. 5.5 Time diagram of the CRFTPC method implementation.

Thus the cost function is defined as

$$P = \frac{1}{2}\int_0^{\frac{T_d}{2}} e^T(t+\tau)e(t+\tau)d\tau \tag{5.8}$$

where T_d is the sampling period.

With

$$e(t+\tau) = y_r(t+\tau) - y(t+\tau) = [E \quad M]\begin{bmatrix} e(t) \\ \frac{de(t)}{dt} \end{bmatrix} \tag{5.9}$$

where $e(t) = y_r(t) - y(t) = \begin{bmatrix} e_d \\ e_q \\ e_\omega \end{bmatrix} = \begin{bmatrix} i_d^{ref}(t) - i_d(t) \\ i_q^{ref}(t) - i_q(t) \\ \omega^{ref}(t) - \omega(t) \end{bmatrix}$.

Substituting Eq. (5.9) into Eq. (5.8), the cost function can be expressed as

$$\begin{aligned} P &= \frac{1}{2}\int_0^{\frac{T_d}{2}} \begin{bmatrix} e^T(t) & \frac{de^T(t)}{dt} \end{bmatrix} \begin{bmatrix} E^T \\ M^T \end{bmatrix} [E \quad M] \begin{bmatrix} e(t) \\ \frac{de(t)}{dt} \end{bmatrix} d\tau \\ &= e^T(t)P_1 e(t) + \frac{de^T(t)}{dt}P_2 e(t) + e^T(t)P_3\frac{de(t)}{dt} + \frac{de^T(t)}{dt}P_4\frac{de(t)}{dt} \end{aligned} \tag{5.10}$$

where $P_1 = \frac{1}{2}\int_0^{\frac{T_d}{2}} E^T E d\tau$, $P_2 = \frac{1}{2}\int_0^{\frac{T_d}{2}} M^T E d\tau$, $P_3 = \frac{1}{2}\int_0^{\frac{T_d}{2}} E^T M d\tau$, $P_4 = \frac{1}{2}\int_0^{\frac{T_d}{2}} M^T M d\tau$.

The necessary condition for the optimal control law is given by

$$\frac{\partial P}{\partial u(t)} = 0 \tag{5.11}$$

Substituting Eq. (5.10) into Eq. (5.11), we obtain

$$-\frac{\partial \frac{d\boldsymbol{y}^{T}(t)}{dt}}{\partial \boldsymbol{u}(t)}\boldsymbol{P}_{2}\boldsymbol{e}(t)-\boldsymbol{e}^{T}(t)\boldsymbol{P}_{3}\frac{\partial \frac{d\boldsymbol{y}(t)}{dt}}{\partial \boldsymbol{u}(t)}+\frac{\partial \frac{d\boldsymbol{y}^{T}(t)}{dt}}{\partial \boldsymbol{u}(t)}\boldsymbol{P}_{4}\frac{d\boldsymbol{y}(t)}{dt}+\frac{d\boldsymbol{y}^{T}(t)}{dt}\boldsymbol{P}_{4}\frac{\partial \frac{d\boldsymbol{y}(t)}{dt}}{\partial \boldsymbol{u}(t)}=0 \quad (5.12)$$

Thus the optimal control law can be obtained according to Eq. (5.12).

$$\boldsymbol{u}(t)=(\boldsymbol{B})^{-1}\left[\frac{\boldsymbol{P}_{2}}{\boldsymbol{P}_{4}}\boldsymbol{e}(t)-\boldsymbol{A}\boldsymbol{x}(t)-\boldsymbol{D}\boldsymbol{f}_{\psi}+\boldsymbol{G}\delta(t)\right] \quad (5.13)$$

To enhance the robustness of the predictive control system, the IT-SMO is designed to observe external disturbances. The disturbance observed value is used as the feedback input for the predictive control system. According to Eq. (5.13), the optimal control law can be expressed as

$$\boldsymbol{u}(t)=(\boldsymbol{B})^{-1}\left[\frac{\boldsymbol{P}_{2}}{\boldsymbol{P}_{4}}\boldsymbol{e}(t)-\boldsymbol{A}\boldsymbol{x}(t)-\boldsymbol{D}\boldsymbol{f}_{\psi}+\boldsymbol{G}\hat{\delta}(t)\right] \quad (5.14)$$

where $\hat{\delta}$ are the observed values of δ.

According to Eq. (5.14), the error equation of the predictive closed-loop system can be expressed as

$$\begin{aligned}\frac{d\boldsymbol{e}(t)}{dt}&=\frac{d\boldsymbol{y}_{r}(t)}{dt}-\frac{d\boldsymbol{y}(t)}{dt}=-\frac{d\boldsymbol{y}(t)}{dt}=-\boldsymbol{E}\frac{d\boldsymbol{x}(t)}{dt}\\&=-\boldsymbol{A}\boldsymbol{x}(t)-\boldsymbol{B}(\boldsymbol{B})^{-1}\left[\frac{\boldsymbol{P}_{2}}{\boldsymbol{P}_{4}}\boldsymbol{e}(t)-\boldsymbol{A}\boldsymbol{x}(t)-\boldsymbol{D}\boldsymbol{f}_{\psi}+\boldsymbol{G}\hat{\delta}(t)\right]\\&\quad-\boldsymbol{D}\boldsymbol{f}_{\psi}+\boldsymbol{G}\delta(t)\\&=-\frac{\boldsymbol{P}_{2}}{\boldsymbol{P}_{4}}\boldsymbol{e}(t)+\boldsymbol{G}\boldsymbol{e}_{\delta}(t)\end{aligned} \quad (5.15)$$

where $\mathbf{e}_{\delta}(t)=\left(\delta(t)-\hat{\delta}(t)\right)$.

In Section V, when the proposed IT-SMO is convergent, where $\mathbf{e}_{\delta}(t)=0$, then $\frac{d\boldsymbol{e}(t)}{dt}=-\frac{\boldsymbol{P}_{2}}{\boldsymbol{P}_{4}}\boldsymbol{e}(t)$. The Lyapunov function is defined as

$$V_{1}=\frac{1}{2}\boldsymbol{e}^{T}(t)\boldsymbol{e}(t) \quad (5.16)$$

By differentiating the Lyapunov function in Eq. (5.16), we obtain:

$$\begin{aligned}\frac{dV_1}{dt} &= \boldsymbol{e}(t)\frac{d\boldsymbol{e}(t)}{dt} = \boldsymbol{e}^T(t)\left(-\frac{\boldsymbol{P}_2}{\boldsymbol{P}_4}\boldsymbol{e}(t)\right) \\ &= -\frac{3}{T_d}e_d^2 - \frac{3}{T_d}e_q^2 - \frac{3}{T_d}e_\omega^2 \leq 0\end{aligned} \tag{5.17}$$

From Eq. (5.17), the predictive closed-loop system is locally asymptotically stable. This completes the proof.

The predictive speed controller is used to achieve speed control in this chapter. The IT-SMO is designed to observe external disturbances to improve robustness against load perturbations and permanent magnet demagnetization.

According to Eq. (5.14), the predictive speed controller can be designed as

$$i_q^{ref} = i_q = \frac{J}{1.5n_p^2\psi_{ro}}\left[\frac{3}{T_d}e_\omega + \frac{B}{J}\omega + \frac{n_p}{J}\hat{\delta}_\omega\right] \tag{5.18}$$

where $\hat{\delta}_\omega$ is the observed value of δ_ω. i_q^{ref} is the q-axis current reference output by predictive speed controller.

The main objective of the current control is to efficiently control motor currents with high accuracy. Parameter perturbation and permanent magnet demagnetization deteriorate current control performance if these factors are not considered in the design of the controller.

According to Eq. (5.18), the predictive speed controller can be designed as

$$\begin{cases} u_d^{ref} = u_d = \left[\dfrac{3L_{do}}{T_d}e_d + R_o i_d - L_{qo} i_q \omega\right] + \hat{\delta}_d \\ u_q^{ref} = u_q = \left[\dfrac{3L_{qo}}{T_d}e_q + L_{do} i_d \omega + R_o i_q + \psi_{ro}\omega\right] + \hat{\delta}_q \end{cases} \tag{5.19}$$

where $\hat{\delta}_d$, $\hat{\delta}_q$ are the observed values of δ_d, δ_q, u_d^{ref}, and u_q^{ref} are the d- and q-axis voltage reference output by predictive current controller. The block diagram of the PMSM drive system with CRFTPC method is shown in Fig. 5.6.

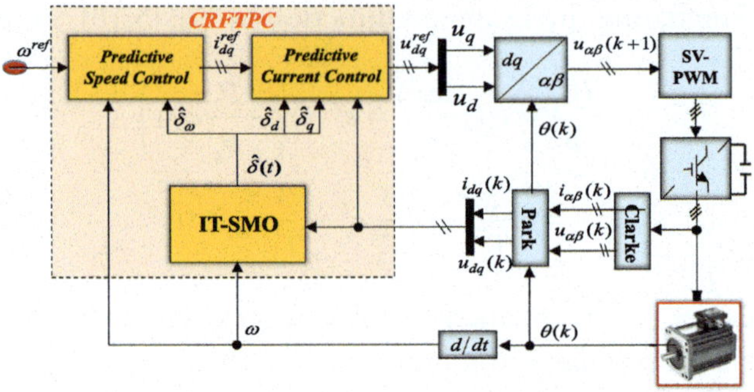

Fig. 5.6 The block diagram of the CRFTPC method with IT-SMO for PMSM.

Design of IT-SMO

According to Eq. (5.15), the IT-SMO can be designed as

$$\frac{d\hat{\boldsymbol{x}}(t)}{dt} = \boldsymbol{A}\hat{\boldsymbol{x}}(t) + \boldsymbol{B}\boldsymbol{u}(t) + \boldsymbol{D}\boldsymbol{f}_{\psi} - \boldsymbol{U}_{o} \tag{5.20}$$

where $\hat{\boldsymbol{x}}$ is the observed value of $\boldsymbol{x}$; $\boldsymbol{U}_o = \begin{bmatrix} U_{od} & U_{oq} & U_{o\omega} \end{bmatrix}^T$ is the sliding mode control function.

Considering the following integral terminal sliding surface vector,

$$\boldsymbol{s}_o = \boldsymbol{e}_o + \lambda \int_0^t sgn(\boldsymbol{e}_o)\, d\tau \tag{5.21}$$

where $\boldsymbol{s}_o = \begin{bmatrix} s_{od} & s_{oq} & s_{o\omega} \end{bmatrix}^T$, $\lambda > 0$, $\boldsymbol{sgn}(\cdot)$ is the sign function, the $\boldsymbol{e}_o$ is defined as

$$\boldsymbol{e}_o = \hat{\boldsymbol{x}}(t) - \boldsymbol{x}(t) = \begin{bmatrix} e_{od} \\ e_{oq} \\ e_{o\omega} \end{bmatrix} = \begin{bmatrix} \hat{i}_d - i_d \\ \hat{i}_q - i_q \\ \hat{\omega} - \omega \end{bmatrix}$$

In practical applications, chattering suppression is important when designing observers, as the chattering of the sign function can easily result in the chattering of the system. For this purpose, the sign function is replaced with a hyperbolic tangent function, which provides smooth continuity. The hyperbolic tangent function can be expressed as:

$$sgn(v) = tanh(v) = \frac{e^{2v} - 1}{e^{2v} + 1} \tag{5.22}$$

Taking the time derivative of Eq. (5.21) and combining Eq. (5.15) with Eq. (5.20), the error equation of the TI-SMO can be obtained as

$$\frac{\boldsymbol{s}_o}{dt} = \boldsymbol{A}\boldsymbol{e}_o + \boldsymbol{G}\delta(t) + \lambda sgn(\boldsymbol{e}_o) - \boldsymbol{U}_o \tag{5.23}$$

Consider the integral terminal sliding surface Eq. (5.21) with the error equation Eq. (5.23). When the sliding mode control function of formula Eq. (5.24) is designed, the error $\boldsymbol{e}_o$ convergence can be guaranteed.

$$\boldsymbol{U}_o = \boldsymbol{A}\boldsymbol{e}_o + \lambda sgn(\boldsymbol{e}_o) + \boldsymbol{k}\boldsymbol{s}_o + \boldsymbol{k}_s sgn(\boldsymbol{s}_o) \tag{5.24}$$

where $\boldsymbol{k} = \begin{bmatrix} k_1 & 0 & 0 \\ 0 & k_2 & 0 \\ 0 & 0 & k_3 \end{bmatrix}$ is a positive diagonal matrix, $\boldsymbol{k}_s = \begin{bmatrix} k_{s1} & 0 & 0 \\ 0 & k_{s2} & 0 \\ 0 & 0 & k_{s3} \end{bmatrix}$ is the observer sliding gain.

Consider the following Lyapunov function candidate

$$V_2 = \frac{1}{2}\boldsymbol{s}_o^T \boldsymbol{s}_o \tag{5.25}$$

Differentiating the Lyapunov function in Eq. (5.25) and combining the obtained result with the result in Eq. (5.24), yields

$$\frac{dV_2}{dt} = \boldsymbol{s}_o^T\left(\boldsymbol{A}\boldsymbol{e}_o + \boldsymbol{G}\delta(t) + \lambda sgn(\boldsymbol{e}_o) - \boldsymbol{U}_o\right) \tag{5.26}$$

By applying the sliding mode control function (5.24), it yields

$$\begin{aligned} \frac{dV_2}{dt} &= \boldsymbol{s}_o^T\left(\boldsymbol{G}\delta(t) - \boldsymbol{k}\boldsymbol{s}_o - \boldsymbol{k}_s sgn(\boldsymbol{s}_o)\right) \\ &= -k_1 s_{od}^2 - k_2 s_{oq}^2 - k_3 s_{o\omega}^2 + \boldsymbol{s}_o^T \boldsymbol{G}\delta(t) - \boldsymbol{s}_o^T \boldsymbol{k}_s sgn(\boldsymbol{s}_o) \\ &\le \boldsymbol{s}_o^T \boldsymbol{G}\delta(t) - k_{s1} \| s_{od} \| - k_{s2} \| s_{oq} \| - k_{s3} \| s_{o\omega} \| \\ &\le \| \boldsymbol{s}_o \| \boldsymbol{G} \| \| \delta(t) \| - k_{s4} \| \boldsymbol{s}_o \| \\ &= \| \boldsymbol{s}_o \| \left(\| \boldsymbol{G} \| \| \delta(t) \| - k_{s4} \right) \end{aligned} \tag{5.27}$$

where $k_{s4} = \min\{k_{s1}, k_{s2}, k_{s3}\}$.

According to an engineering point of view, the disturbances $\delta(t)$ should be bounded, that is, there exists a normal value N satisfying $\| \delta(t) \| \le N$.

According to Eq. (5.27), if k_{s4} satisfies the condition of $k_{s4} \geq \|\mathbf{G}\| N + \eta$, where $\eta > 0$, then $\frac{dV_2}{dt} \leq -\eta \| \mathbf{s}_o \|$. Therefore, the stability and convergence of the IT-SMO is guaranteed based on the sliding-mode control theory. This completes the proof.

According to the sliding mode equivalent principle, when the system reaches the sliding-mode surface that is $\frac{d\mathbf{e}_o}{dt} = \mathbf{e}_o = 0$, then the error equation Eq. (5.23) can be simplified as

$$\mathbf{G}\delta(t) = \mathbf{U}_o \tag{5.28}$$

From Eq. (5.28), the estimated disturbances $\hat{\delta}(t)$ can be represented as

$$\begin{cases} \hat{\delta}_d = L_{do} U_{od} \\ \hat{\delta}_q = L_{qo} U_{oq} \\ \hat{\delta}_\omega = J / n_p U_{o\omega} \end{cases} \tag{5.29}$$

5.5 Simulation results

Simulations are established in MATLAB/Simulink, the DC voltage is 1500 V, and the amplitude value of the stator current is 350 A. The proposed IT-SMO parameters are $\lambda = 30$, $K_1 = K_2 = K_3 = 5000$ and $K_{s1} = K_{s2} = K_{s3} = 100$. To realize the inductance parameter perturbation and permanent magnet demagnetization in MATLAB/Simulink, we reconstruct the PMSM module by using the step module to simulate the inductance parameter perturbation and permanent magnet demagnetization. The main parameters of PMSM used in the simulation are shown in Table 5.1.

Control performance of the proposed CRFTPC method under inductance parameter perturbations

Fig. 5.7 illustrates the simulation results of the impact of inductance parameter perturbation on the proposed CRFTPC method under torque mutation conditions. From Fig. 5.7A, it can be observed that the proposed CRFTPC method effectively achieves efficient control under torque mutation conditions, even when the inductance parameters are mismatched. In addition, Fig. 5.7A demonstrates that the proposed CRFTPC method utilizes a speed-current cascade control structure, which effectively suppresses

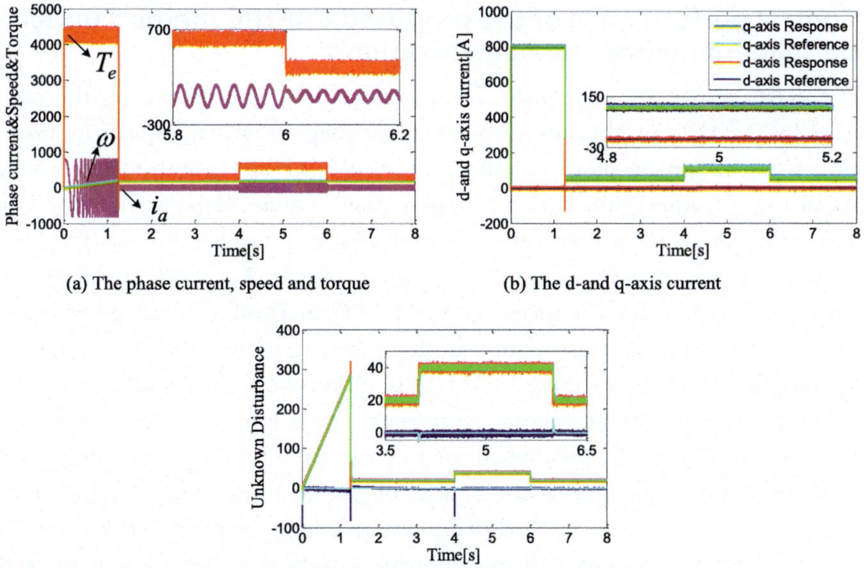

(a) The phase current, speed and torque (b) The d-and q-axis current

(c) Unknown disturbance observation

Fig. 5.7 Simulation results of the proposed CRFTPC method under inductance parameter perturbation. (a) The phase current, speed, and torque. (b) The *d*- and *q*-axis currents. (c) Unknown disturbance observation.

the influence of inductance parameter mismatch on speed and torque. In Fig. 5.7B, it is clearly shown that the proposed CRFTPC method effectively eliminates the deviation of *d*- and *q*-axis currents, allowing the *d*- and *q*-axis response current values to accurately track the reference values under inductance parameter perturbation. The primary reason for eliminating *d*- and *q*-axis current deviation is that the designed IT-SMO accurately compensates for speed and current controllers under inductance parameter mismatch. It is worth noting that a distinct peak occurs when the motor torque stabilizes, which is caused by the compensation provided by the designed IT-SMO. Fig. 5.7C presents the observation results of unknown disturbances caused by inductance parameter perturbation in the designed IT-SMO. The designed IT-SMO accurately observes *d*- and *q*-axis unknown disturbances caused by inductance parameter perturbation, thereby effectively compensating for speed and torque, as clearly shown in Fig. 5.7C. It is important to note that the designed IT-SMO exhibits noticeable overshoot in cases of sudden torque changes, which is attributed to observer modulation, as clearly depicted in Fig. 5.7C.

Control performance of the proposed CRFTPC method under permanent magnet demagnetization

Fig. 5.8A shows the simulation results of speed and torque using the proposed CRFTPC method under permanent magnet demagnetization. From Fig. 5.8A, it can be seen that the proposed CRFTPC method effectively suppresses torque ripple caused by permanent magnet demagnetization. The significant increase in stator current after adopting the proposed CRFTPC method is due to maintaining the torque constant. The conclusion drawn from Fig. 5.8A is that the proposed CRFTPC method achieves good speed and torque control performance under permanent magnet demagnetization. In addition, the proposed CRFTPC method effectively eliminates the influence of permanent magnet demagnetization and enhances the control robustness of torque and speed during torque increases and decreases. Fig. 5.8B shows the simulation results of the *d*- and *q*-axis current tracking performance of the PMSM drive system using the proposed CRFTPC method under permanent magnet demagnetization. In the case of permanent magnet demagnetization, the *d*- and *q*-axis current response values

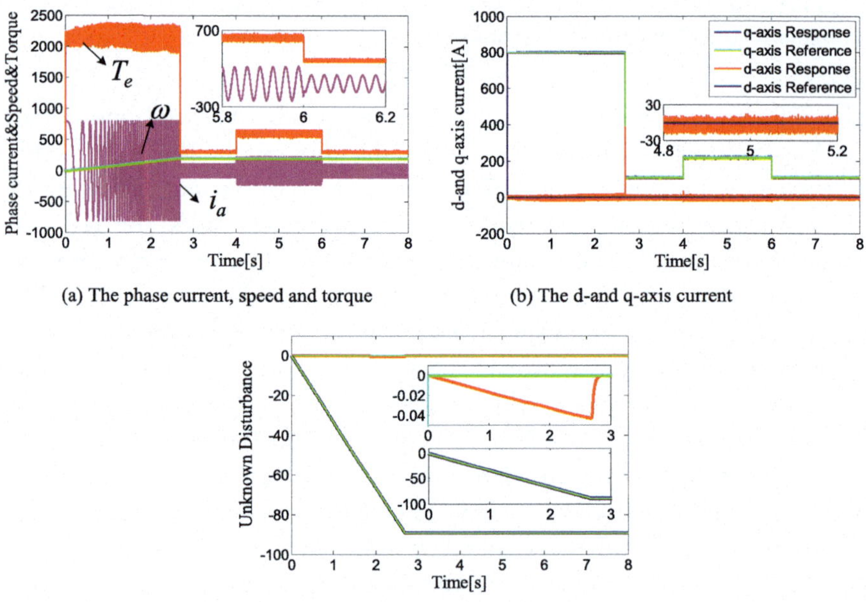

(a) The phase current, speed and torque

(b) The d-and q-axis current

(c) Unknown disturbance observation

Fig. 5.8 Simulation results of the proposed CRFTPC method under permanent magnet demagnetization. (a) The phase current, speed, and torque. (b) The *d*- and *q*-axis currents. (c) Unknown disturbance observation.

accurately track the reference values by adopting the proposed CRFTPC method, as clearly shown in Fig. 5.8B. It is worth noting that the current tracking performance of the PMSM drive system remains good even under sudden torque changes. Note that there is a surge when the load torque is changed suddenly, which is caused by the designed IT-SMO compensating for the speed and current controllers under permanent magnet demagnetization. Fig. 5.8C presents the observation results of the designed IT-SMO for unknown disturbances caused by permanent magnet demagnetization. From Fig. 5.8C, it can be observed that the impact of permanent magnet demagnetization on the speed and current control performance of the PMSM drive system is effectively eliminated by accurately compensating for unknown disturbances.

5.6 Conclusion

In this chapter, a CRFTPC strategy based on IT-SMO for the speed and current control of the PMSM, without the conventional speed PI controller, has been developed. The proposed CRFTPC can achieve a fast dynamic response, which is of significant importance for high-performance FOC-based PMSM drives. The steady-state current tracking error caused by parameter perturbation and permanent magnet demagnetization is eliminated by introducing the disturbance term into the proposed control law. Moreover, the proposed CRFTPC is able to effectively guarantee the control precision of the speed loop regardless of load perturbation and permanent magnet demagnetization. The proposed scheme has been verified through simulations and experimental tests, and the results are satisfactory. Compared with the conventional speed PI controller, the proposed scheme demonstrates its superiority in control precision and disturbance rejection. In addition, the proposed scheme exhibits strong robustness and excellent dynamic response in the case of parameter perturbation and permanent magnet demagnetization, where the weaknesses of the conventional PCC are successfully overcome.

References

[1] A. Consoli, G. Scarcella, et al, Slip-Frequency Detection for Indirect Field-Oriented Control Drives, IEEE Trans. Ind. Electron. 40 (01) (2004) 194–201.
[2] W. Qian, S.K. Panda, et al, Torque ripple minimization in PM synchronous motors using iterative learning control, IEEE Trans. Power Electron. 19 (02) (2004) 272–279.
[3] B. Wang, X. Chen, et al, Robust Predictive Current Control With Online Disturbance Estimation for Induction Machine Drives, IEEE Trans. Power Electron. 32 (06) (2017) 4663–4674.

[4] S.D. M, M. Milinkovic, et al, Stationary frame induction motor feed forward current controller with back EMF compensation, IEEE Trans. Power Electron. 30 (04) (2015) 1356–1366.
[5] H. Guzman, D.M. J, et al, Comparative study of predictive and resonant controllers in fault-tolerant five-phase induction motor drives, IEEE Trans. Ind. Electron. 63 (01) (2016) 606–617.
[6] K. Belda, D. Vošmik, Explicit generalized predictive control of speed and position of PMSM drives, IEEE Trans. Ind. Electron. 63 (06) (2016) 3889–3896.
[7] M. Zhu, H. Wensong, et al, Torque-Ripple-Based Interior Permanent-Magnet Synchronous Machine Rotor Demagnetization Fault Detection and Current Regulation, IEEE Trans. Ind. Appl. 53 (03) (2017) 2795–2804.
[8] A.H. J, L.D. M, A New Bumpless Rotor-Flux Position Estimation Scheme for Vector-Controlled Washing Machine, IEEE Trans. Ind. Inform. 12 (02) (2016) 466–473.
[9] L. Shihua, Z. Liu, Adaptive speed control for permanent-magnet synchronous motor system with variations of load inertia, IEEE Trans. Ind. Electron. 56 (08) (2009) 3050–3059.
[10] S.-C. Carpiuc, et al, Fast real-time constrained predictive current control in permanent magnet synchronous machine-based automotive traction drives, IEEE Trans. Transp. Electr. 1 (04) (2015) 326–335.
[11] P. Cortes, J. Rodriguez, et al, Delay Compensation in Model Predictive Current Control of a Three-Phase Inverter, IEEE Trans. Ind. Electron. 59 (02) (2012) 1323–1325.
[12] C.S. Lim, E. Levi, et al, FCSMPC- Based Current Control of a Five-Phase Induction Motor and its Comparison with PI-PWM Control, IEEE Trans. Ind. Electron. 61 (01) (2014) 149–163.
[13] M. Siami, K.D. A, et al, Robustness improvement of predictive current control using prediction error correction for permanent-magnet synchronous machines, IEEE Trans. Ind. Electron. 63 (06) (2016) 3458–3466.
[14] M. Yang, X. Lang, et al, A Flux Immunity Robust Predictive Current Control With Incremental Model and Extended State Observer for PMSM Drive, IEEE Trans. Power Electron. 32 (12) (2017) 9267–9279.
[15] Z. Xiaoguang, H. Benshuai, et al, Deadbeat Predictive Current Control of Permanent-Magnet Synchronous Motors with Stator Current and Disturbance Observer, IEEE Trans. Power Electron. 32 (05) (2017) 3818–3834.
[16] T. Türker, et al, A robust predictive current controller for PMSM drives, IEEE Trans. Ind. Electron. 63 (06) (2016) 3906–3914.
[17] S. Chai, W. Liuping, et al, A cascade MPC control structure for a PMSM with speed ripple minimization, IEEE Trans. Ind. Electron. 60 (08) (2013) 2978–2987.
[18] J. Sawma, F. Khatounian, et al, Cascaded dual-model-predictive control of an active front-end rectifier, IEEE Trans. Ind. Electron. 63 (07) (2016) 4604–4614.
[19] E.J. Fuentes, C.A. Silva, et al, Predictive Speed Control of a Two-Mass System Driven by a Permanent Magnet Synchronous Motor, IEEE Trans. Ind. Electron. 59 (07) (2012) 2840–2848.
[20] E. Fuentes, D. Kalise, et al, Cascade-Free Predictive Speed Control for Electrical Drives, IEEE Trans. Ind. Electron. 61 (05) (2014) 2176–2184.
[21] C. Garcia, et al, Full predictive cascaded speed and current control of an induction machine, IEEE Trans. Energy Convers. 31 (03) (2016) 1059–1067.
[22] A.D. Alexandrou, et al, Development of a constant switching frequency deadbeat predictive control technique for field-oriented synchronous permanent-magnet motor drive, IEEE Trans. Ind. Electron. 63 (08) (2016) 5167–5175.

CHAPTER 6

Advanced robust fault-tolerant predictive control strategies for PMSM drives under parameter disturbances

6.1 Introduction

Permanent magnet synchronous motors (PMSMs) have been widely applied in modern industry and transportation as a crucial power source due to their advantages of high efficiency, power density, and torque density [1,2]. Due to their benefits of simple structure, convenient parameter adjustment, and strong dynamic response, the model predictive control (MPC) is highly suitable for multiobjective coordination and nonlinear control systems and has been widely used in PMSM drive systems [3–5]. The MPC method heavily relies on the accurate mathematical model of the PMSM drive system, so its control performance largely depends on the accuracy of the model parameters [6,7]. However, the PMSM drive system may experience parameter disturbance faults during actual operation, where the parameters used by MPC do not match the actual parameters of the motor. These parameter disturbance faults can lead to inaccurate voltage calculations, thereby deteriorating the performance of MPC and increasing stator flux/torque ripple.

The PMSM drive system is a dual-loop control system, including a speed external loop and a current internal loop. Many improved methods have been proposed to address the issue of parameter disturbance faults in the current internal loop [6,8–16]. Wang et al. [6] proposed a deadbeat predictive current control method that combines high-frequency current injection, which can effectively improve the parameter disturbance identification accuracy of interior PMSMs in electric vehicles. An et al. [8,9] proposed an adaptive mechanism based on a fuzzy logic controller to adjust the continuously updated inductance parameter, which can improve the accuracy of inductance parameter disturbance fault diagnosis in the dq-axis while suppressing current distortion. Xie et al. [10] proposed a method for

Robust Model Predictive Control of Permanent Magnet Synchronous Motor Under Parameter Mismatch
https://doi.org/10.1016/B978-0-443-44209-4.00011-7

motor parameter disturbance fault diagnosis by using an online optimized particle swarm algorithm, which offers advantages such as fast convergence, good stability, and ease of implementation. Bai et al. [11] introduced a multimodel adaptive disturbance observer based on the single-model adaptive principle, which can accelerate the speed of disturbance fault diagnosis and improve transient performance. Yu et al. [12] proposed a composite observer involving generalized proportional-integral and sliding mode observers, which can achieve zero steady-state error and improve dynamic response performance. Yang et al. [13] proposed a Luenberger disturbance observer without involving motor parameters, which can achieve delay compensation and significantly improve current tracking performance. Zhang et al. [14] proposed a nonparametric predictive model based on current prediction errors, which can effectively update the predictive model in real time to significantly enhance the accuracy of parameter disturbance fault diagnosis. In [6,8–16], the proposed improved methods effectively address the issue of current internal loop parameter disturbance faults in the PMSM drive system. However, the parameter disturbance faults of the speed external loop can also affect the control performance of the PMSM drive system, thereby restricting the practical engineering application of PMSM drives.

Many improved predictive control methods have been proposed to overcome the impact of parameter disturbance faults on the speed external loop of PMSM drive systems. Xu et al. [17] proposed a sliding mode observer based on a sine saturation function and a speed external loop with deadbeat control, which can achieve accurate parameter disturbance fault diagnosis and significantly improve speed tracking performance. Wu et al. [18] proposed an improved sliding mode approach rate and established a rotating speed flux cascade sliding mode control structure, which effectively resists disturbance performance under system parameter disturbance faults. Chen et al. [19] proposed an extended state observer with a resonant model and decoupled feedback controllers applied to the internal and external control loops, which can suppress external parameter disturbance faults and maintain high tracking accuracy. Yu et al. [20] developed a cascaded deadbeat control strategy for speed and current using generalized proportional-integral and sliding mode observations, which can mitigate the intrinsic time-delay problem in the speed observer and maintain strong robustness against load disturbances and parameter disturbances. Yang et al. [12] developed a control framework in which the speed external loop implements sliding mode control integrating motor parameters into the control rate, which can provide fast fault diagnosis and robust disturbance rejection.

Li et al. [21] introduced a speed-current dual second-order sliding mode disturbance observer integrated with deadbeat predictive control, which can completely correct control errors and effectively suppress sliding mode chattering. In [12,17–22], the proposed improved methods can effectively eliminate the influence of parameter disturbance faults on the speed external loop of PMSM drive systems. However, the PMSM drive system is a dual-loop structure, and the improved control method of an independent speed external loop or current internal loop cannot effectively solve the problem of parameter disturbance fault diagnosis.

This chapter is organized as follows. The PMSM mathematical model under parameter disturbances is introduced in Section 6.2, providing the theoretical foundation for control system analysis. Building on this foundation, the RS-SFTPC strategy with a composite observer is proposed in Section 6.3, presenting the core methodology of this research. The simulation results validating the proposed method's effectiveness are presented in Section 6.4. The conclusion is presented in Section 6.5.

6.2 Parameters disturbance faults analysis of PMSM drives

Under normal conditions, the stator flux and voltage equations of the PMSM drive system can be expressed as

$$\begin{cases} u_d = R_o i_d - \omega_e \psi_q + \dfrac{d\psi_d}{dt} \\ u_q = R_o i_q + \omega_e \psi_d + \dfrac{d\psi_q}{dt} \end{cases} \tag{6.1}$$

with

$$\begin{cases} \psi_d = L_o i_d + \psi_o \\ \psi_q = L_o i_q \end{cases} \tag{6.2}$$

Under normal conditions, the motion equation of the PMSM drive system can be expressed as

$$\frac{d\omega_e}{dt} = \frac{p}{J}(T_e - T_L) = \frac{3p^2\psi_o}{2J} i_q - \frac{pT_L}{J} \tag{6.3}$$

where u_d and u_q represent the d-axis and q-axis voltage components, respectively; i_d and i_q represent the d-axis and q-axis current components;

ψ_d and ψ_q represent the d-axis and q-axis stator flux components; R_o and ψ_o represent the stator resistance and rotor magnetic flux, respectively. L_o represents the nominal values of the stator inductance; p is the number of pole pairs of the motor; J is the moment of inertia; T_L is the load torque; T_e is the electromagnetic torque.

To analyze the model parameters disturbance, introducing $\psi_\delta = \Delta\psi + \psi_o$, $L_\delta = \Delta L + L_o$, where ΔL and $\Delta\psi$ are uncertain components caused by the parameter disturbance faults. According to Eq. (6.1)–Eq. (6.3), the stator flux and voltage equations of the PMSM drives system can be rewritten as

$$\frac{d\boldsymbol{x}}{dt} = \boldsymbol{A}\boldsymbol{x} + \boldsymbol{B}\boldsymbol{u} + \boldsymbol{C}\boldsymbol{f}_a + \boldsymbol{D}\boldsymbol{f}_b \tag{6.4}$$

where

$$\boldsymbol{x} = \begin{bmatrix} \psi_d \\ \psi_q \\ \omega_e \end{bmatrix}, \quad \boldsymbol{u} = \begin{bmatrix} u_d \\ u_q \\ \psi_q \end{bmatrix}, \quad \boldsymbol{f}_a = \begin{bmatrix} f_d \\ f_q \\ f_\omega \end{bmatrix}, \quad \boldsymbol{f}_b = \begin{bmatrix} 0 \\ 0 \\ T_L \end{bmatrix}, \quad \boldsymbol{A} = \begin{bmatrix} \frac{-R_o}{L_o} & \omega_e & 0 \\ -\omega_e & \frac{-R_o}{L_o} & 0 \\ 0 & 0 & 0 \end{bmatrix},$$

$$\boldsymbol{B} = \begin{bmatrix} 1 & 0 & 0 \\ 0 & 1 & 0 \\ 0 & 0 & \frac{3p^2\psi_{fo}}{2JL_o} \end{bmatrix}, \boldsymbol{C} = \begin{bmatrix} 1 & 0 & 0 \\ 0 & 1 & 0 \\ 0 & 0 & 1 \end{bmatrix}, \boldsymbol{D} = \begin{bmatrix} 0 & 0 & -\frac{n}{J} \end{bmatrix}^T,$$

with

$$\begin{cases} f_d = \dfrac{R_o}{L_o}\psi_o + \dfrac{\Delta L R_o}{L_o + \Delta L} i_d - \dfrac{\Delta L}{L_o + \Delta L} u_d \\ f_q = \dfrac{\Delta L}{L_o + \Delta L}(R_o i_q - u_q + \psi_o \omega_e) - \dfrac{L_o}{L_o + \Delta L}\Delta\psi\omega_e \\ f_\omega = \dfrac{3p^2\Delta\psi}{2J} i_q \end{cases} \tag{6.5}$$

where f_d and f_q are disturbance terms in the flux linkage loop; f_ω is disturbance terms in the speed loop.

According to Eqs. (6.1) and (6.2), the voltage equations of the PMSM under steady-state conditions can be obtained as

$$\begin{cases} u_d' = R_o i_d - L_o \omega_e i_q \\ u_q' = R_o i_q + L_o \omega_e i_d + \omega_e \psi_o \end{cases} \tag{6.6}$$

where u_d' is the d-axis voltage component in steady state; u_q' is the q-axis voltage component in steady state.

Introducing $\delta_1 = \Delta\psi / \psi_o$, $\delta_2 = \Delta L / L_o$, where δ_1 and δ_2 are the parameters disturbance rates of flux linkage and inductance, respectively. Under $i_d^{ref} = 0$; combining Eq. (6.4), (6.5) and (6.6), yields

$$\begin{cases} \Delta E_d = -\dfrac{R_o}{L_o}\psi_o - \dfrac{\delta_2}{1+\delta_2}\left(\omega_e L_o \dfrac{2T_e}{3p\psi_o}\right) \\ \Delta E_q = -\dfrac{\delta_1}{1+\delta_2}\omega_e \psi_o \\ \Delta E_\omega = -\delta_1 \dfrac{p}{J} T_e \end{cases} \tag{6.7}$$

where ΔE_d, ΔE_q, and ΔE_ω denote the absolute values of the d-axis disturbance term, the q-axis disturbance term, and the rotational speed disturbance term, respectively.

According to Eq. (6.7), parameter disturbance faults can be quantitatively analyzed under various operating conditions. Fig. 6.1 illustrates the impact of parameter disturbances on the d-axis stator flux loop. The

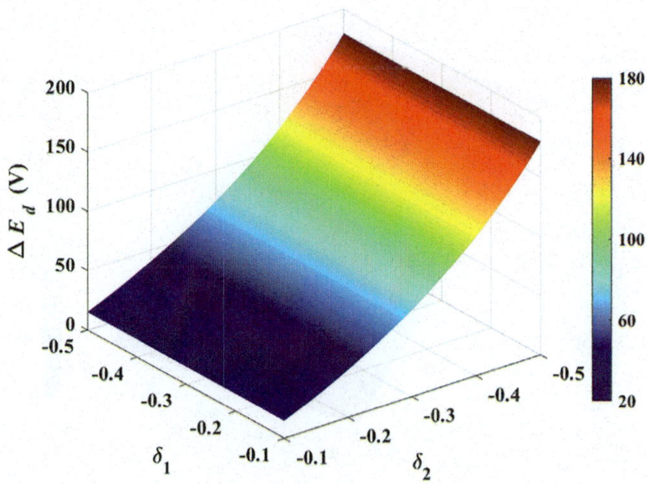

Fig. 6.1 The d-axis flux linkage loop disturbance under various levels of parameter deviations.

results show that the d-axis stator flux disturbance increases with higher inductance disturbance rates, reaching a peak value of 179 V at 300 rad/s. Fig. 6.2 demonstrates the influence of parameter disturbances on the q-axis stator flux loop. It is observed that the q-axis flux linkage disturbance grows with increasing inductance and flux linkage disturbance rates, peaking at 234 V when the speed is 300 rad/s. Fig. 6.3 presents the effect of parameter disturbances on the speed loop under different

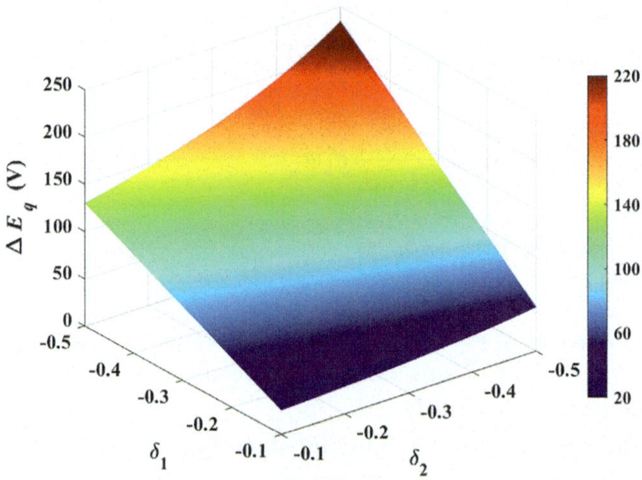

Fig. 6.2 The *q*-axis flux linkage loop disturbance under various levels of parameter deviations.

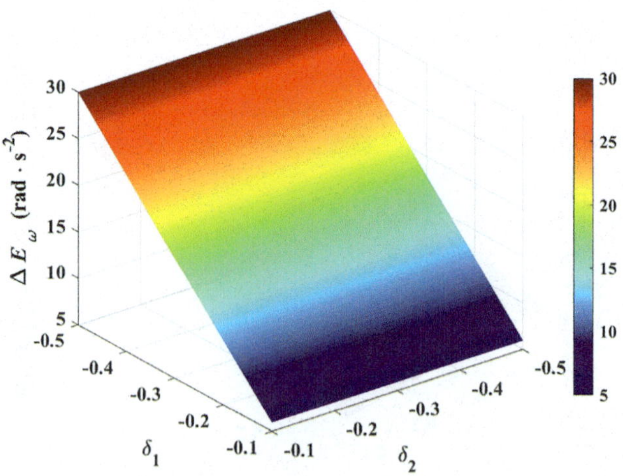

Fig. 6.3 Speed loop disturbance under various levels of parameter deviations.

torque conditions. At a torque of 1500 N · m, the speed disturbance term increases with the flux linkage disturbance rate and reaches a maximum value of 30 rad/s^2. The results clearly indicate that higher torque levels lead to larger disturbance terms under the same parameter disturbance conditions. In summary, parameter disturbance faults can significantly degrade the control accuracy of the system. Notably, the disturbance term escalates with increasing parameter disturbance rates, which may lead to system failure in severe cases.

6.3 RS-SFTPC method with composite discrete ESO

Parameter disturbance faults and digital control delays can significantly degrade the control accuracy of PMSM drive systems, particularly when conventional compensation strategies fail to effectively address these issues. To this end, this section proposes an RS-SFTPC strategy based on an ultra-local model. By integrating a composite discrete ESO, this strategy reconstructs the predictive control architecture while achieving real-time diagnosis of multiloop disturbances and delay compensation.

RS-SFTPC–based on composite discrete ESO

For any given single-input, single-output system, it is possible to express it using an nth-order differential equation, which is presented in the form of Eq. (6.8).

$$a_0 y^{(n)} + a_1 y^{(n-1)} + \ldots + a_{n-1} y' + a_n y = b_0 u^m + b_1 u^{m-1} + \ldots + b_{m-1} u' + b_m u \tag{6.8}$$

where u represents the input of the system; y denotes the output of the system; a_i ($i = 1,2, \ldots, n$) and b_j ($j = 1,2, \ldots, m$) are the coefficients of variables y and u, respectively.

According to Eq. (6.8) and model-free control [22], Eq. (6.8) can be represented by the ultra-local model, as shown in Eq. (6.9).

$$y^{(v)} = \alpha u + F \tag{6.9}$$

where y is the control output; u is the control input; α is the designed input gain; v represents the order of the system; and F denotes the total disturbance of the system, which encompasses the sum of disturbances and unmodeled parts.

When $v = 1$, the ultra-local model expression for a first-order nonlinear system can be given as

$$\frac{dy}{dt} = \alpha u + F \tag{6.10}$$

Combining with Eq. (6.4) and Eq. (6.10), one obtains

$$\frac{d\omega_e}{dt} = \alpha \psi_q + d \tag{6.11}$$

where α is the gain of the speed controller to be designed; d represents the lumped disturbances in the speed loop.

Using the forward Euler discretization method, discretizing Eq. (6.11) yields

$$\omega_e(k+1) = \omega_e(k) + T_s\left(\alpha \psi_q(k) + d(k)\right) \tag{6.12}$$

where T_s represents the sampling time.

According to Eq. (6.12), the q-axis flux linkage command value can be obtained as

$$\psi_q^{ref}(k) = \frac{\omega_e^{ref}(k)}{\alpha T_s} - \frac{\hat{\omega}_e(k)}{\alpha T_s} - \frac{\hat{d}(k)}{\alpha} \tag{6.13}$$

where ψ_q^{ref} is the command value of the q-axis flux linkage; ω_e^{ref} is the command value of the speed; $\hat{\omega}_e$ is the observed value of ω_e; $\hat{d}$ is the observed value of d.

To address the one-sample delay issue present in actual digital control systems, $\hat{\omega}_e(k)$ and $\hat{d}(k)$ are replaced with $\hat{\omega}_e(k+1)$ and $\hat{d}(k+1)$, respectively. Ultimately, the command value for the q-axis flux linkage can be obtained as

$$\psi_q^{ref}(k) = \frac{\omega_e^{ref}(k+1)}{\alpha T_s} - \frac{\hat{\omega}_e(k+1)}{\alpha T_s} - \frac{\hat{d}(k+1)}{\alpha} \tag{6.14}$$

where $\hat{\omega}(k+1)$ and $\hat{d}(k+1)$ are the predicted values of speed and disturbance, respectively.

The three known time points are $t_{k-2} = (k-2)T_s$, $t_{k-1} = (k-1)T_s$ and $t_k = kT_s$. Based on the Lagrange interpolation method, the q-axis stator flux command value at the time (k + 1) can be deduced. The Lagrange basis polynomial $L_i(t)$ is given as

$$\begin{cases} L_0(t) = \dfrac{(t - t_{k-1})(t - t_k)}{(t_{k-2} - t_{k-1})(t_{k-2} - t_k)} = \dfrac{(t - t_{k-1})(t - t_k)}{2T_s^2} \\ L_1(t) = \dfrac{(t - t_{k-2})(t - t_k)}{(t_{k-1} - t_{k-2})(t_{k-1} - t_k)} = \dfrac{(t - t_{k-2})(t - t_k)}{-T_s^2} \\ L_2(t) = \dfrac{(t - t_{k-2})(t - t_{k-1})}{(t_k - t_{k-2})(t_k - t_{k-1})} = \dfrac{(t - t_{k-2})(t - t_{k-1})}{2T_s^2} \end{cases} \tag{6.15}$$

Upon substituting the target time $t = (k+1)T_s$ into Eq. (6.15) can be derived as

$$\psi_q^{ref}(k+1) = \psi_q^{ref}(k-2) \cdot L_0(t_{k+1}) + \psi_q^{ref}(k-1) \cdot L_1(t_{k+1}) + \psi_q^{ref}(k) \cdot L_2(t_{k+1}) \tag{6.16}$$

Through calculations, the values of the coefficients are determined to be $L_0(t_{k+1}) = 1$, $L_1(t_{k+1}) = -3$, and $L_2(t_{k+1}) = 3$, respectively. Consequently, the final prediction formula is obtained as

$$\psi_q^{ref}(k+1) = 3\psi_q^{ref}(k) - 3\psi_q^{ref}(k-1) + \psi_q^{ref}(k-2) \tag{6.17}$$

According to Eq. (6.4), the state equation for the stator flux can be obtained as

$$\begin{bmatrix} \psi_d(k+1) \\ \psi_q(k+1) \end{bmatrix} = \underbrace{\begin{bmatrix} 1 - \dfrac{T_s R_o}{L_o} & T_s \omega_e \\ -T_s \omega_e & 1 - \dfrac{T_s R_o}{L_o} \end{bmatrix}}_{E_s} \underbrace{\begin{bmatrix} \psi_d(k) \\ \psi_q(k) \end{bmatrix}}_{\psi_{s(k)}} + T_s \underbrace{\begin{bmatrix} u_d(k) \\ u_q(k) \end{bmatrix}}_{U_s(k)} + T_s \underbrace{\begin{bmatrix} f_d(k) \\ f_q(k) \end{bmatrix}}_{f_s(k)} \tag{6.18}$$

According to Eq. (6.10), the ultra-local model of the stator flux inner loop is designed as

$$\begin{bmatrix} \psi_d(k+1) \\ \psi_q(k+1) \end{bmatrix} = \begin{bmatrix} 1 & 0 \\ 0 & 1 \end{bmatrix} \begin{bmatrix} \psi_d(k) \\ \psi_q(k) \end{bmatrix} + \underbrace{\begin{bmatrix} \beta_1 & 0 \\ 0 & \beta_2 \end{bmatrix}}_{\beta} \begin{bmatrix} u_d \\ u_q \end{bmatrix} + T_s \underbrace{\begin{bmatrix} F_d(k) \\ F_q(k) \end{bmatrix}}_{F_s(k)} \tag{6.19}$$

with

$$\begin{bmatrix} F_d(k) \\ F_q(k) \end{bmatrix} = \begin{bmatrix} -\dfrac{R_o}{L_o} & \omega_e \\ -\omega_e & -\dfrac{R_o}{L_o} \end{bmatrix} \begin{bmatrix} \psi_d(k) \\ \psi_q(k) \end{bmatrix} + \begin{bmatrix} f_d(k) \\ f_q(k) \end{bmatrix} \tag{6.20}$$

where $\beta_1, \beta_2 \in \Re$ represents the gain matrix that needs to be designed for the stator flux loop.

According to Eq. (6.19), under parameter disturbance faults, the required dq-axis voltage command values can be obtained as follows.

$$\begin{cases} u_d^{ref}(k) = \beta_1^{-1}\left[\dfrac{\psi_d^{ref}(k+1)}{T_s} - \dfrac{\widehat{\psi}_d(k+1)}{T_s}\right] - \beta_1^{-1}\widehat{F}_d(k+1) \\ u_q^{ref}(k) = \beta_2^{-1}\left[\dfrac{\psi_q^{ref}(k+1)}{T_s} - \dfrac{\widehat{\psi}_q(k+1)}{T_s}\right] - \beta_2^{-1}\widehat{F}_q(k+1) \end{cases} \tag{6.21}$$

Combining Eq. (6.14) and Eq. (6.21), the output of the robust speed-stator flux-tolerant predictive controller in discrete-time settings can be obtained as

$$\boldsymbol{U}_o^{ref}(k) = \mathbf{G}\left[\boldsymbol{x}^{ref}(k+1) - \hat{\boldsymbol{x}}(k+1)\right] - \boldsymbol{H}\widehat{\boldsymbol{F}}_o(k+1) \tag{6.22}$$

where $\boldsymbol{U}_o^{ref} = \begin{bmatrix} u_d^{ref} \\ u_q^{ref} \\ \psi_q^{ref} \end{bmatrix}$, $\boldsymbol{x}^{ref} = \begin{bmatrix} \psi_d^{ref} \\ \psi_q^{ref} \\ \omega_e^{ref} \end{bmatrix}$, $\hat{\boldsymbol{x}} = \begin{bmatrix} \hat{\psi}_d \\ \hat{\psi}_q \\ \hat{\omega}_e \end{bmatrix}$, $\hat{\boldsymbol{F}}_o = \begin{bmatrix} \hat{F}_d \\ \hat{F}_d \\ \hat{d} \end{bmatrix}$, $\mathbf{G} = T_s^{-1} \cdot \boldsymbol{H}$,

$$\boldsymbol{H} = \begin{bmatrix} \beta_1^{-1} & 0 & 0 \\ 0 & \beta_2 & 0 \\ 0 & 0 & \alpha^{-1} \end{bmatrix}.$$

Fig. 6.4 shows the RS-SFTPC strategy of the PMSM drive system based on a composite discrete ESO. The core components of the control schematic focus on the speed external loop and the stator flux internal loop

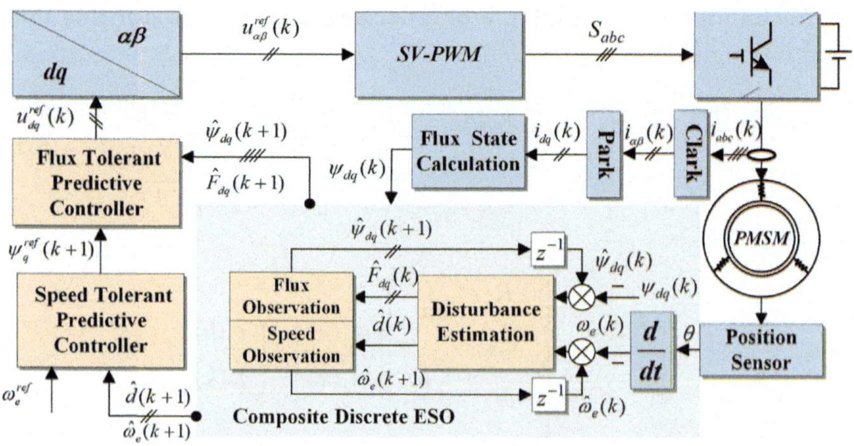

Fig. 6.4 Structural diagram of the PMSM drive system with RS-SFTPC.

control, both of which adopt a tolerant predictive controller structure based on an ultra-local model. The composite discrete ESO is used for parameter disturbance fault diagnosis, and its estimated values are input into the corresponding tolerant predictive controller so that the controller can generate the required voltage command values to drive the PMSM.

Design of composite discrete ESO

The composite discrete ESO is designed to diagnose parameter disturbance faults, in which both the speed loop and the stator flux loop are addressed using a second-order ESO.

$$\begin{cases} e_\omega = \hat{\omega}_e - \omega_e \\ \dfrac{d\hat{\omega}_e}{dt} = \alpha\psi_q - \gamma_1 e_\omega + \hat{d} \\ \dfrac{d\hat{d}}{dt} = -\gamma_2 e_\omega \end{cases} \tag{6.23}$$

where γ_1 and γ_2 are the observer gains to be designed, $\gamma_1, \gamma_2 \in \Re$.

According to Eq. (6.23), the characteristic equation of the speed ESO can be derived as

$$s^2 + \gamma_1 s + \gamma_2 = 0 \tag{6.24}$$

To position the roots of the characteristic equation, wit ω_0 representing the system bandwidth, the following definition can be set as

$$\begin{cases} \gamma_1 = 2\omega_0 \\ \gamma_2 = \omega_0{}^2 \end{cases} \tag{6.25}$$

According to Eq. (6.19), define $\boldsymbol{U}_s$ and $\boldsymbol{F}_s$ as the state variables of the flux linkage control system. The flux linkage observer can be constructed as

$$\begin{cases} \boldsymbol{e}_s = \hat{\psi}_s - \psi_s \\ \dfrac{d\hat{\psi}_s}{dt} = \beta\boldsymbol{U}_s - \zeta_1\boldsymbol{e}_s + \hat{\boldsymbol{F}}_s \\ \dfrac{d\hat{\boldsymbol{F}}_s}{dt} = -\zeta_2\boldsymbol{e}_s \end{cases} \tag{6.26}$$

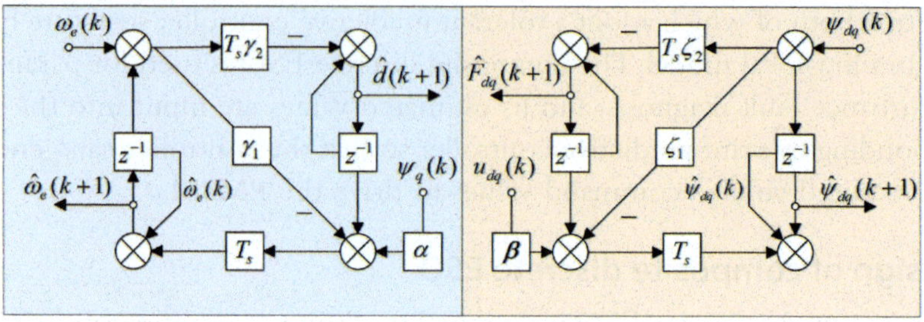

Fig. 6.5 Block diagram based on the composite discrete ESO.

where $\boldsymbol{e}_s = \begin{bmatrix} e_d \\ e_q \end{bmatrix} = \begin{bmatrix} \hat{\psi}_d - \psi_d \\ \hat{\psi}_q - \psi_q \end{bmatrix}$; ζ_1 and ζ_2 are the observer gains to be designed, $\zeta_1 = \zeta_1 \cdot \boldsymbol{I}$, $\zeta_2 = \zeta_2 \cdot \boldsymbol{I}$, $\boldsymbol{I}$ is the identity matrix, $[\zeta_1 \quad \zeta_2] = [2\omega_0 \quad \omega_0^2]$. In a general sense, when ω_0 is excessively small, the observer's dynamic performance will degrade. On the other hand, if ω_0 is set too high, the system's robustness will be adversely affected, which may result in system instability. Therefore, under the premise of maintaining system stability, ω_0 should be taken as large a value as possible (Fig. 6.5).

According to Eqs. (6.23) and (6.26), the composite discrete ESO can be obtained as

$$\begin{cases} \boldsymbol{e}_o(k) = \hat{\boldsymbol{x}}(k) - \boldsymbol{x}(k) \\ \hat{\boldsymbol{x}}(k+1) = \hat{\boldsymbol{x}}(k) + \boldsymbol{M}\boldsymbol{U}_o(k) + \boldsymbol{N}\widehat{\boldsymbol{F}}_o(k) - \kappa_1 \boldsymbol{e}_o(k) \\ \widehat{\boldsymbol{F}}_o(k+1) = \widehat{\boldsymbol{F}}_o(k) - \kappa_2 \boldsymbol{e}_o(k) \end{cases} \tag{6.27}$$

where $\boldsymbol{e}_o = \begin{bmatrix} e_d \\ e_q \\ e_\omega \end{bmatrix}$, $\boldsymbol{M} = T_s \begin{bmatrix} \beta_1 & 0 & 0 \\ 0 & \beta_2 & 0 \\ 0 & 0 & \alpha \end{bmatrix}$, $\boldsymbol{N} = T_s \boldsymbol{I}_{3\times 3}$,

$$\kappa_1 = T_s \begin{bmatrix} \zeta_1 & 0 & 0 \\ 0 & \zeta_1 & 0 \\ 0 & 0 & \gamma_1 \end{bmatrix}, \kappa_2 = T_s \begin{bmatrix} \zeta_2 & 0 & 0 \\ 0 & \zeta_2 & 0 \\ 0 & 0 & \gamma_2 \end{bmatrix}.$$

6.4 Simulation results

To validate the feasibility of the proposed RS-SFTPC strategy with parameter disturbance fault diagnosis, a simulation model is constructed on the MATLAB/Simulink platform. A comparative analysis is conducted

Table 6.1 Main parameters of PMSM.

Parameter	Value	Parameter	Value
Inductance (L_o)	2.892/mH	Stator phase resistance (R_o)	0.02/Ω
Flux linkage of PM (Ψ_o)	0.782/Wb	Rotational inertia (T_n)	1200/ N·m

on the simulation results of the conventional PI-MPC method and the proposed RS-SFTPC method under the same operating conditions. The simulation parameters are detailed in Table 6.1. In the simulation, the speed command for both control strategies is set to 80 rad/s, and the load torque is maintained constant at 1200 N · m. The bandwidth is set to 8000 rad/s at this frequency to guarantee the convergence of the ESO.

Simulation for PMSM under inductance parameter disturbance fault

Under both the conventional PI-MPC method and the proposed RS-SFTPC method, configure the system is configured such that, at the 0.6-second mark, the inductance parameter experiences a sudden decrease to half of its original value, transitioning abruptly from 2.892 mH to 1.446 mH. Comparative simulation results of the conventional PI-MPC strategy and the proposed RS-SFTPC strategy under inductance parameter disturbance are shown in Figs. 6.6–6.9. Fig. 6.6 presents a comparative simulation of stator flux linkage tracking between the conventional PI-MPC method and the proposed RS-SFTPC method. During the period from 0 to 0.6 s under normal operating conditions, both control strategies demonstrate satisfactory stator flux tracking performance. However, significant differences emerge when an inductance parameter disturbance occurs. As shown in Fig. 6.6A, under the conventional PI-MPC method, the actual stator flux linkage fails to accurately follow its reference value, exhibiting a noticeable deviation along with considerable flux fluctuation, and the steady-state error reaches as high as 0.4 Wb. In contrast, Fig. 6.6B illustrates that, with the proposed RS-SFTPC method, the stator flux tracking error is significantly reduced to 0.002 Wb, representing a 99.5% reduction compared to the conventional PI-MPC approach, which highlights its superior robustness and tracking accuracy under inductance parameter disturbance conditions.

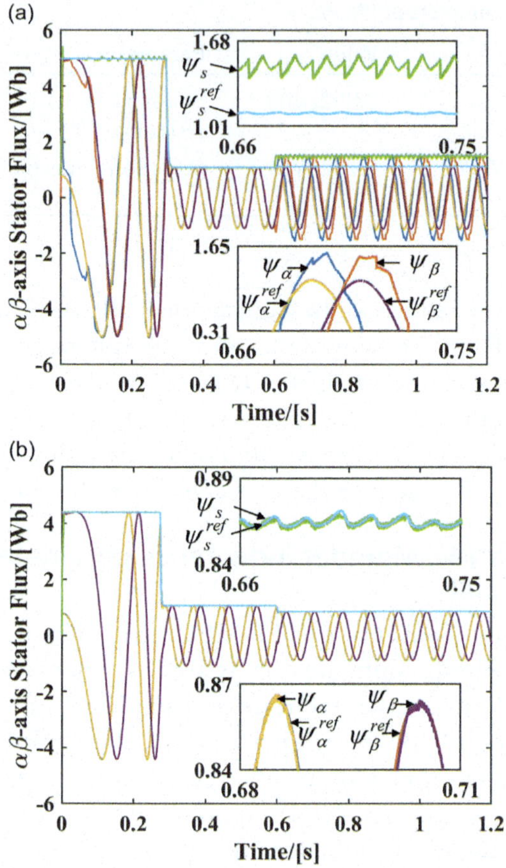

Fig. 6.6 Simulation results of the stator flux linkage under an inductance parameter disturbance fault. (a) Conventional PI-MPC method. (b) Proposed RS-SFTPC method.

Fig. 6.7 presents a comparative simulation of the three-dimensional trajectory of stator flux for the conventional PI-MPC and the proposed RS-SFTPC method. Fig. 6.7A intuitively demonstrates that when the system is subjected to an inductance parameter disturbance fault, the stator flux trajectory under the conventional PI-MPC method exhibits a significant deviation, with a nonnegligible tracking error and noticeable distortion in its shape. In contrast, as shown in Fig. 6.7B, after implementing the proposed RS-SFTPC meth permanent magnet synchronous motor the oscillation amplitude and tracking error of the flux trajectory under the same fault condition are substantially suppressed, and the trajectory shape is well maintained, thanks to the effective compensation

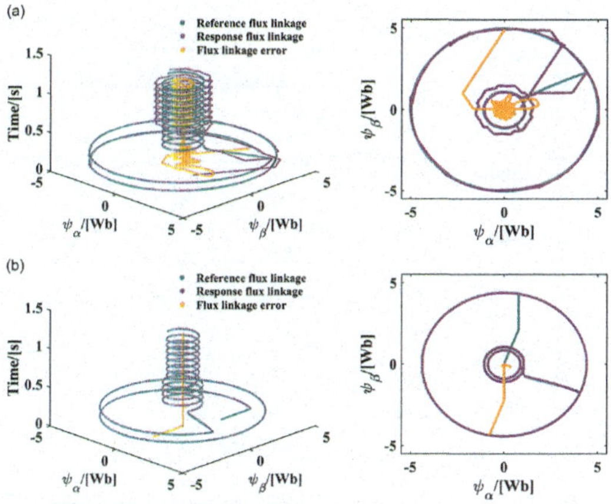

Fig. 6.7 Simulation results of the three-dimensional trajectory of the flux under an inductance parameter disturbance fault. (1) Conventional PI-MPC method. (2) Proposed RS-SFTPC method.

provided by the integrated observer for the disturbance. The comparative results indicate that the RS-SFTPC method can effectively enhance the system's robustness against parameter disturbances, significantly improving the tracking accuracy of the stator flux trajectory and suppressing its oscillations.

Fig. 6.8 presents a comparative simulation of the phase current and torque for the conventional PI-MPC and the proposed RS-SFTPC method. From the analysis of Fig. 6.8, it can be observed that under normal operating conditions, with no inductance parameter disturbance fault, both methods exhibit excellent control performance. In Fig. 6.8A, the conventional PI-MPC method experiences a torque ripple of 98 N · m after 0.6 s, with the current waveform showing significant distortion and a high degree of waveform irregularity under inductance parameter disturbance fault. In Fig. 6.8B, the torque ripple for the proposed RS-SFTPC method is reduced to 35 N · m, and the current distortion rate is significantly lower compared to the conventional PI-MPC method. In summary, while inductance parameter disturbance faults adversely affect the control performance of PMSM drives, the proposed RS-SFTPC method can effectively suppress the impact of such faults on control performance. The simulation results for phase current and torque demonstrate that the proposed RS-SFTPC method can effectively suppress torque pulsation and current waveform distortion.

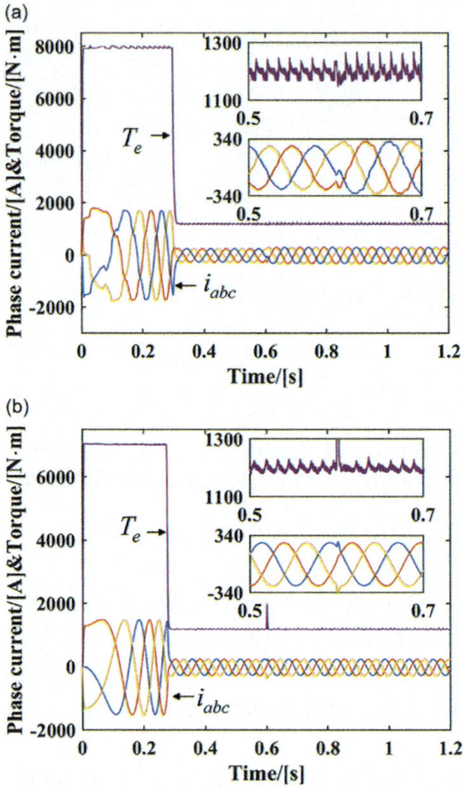

Fig. 6.8 Simulation results of phase current and torque under an inductance parameter disturbance fault. (a) Conventional PI-MPC method. (b) Proposed RS-SFTPC method.

Fig. 6.9 presents the observer results of rotor speed and stator flux state by using the proposed RS-SFTPC method in the case of inductance parameter disturbance fault. From Fig. 6.9A, it can be observed that the estimated rotor speed obtained by the composite discrete ESO is very close to the actual rotor speed state of the motor, with the observation errors falling within an acceptable range. In Fig. 6.9B, the composite discrete ESO observed estimates of the $\alpha\beta$ -axis stator flux components align nearly perfectly with the actual flux state. Both the adjustment duration and steady-state error are insignificant in the event of inductance parameter disturbance fault. The simulation analysis of the observed values for rotor speed and stator flux indicates that the composite discrete ESO can accurately observe both the rotor speed and flux, and effectively diagnose inductance parameter disturbance fault.

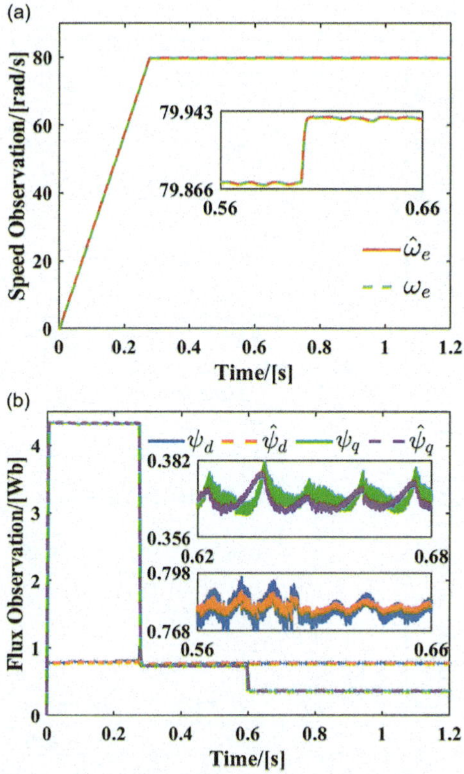

Fig. 6.9 Simulation results of speed and flux linkage observation under inductance parameter disturbance fault: (a) Speed observation, (b) Flux linkage observation.

Simulation for PMSM underflux linkage parameter disturbance fault

Under both the conventional PI-MPC method and the proposed RS-SFTPC method, the system is configured such that, at the 0.6-second mark, the flux linkage parameter experiences a sudden decrease to half of its original value, transitioning abruptly from 0.782 Wb to 0.391 Wb. Comparative simulation results of the conventional PI-MPC strategy and the proposed RS-SFTPC strategy under flux linkage parameter disturbance are shown in Figs. 6.10–6.13.

Fig. 6.10 illustrates a simulation comparison of stator flux between the conventional PI-MPC and the proposed RS-SFTPC method under flux linkage parameter disturbance fault. In Fig. 6.10, it can be observed that both methods achieve satisfactory control results and accurately track the

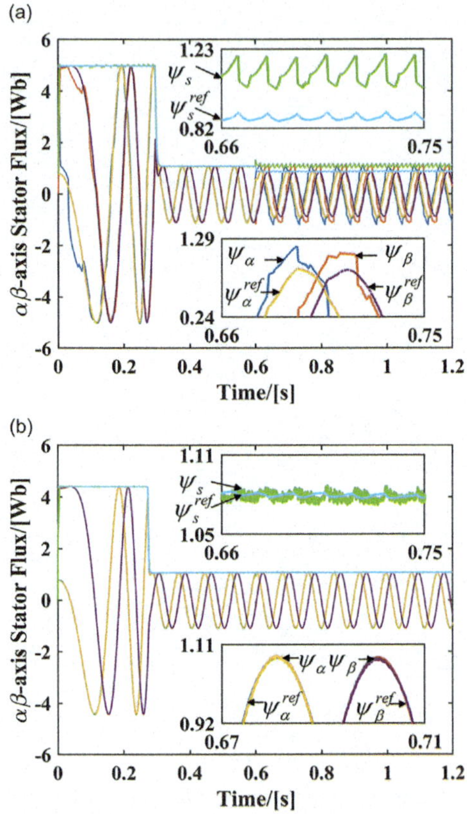

Fig. 6.10 Simulation results of the stator flux linkage under a flux linkage parameter disturbance fault. (a) Conventional PI-MPC method. (b) Proposed RS-SFTPC method.

stator flux under normal operating conditions. As shown in Fig. 6.10A, when using the conventional PI-MPC method, the actual stator flux value fails to fully track the commanded value, resulting in a static error as high as 0.4 Wb. This indicates the limited disturbance rejection capability of the conventional method under parameter perturbation conditions. In contrast, Fig. 6.10B clearly demonstrates that with the proposed RS-SFTPC method, the actual flux value can track the command almost perfectly, with only a minimal static error of 0.005 Wb. It is particularly noteworthy that when using the RS-SFTPC method, all components of the stator flux in the *d-q* reference frame exhibit very small pulsation amplitudes, further confirming the significant advantage of this method in suppressing flux fluctuations. Comparative results indicate that the RS-SFTPC method exhibits stronger robustness and superior tracking accuracy under flux linkage parameter disturbance conditions.

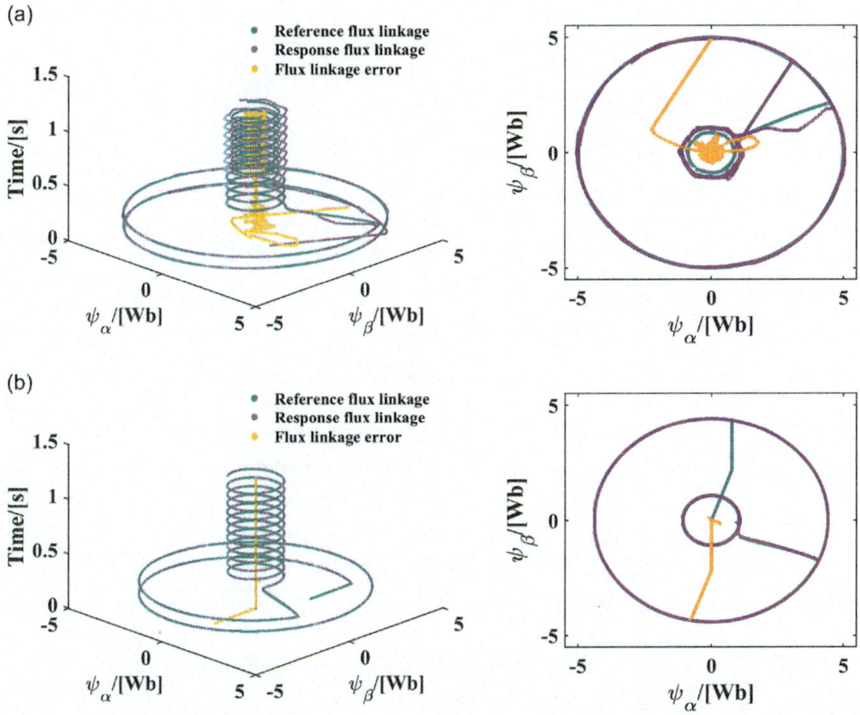

Fig. 6.11 Simulation results of the three-dimensional trajectory of the flux under a flux linkage parameter disturbance fault. (a) Conventional PI-MPC method. (b) Proposed RS-SFTPC method.

Fig. 6.11 presents a comparative simulation of the three-dimensional trajectory of stator flux for the conventional PI-MPC and the proposed RS-SFTPC method. As shown in Fig. 6.11A, when the system experiences a flux linkage parameter disturbance fault, the conventional PI-MPC method exhibits a noticeable tracking error in the steady-state phase. In stark contrast, Fig. 6.11B clearly demonstrates that, even under the same fault condition, the proposed RS-SFTPC method is capable of achieving precise tracking control and effectively confining the steady-state error within permissible limits. Through a comprehensive analysis of the flux response under both control strategies, it is evident that the RS-SFTPC method not only significantly suppresses stator flux pulsation but also demonstrates superior steady-state control performance under flux linkage parameter disturbance conditions.

Fig. 6.12 presents a comparison of simulated phase current and torque responses between the conventional PI-MPC and the proposed RS-SFTPC method in the presence of a flux linkage parameter disturbance fault.

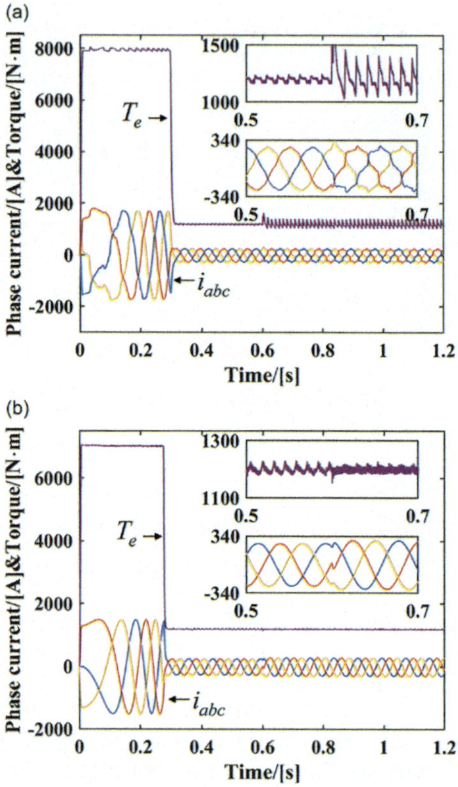

Fig. 6.12 Simulation results of phase current and torque under flux linkage parameter disturbance fault: (a) Conventional PI-MPC method, (b) Proposed RS-SFTPC method.

Through detailed analysis of Fig. 6.12A, it can be observed that when using the conventional PI-MPC method, the system output torque exhibits noticeable ripple with an amplitude reaching 330 N · m, accompanied by significant distortion in the current waveform. In Fig. 6.12B, however, the proposed RS-SFTPC method demonstrates superior control performance by effectively reducing the torque ripple amplitude to 30 N · m, achieving a substantial reduction of 90.91%, while maintaining stable current waveform characteristics with low distortion levels throughout the operation. Comprehensive analysis of both torque and current response characteristics clearly indicates that the RS-SFTPC method can effectively suppress torque ripple caused by flux linkage parameter disturbances while significantly reducing the distortion level of current waveforms.

Fig. 6.13 presents the observed results of speed and flux states under a flux linkage parameter disturbance fault. Fig. 6.13A shows that the speed observation error is less than 0.001 rad/s, with minimal steady-state error.

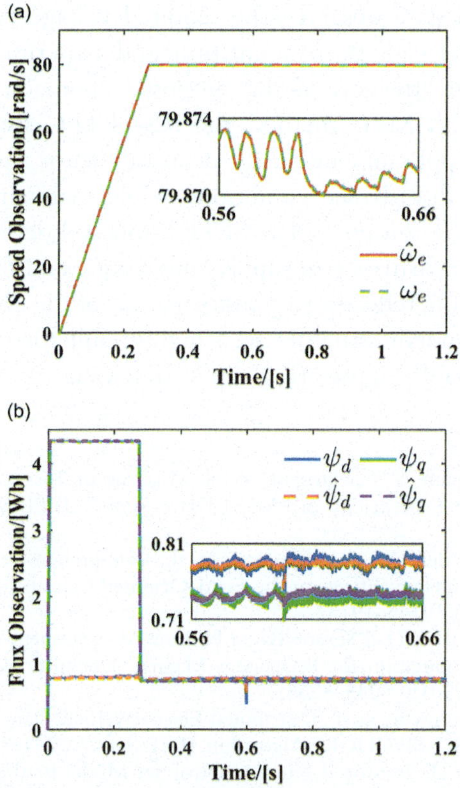

Fig. 6.13 Simulation results of speed and flux linkage observation under a flux linkage parameter disturbance fault. (a) Speed observation. (b) Flux linkage observation.

In Fig. 6.13B, it can be seen that the observation error of the dq-axis flux components is minimal, with a quick adjustment time following the parameter disturbance fault. From the analysis of the simulation under the flux linkage parameter disturbance fault, it can be determined that the composite discrete extended state observer accurately monitors speed and flux state values with a high degree of robustness.

6.5 Conclusion

This chapter introduces a robust self-sensing fault-tolerant predictive control (RS-SFTPC) strategy. The key advantage of this strategy lies in its independence from motor model parameters, thereby effectively mitigating control performance degradation caused by parameter disturbance faults. By integrating robust fault-tolerant predictive controllers with a composite discrete extended state observer, the required voltage vectors for driving the motor are derived. Both simulation and experimental results have confirmed the effectiveness of the proposed approach. Compared with the conventional PI-MPC method, the RS-SFTPC strategy significantly reduces torque ripple and markedly decreases stator flux tracking errors under parameter disturbance conditions, while maintaining smooth current waveforms. The proposed RS-SFTPC method demonstrates superior comprehensive performance in suppressing torque and flux ripples, as well as current distortion, reflecting a transformation of the control system from a parameter-dependent structure to a highly fault-tolerant one, and thus possesses considerable engineering application value.

References

[1] M.S. Khajueezadeh, S. FeizHoseini, et al., Analysis of Torsional Vibrations on the Resolver Under Eccentricity in PMSM Drive System, IEEE Sens. J. 22 (22) (2022) 21592–21599.
[2] M.S. KhajueeZadeh, M. Emadaleslami, et al., Comprehensive Investigation of the Resolver's Eccentricity Effect on the Field-Oriented Control of PMSM, IEEE Sens. J. 23 (17) (2023) 19145–19152.
[3] S. Du, J. Hu, et al., A Hall Sensor-Based Position Measurement With On-Line Model Parameters Computation for Permanent Magnet Synchronous Linear Motor, IEEE Sens. J. 18 (13) (2018) 5245–5255.
[4] L. Yaming, Z. Rongyun, et al., Distributed Electric Vehicle State Parameter Estimation Based on the ASO-SRGHCKF Algorithm, IEEE Sens. J. 22 (19) (2022) 18780–18792.
[5] T. Li, X. Sun, et al., Improved Finite Control Set Model Predictive Current Control for Permanent Magnet Synchronous Motor With Sliding Mode Observer, IEEE Trans. Transp. Electrific. 10 (01) (2024) 699–710.
[6] Y. Wang, et al., A Robust DPCC for IPMSM Based on a Full Parameter Identification Method, IEEE Trans. Ind. Electron. 70 (08) (2023) 7695–7705.
[7] S. Wang, Y. Hu, et al., Parameter Robustness Improvement of Predictive Current Control for Permanent-Magnet Synchronous Motors, IEEE J. Emerg. Sel. Top. Power Electron. 10 (02) (2022) 1619–1626.
[8] X. An, G. Liu, et al., Adjustable Model Predictive Control for IPMSM Drives Based on Online Stator Inductance Identification, IEEE Trans. Ind. Electron. 69 (04) (2022) 3368–3381.
[9] X. An, G. Liu, et al., Robust Predictive Current Control for Fault-Tolerant Operation of Five-Phase PM Motors Based on Online Stator Inductance Identification, IEEE Trans. Power Electron. 36 (11) (2021) 13162–13175.
[10] C. Xie, S. Zhang, et al., Parameter Identification for SPMSM With Deadbeat Predictive Current Control Using Online PSO, IEEE Trans. Transp. Electrific. 10 (02) (2024) 4055–4064.
[11] C. Bai, Z. Yin, et al., Multiple-Models Adaptive Disturbance Observer-Based Predictive Control for Linear Permanent-Magnet Synchronous Motor Vector Drive, IEEE Trans. Power Electron. 37 (08) (2022) 9596–9611.
[12] K. Yu, Z. Wang, Improved Deadbeat Predictive Current Control of Dual Three-Phase Variable-Flux PMSM Drives With Composite Disturbance Observer, IEEE Trans. Power Electron. 37 (07) (2022) 8310–8321.

[13] N. Yang, S. Zhang, et al., A New Model-Free Deadbeat Predictive Current Control for PMSM Using Parameter-Free Luenberger Disturbance Observer, IEEE J. Emerg. Sel. Top. Power Electron. 11 (01) (2023) 407–417.
[14] X. Zhang, Z. Liu, et al., Model Predictive Current Control for PMSM Drives Based on Nonparametric Prediction Model, IEEE Trans. Transp. Electrific. 10 (01) (2024) 711–719.
[15] H. Liu, W. Lin, et al., Model Predictive Current Control With Model-Aid Extended State Observer Compensation for PMSM Drive, IEEE Trans. Power Electron. 38 (03) (2023) 3152–3162.
[16] C. Li, Y. Yan, et al., Model Predictive Current Control With Variable Gain Adaptive Observer Based on Current Augmenter Prediction Model for IPMSM Drives, IEEE Trans. Veh. Technol. 71 (06) (2022) 6131–6144.
[17] Y. Xu, Z. Yan, et al., Model Predictive Current Control of Permanent Magnet Synchronous Motor Based on Sliding Mode Observer With Enhanced Current and Speed Tracking Ability Under Disturbance, IEEE Trans. Energy Convers. 38 (02) (2023) 948–958.
[18] G. Wu, H. Xiao, et al., Robust Cascaded Model-Free Tolerant Control With Parameters Disturbance Fault Diagnosis for PMSM Drive System, IEEE Sens. J. 25 (01) (2025) 1144–1154.
[19] P. Chen, Y. Luo, et al., A Current- and Speed-Loop Decoupling Controller for SPMSM Under Periodic Disturbances, IEEE Trans. Power Electron. 39 (06) (2024) 6889–6902.
[20] M. Yang, Enhanced Model-Free Predictive Speed Control Without Weighting Factors for PMSM Based on Newly Designed Cost Function, IEEE Trans. Transp. Electrific. 10 (03) (2024) 5703–5714.
[21] S. Li, Y. Xu, et al., Robust Deadbeat Predictive Direct Speed Control for PMSM With Dual Second-Order Sliding-Mode Disturbance Observers and Sensitivity Analysis, IEEE Trans. Power Electron. 38 (07) (2023) 8310–8326.
[22] Z. Yang, C. Miao, et al., Model Predictive Current Control for IPMSM Drives With Extended-State-Observer-Based Sliding Mode Speed Controller, IEEE Trans. Energy Convers. 38 (02) (2023) 1471–1480.

CHAPTER 7

Robust cascaded model-free stator flux sliding mode control for PMSMs considering parameters mismatch

7.1 Introduction

Permanent magnet synchronous motors (PMSMs) have been widely used in new energy electric vehicles due to their advantages, such as high power density and high efficiency [1–3]. The model predictive control (MPC) offers the benefits of high control accuracy and fast dynamic response, making it particularly suitable for PMSM drive systems in electric vehicles [4,5]. The MPC method strictly relies on the mathematical model of the controlled object (PMSM), which depends on the accuracy of the flux linkage parameter and the inductance parameter [6]. However, the flux linkage parameter and inductance parameter utilized in MPC may not match the actual motor parameters due to the possibility of time-varying parameters during motor operation. As the operating time of electric vehicles increases, it is inevitable that a mismatch will occur between the PMSM parameters and the MPC parameters. Therefore parameter mismatch faults have become the most common and widespread type of fault in PMSM drive systems.

Online parameter identification is the most effective method for diagnosing parameter mismatch faults [7–15]. An et al. [7] proposed an adjustable MPC strategy for the PMSM drive system, which can automatically identify online parameters to suppress disturbances caused by parameter mismatch faults. Li et al. [8] proposed a novel state parameter estimation method based on sensorless control of the PMSM drive system in distributed drive electric vehicles, which can effectively improve the observation accuracy of important state parameters. Liu et al. [9] proposed a square root generalized high-order cubature Kalman filter estimation algorithm based

Robust Model Predictive Control of Permanent Magnet Synchronous Motor Under Parameter Mismatch
https://doi.org/10.1016/B978-0-443-44209-4.00009-9

on the atomic search optimization algorithm, which can precisely obtain state parameter information in distributed drive electric vehicles. Zhang et al. [10,11] proposed a current prediction error model that considers uncertainties in model parameters, which can simultaneously identify stator inductance and flux linkage under parameter mismatch faults. Dang et al. [12] proposed an online parameter estimation method based on a discrete-time dynamic model for the PMSM drive system, which can improve robustness to electrical parameter changes and load disturbances under parameter mismatch faults. Uddin et al. [13] presented an online parameter-estimation-based high-speed control method for the PMSM drive system, which can effectively eliminate the impact of parameter mismatch faults on the control system. Zhou et al. [14] proposed a current prediction error-based parameter identification method for the PMSM drive system, which can effectively reduce current fluctuations and current tracking errors caused by parameter mismatch faults. Wang et al. [15] presented a robust deadbeat predictive current control method for the PMSM drive system based on a full parameter identification method, which can improve the robustness of the control system under parameter mismatch faults. The online parameter identification method can accurately identify important parameter information in the drive system. However, the online parameter identification method relies heavily on information from the motor drive system, resulting in a large computational load when diagnosing parameter mismatch faults.

The design of an unknown disturbance observer is another effective method for diagnosing parameter mismatch faults [16–22]. Han et al. [16] presented a novel sliding mode disturbance observer-based technique for diagnosing demagnetization faults in PMSM drive systems, which can identify the overall disturbance caused by parameter mismatch faults. Chen et al. [17] proposed a unified parameter-sensitive method with an adaptive full-order observer for estimating rotor-speed errors caused by all four kinds of parameter mismatch faults. Bai et al. [18] proposed a robust continuous-time predictive control method based on an augmented internal model disturbance observer for estimating the lumped disturbances of parameter mismatch faults. Yuan et al. [19] proposed a novel model-free predictive current control method using an extended state observer for the PMSM drive system, which can effectively suppress the disturbances caused by parameter mismatch faults. Sun et al. [20] proposed an adaptive model-free predictive current control method with optimal virtual vector modulation for PMSM drive systems, which can enhance the parameter robustness of drive systems under parameter mismatch faults. Luo et al. [21] designed a robust model predictive controller with a Luenberger observer for a six-phase PMSM drive system, which can compensate for the localization error caused by parameter mismatch faults. Babayomi et al. [22] proposed a novel model-free predictive control method with cascade-parallel extended state observers, which can achieve better disturbance rejection under parameter mismatch faults. The parameter mismatch fault can be effectively diagnosed by constructing the unknown disturbance observer, but there is a drawback in that the controller of the drive system relies on the accuracy of the disturbance observer. Due to the fact that the controller only relies on simple compensation from the disturbance observer, the controller of the drive system cannot handle parameter mismatch faults when the disturbance observer diagnosis is inaccurate.

In summary, to improve the robustness of the predictive control system against parameter mismatch faults, a novel robust cascaded model-free tolerant control (RCM-FTC) method with parameter mismatch fault diagnosis is proposed for the PMSM drive system. The multivariable integral terminal sliding mode observer (MIT-SMO) is designed by introducing an improved sliding mode reaching law, which can accurately estimate unknown disturbances caused by parameter mismatch faults. The novel ultralocal model of the speed loop and *dq*-axis stator flux loop is established in the RCM-FTC method under parameter mismatch faults. The proposed RCM-FTC method can effectively reduce reliance on the nominal motor model and enhance disturbance rejection capability under parameter mismatch faults. The proposed RCM-FTC method is significantly better than the conventional MPC method in terms of stator flux tracking accuracy and torque/flux ripple under parameter mismatch faults.

This chapter is organized as follows: The mathematical model of the PMSM drive system under parameter mismatch faults is introduced in Section 7.2. The novel ultralocal model for the PMSM drive system is presented in Section 7.3. The design of the MIT-SMO for diagnosing parameter mismatch faults is detailed in Section 7.4. The design of the robust cascaded model-free fault-tolerant controller is presented in Section 7.5. Simulation results validating the proposed RCM-FTC method are provided in Section 7.6. Finally, the conclusion is drawn in Section 7.7.

7.2 Parameter mismatch fault sensitivity analysis of PMSM drives

The normal stator flux linkage equations of the PMSM drive system can be expressed as

$$\begin{cases} \dfrac{d\psi_{do}}{dt} = -\dfrac{R_o}{L_{so}}\psi_{do} - \omega_e \psi_{qo} + u_{do} \\ \dfrac{d\psi_{qo}}{dt} = -\dfrac{R_o}{L_{so}}\psi_{qo} + \omega_e \psi_{do} + u_{qo} \end{cases} \tag{7.1}$$

with

$$\begin{cases} \psi_{do} = L_{so} i_d + \psi_{ro} \\ \psi_{qo} = L_{so} i_q \end{cases} \tag{7.2}$$

where ψ_{do} and ψ_{qo} represent the d- and q-axis component of the stator flux, respectively; ψ_{ro} denotes the nominal value of the flux linkage; R_o represents the nominal resistance of the motor; L_{so} is the nominal inductance of the motor; ω_e denotes the electrical angular velocity; u_{do} and u_{qo} represent the d- and q-axis voltage component of the stator voltage; i_d and i_q represent the d- and q-axis current component.

The mechanical motion equation of the PMSM drive system is given as

$$\frac{d\omega_e}{dt} = \frac{n_p}{J}(T_e - T_L - B_m \omega_m) \tag{7.3}$$

where n_p represents the number of pole pairs of the motor; T_e denotes the electromagnetic torque; T_L represents the load torque; J represents the moment of inertia; B_m is the damping coefficient; ω_m denotes the mechanical angular velocity.

To analyze the parameters mismatch in the actual operation of motor, introducing $\psi_r = \psi_{ro}(1+\xi_1)$, $L = L_{so}(1+\xi_2)$, that is, $\xi_1 = \Delta\psi_r / \psi_{ro}$, $\xi_2 = \Delta L_s / L_{so}$. where L_s and ψ_s are actual values, ΔL_s and $\Delta\psi_s$ are uncertain components caused by the parameters mismatch fault. According to Eq. (7.1) and Eq. (7.2), the mathematical model of the PMSM drive system under parameters mismatch fault can be rewritten as

$$\frac{d\boldsymbol{x}}{dt} = \boldsymbol{Ax} + \boldsymbol{Bu} + \delta \tag{7.4}$$

where δ is the unknown disturbance caused by parameters mismatch fault, δ_ω represents the unknown disturbance on the speed under parameters mismatch fault; δ_d and δ_q represent the unknown disturbance on the d- and q-axis flux linkage under parameters mismatch fault.

$$\boldsymbol{x} = \begin{bmatrix} \omega_e \\ \psi_d \\ \psi_q \end{bmatrix}, \ \boldsymbol{u} = \begin{bmatrix} i_q \\ u_d \\ u_q \end{bmatrix}, \ \delta = \begin{bmatrix} \delta_\omega \\ \delta_d \\ \delta_q \end{bmatrix}, \ \boldsymbol{A} = \begin{bmatrix} -\frac{B_m}{J} & 0 & 0 \\ 0 & -\frac{R_o}{L_{so}} & \omega_e \\ 0 & -\omega_e & -\frac{R_o}{L_{so}} \end{bmatrix},$$

$$\boldsymbol{B} = \begin{bmatrix} \frac{3n_p^2\psi_{ro}}{2J} & 0 & 0 \\ 0 & 1 & 0 \\ 0 & 0 & 1 \end{bmatrix}, \begin{cases} \delta_\omega = \frac{n_p}{J}\xi_1 T_e \\ \delta_d = \frac{R_o}{L_{so}}\psi_{ro} + \frac{\xi_2}{1+\xi_2}(R_o i_d - u_d) \\ \delta_q = \frac{\xi_2}{1+\xi_2}\left(-u_q + R_o i_q + \frac{\xi_2 - \xi_1}{\xi_2}\omega_e\psi_{ro}\right) \end{cases}.$$

The state equation of the PMSM drive system under steady-state operating conditions can be expressed as

$$\begin{cases} u_d = R_o i_d - \omega_e L_{so} i_q \\ u_q = R_o i_q + \omega_e L_{so} i_d + \omega_e \psi_{ro} \\ T_e = \frac{3n_p}{2}\psi_{ro} i_q \end{cases} \tag{7.5}$$

Combining Eq. (7.4) and Eq. (7.5), the unknown disturbance caused by the parameter mismatch fault can be rewritten as

$$\begin{cases} u_d = R_o i_d - \omega_e L_{so} i_q \\ u_q = R_o i_q + \omega_e L_{so} i_d + \omega_e \psi_{ro} \\ T_e = \frac{3n_p}{2}\psi_{ro} i_q \end{cases} \tag{7.6}$$

According to Eq. (7.6), the unknown disturbance caused by a parameter mismatch fault in the PMSM drive system can be characterized. Fig. 7.1A shows the speed unknown disturbance of the PMSM drive system under a parameter mismatch fault. From Fig. 7.1A, it can be seen that the speed unknown disturbance caused by the parameter mismatch fault can reach up to 39.6 rad/s^2 at a torque of 2000 N · m. It is evident that the parameter mismatch fault has a significant impact on the mechanical motion model of the PMSM drive system. Fig. 7.1B and Fig. 7.1C show the *d*- and *q*-axis flux linkage unknown disturbances of the PMSM drive system under a parameter mismatch fault. From Fig. 7.1B, it can be observed that the *d*-axis flux linkage unknown disturbance caused by the parameter mismatch fault can reach up to 540.3 V at 900 rad/s. From Fig. 7.1C, it can be observed that the *q*-axis flux linkage unknown disturbance of the PMSM drive system reaches up to 703.8 V under a parameter mismatch fault. Through the analysis of the results, it is evident that the mathematical model of the PMSM drive system is highly sensitive to parameter

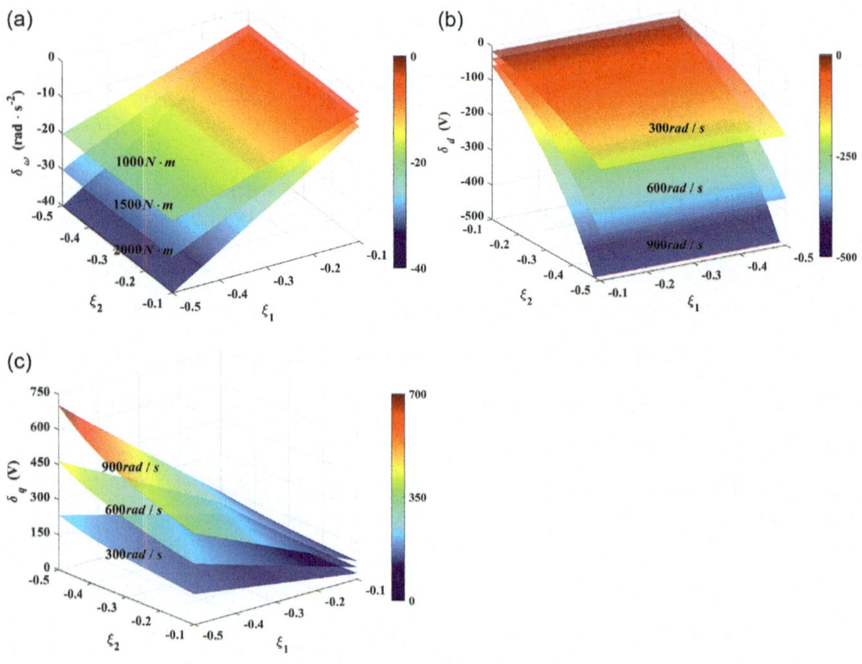

Fig. 7.1 The speed and flux linkage errors under parameter mismatch. (a) The speed error. (b) The *d*-axis flux linkage error. (c) The *q*-axis flux linkage error.

mismatch faults. Parameter mismatch deteriorates the control performance of the drive system if the parameter mismatch fault cannot be accurately diagnosed.

7.3 Ultralocal model of PMSM drives under parameters mismatch fault

The ultralocal model of the nonlinear single-input single-output systems can be given as [23].

$$\frac{dy}{dt} = \alpha u + g(x) \tag{7.7}$$

where y and u represent the system output and control input, respectively; $x \in \Re$ denotes the system state variable; $g : \Re \rightarrow \Re$ is an unknown nonlinear Lipschitz bounded function that depends solely on x (g including bounded disturbance); $\alpha \in \Re$ is a nonzero nonphysical constant gain to be determined, such that αu and x have the same order.

The ultralocal model of the speed loop and flux loop can be derived from the aforementioned model-free control theory.

$$\frac{d\boldsymbol{x}}{dt} = \alpha \boldsymbol{u} + \boldsymbol{g}(\boldsymbol{x}) \tag{7.8}$$

where α represents the state gain matrix of the system; $g(x)$ represents the unknown quantity to be observed.

Expanding Eq. (7.8) yields

$$\frac{d}{dt}\begin{bmatrix} \omega_e \\ \psi_d \\ \psi_q \end{bmatrix} = \begin{bmatrix} \frac{3n_p^2\psi_{ro}}{2J} & 0 & 0 \\ 0 & 1 & 0 \\ 0 & 0 & 1 \end{bmatrix} \begin{bmatrix} i_q \\ u_d \\ u_q \end{bmatrix} + \begin{bmatrix} g_1(x) \\ g_2(x) \\ g_3(x) \end{bmatrix} \tag{7.9}$$

where $\boldsymbol{g}(\boldsymbol{x}) = [g_1(x) \quad g_2(x) \quad g_3(x)]^T$, denotes the parameters mismatch fault disturbance in the ultralocal model.

Based on the novel ultralocal model theory, the parameter mismatch fault disturbance can be decomposed as

$$\boldsymbol{g}(\boldsymbol{x}) = \beta \boldsymbol{x} + \boldsymbol{F} \tag{7.10}$$

where β represents the system gain matrix; $\boldsymbol{F}$ is a nonlinear Lipschitz function. The Eq. (7.10) can be rewritten as

$$g(\boldsymbol{x}) = \begin{bmatrix} -\dfrac{B_{\mathrm{m}}}{J} & 0 & 0 \\ 0 & -\dfrac{R_{\mathrm{o}}}{L_{\mathrm{so}}} & \omega_{\mathrm{e}} \\ 0 & -\omega_{\mathrm{e}} & -\dfrac{R_{\mathrm{o}}}{L_{\mathrm{so}}} \end{bmatrix} \begin{bmatrix} \omega_{\mathrm{e}} \\ \psi_d \\ \psi_q \end{bmatrix} + \begin{bmatrix} F_{\omega} \\ F_d \\ F_q \end{bmatrix} \tag{7.11}$$

where $\boldsymbol{F} = \begin{bmatrix} F_{\omega} & F_d & F_q \end{bmatrix}^T$ denotes the disturbance term caused by the parameters mismatch fault.

Under parameter mismatch fault, the PMSM drive system can be described using the novel ultralocal model as

$$\frac{dx}{dt} = \alpha \boldsymbol{u} + \beta \boldsymbol{x} + \boldsymbol{F} \tag{7.12}$$

where α is the system input gain matrix; β is the system state gain matrix; $\alpha = \begin{bmatrix} \dfrac{3n_{\mathrm{p}}^2\psi_{\mathrm{ro}}}{2J} & 0 & 0 \\ 0 & 1 & 0 \\ 0 & 0 & 1 \end{bmatrix}$, and $\beta = \begin{bmatrix} -\dfrac{B_{\mathrm{m}}}{J} & 0 & 0 \\ 0 & -\dfrac{R_{\mathrm{o}}}{L_{\mathrm{so}}} & \omega_{\mathrm{e}} \\ 0 & -\omega_{\mathrm{e}} & -\dfrac{R_{\mathrm{o}}}{L_{\mathrm{so}}} \end{bmatrix}$.

7.4 MIT-SMO design for diagnosing parameters mismatch fault

In order to diagnose a parameter mismatch fault, according to Eq. (7.12), the MIT-SMO can be designed as

$$\begin{cases} \dfrac{d\boldsymbol{x}}{dt} = \alpha \boldsymbol{u} + \beta \boldsymbol{x} + \boldsymbol{F} \\ \dfrac{d\hat{\boldsymbol{x}}}{dt} = \alpha \boldsymbol{u} + \beta \hat{x} + \omega_{\mathrm{e}} \boldsymbol{P}(\hat{\boldsymbol{x}} - \boldsymbol{x}) + \boldsymbol{Z} \end{cases} \tag{7.13}$$

where $\boldsymbol{P}$ is the sliding mode gain matrix to be designed, and $\boldsymbol{Z} = \begin{bmatrix} Z_\omega & Z_d & Z_q \end{bmatrix}^{\mathrm{T}}$ is the sliding mode control function.

The integral terminal sliding surface is given as

$$\boldsymbol{s}_2 = \boldsymbol{e} + \rho \int_0^t \operatorname{sgn}(\boldsymbol{e}) d\tau \tag{7.14}$$

With

$$\boldsymbol{e} = \begin{bmatrix} e_\omega \\ e_d \\ e_q \end{bmatrix} = \begin{bmatrix} \hat{\omega}_{\mathrm{e}} - \omega_{\mathrm{e}} \\ \hat{\psi}_d - \psi_d \\ \hat{\psi}_q - \psi_q \end{bmatrix} \tag{7.15}$$

where $\boldsymbol{s}_2 = \begin{bmatrix} s_{2\omega} & s_{2d} & s_{2q} \end{bmatrix}^{\mathrm{T}}$; ρ is the parameter of the MIT-SMO to be designed; $\operatorname{sgn}(\varepsilon) = \dfrac{\mathrm{e}^{2\varepsilon} - 1}{\mathrm{e}^{2\varepsilon} + 1}$ is the sign function.

According to Eq. (7.13), one yields

$$\frac{d\boldsymbol{s}_2}{dt} = \beta' \boldsymbol{e} + \omega_{\mathrm{e}} (\beta'' + \boldsymbol{P}) \boldsymbol{e} + \widehat{\boldsymbol{F}} - \boldsymbol{F} + \rho \operatorname{sgn}(\boldsymbol{e}) \tag{7.16}$$

where $\beta' = \begin{bmatrix} -\dfrac{B_{\mathrm{m}}}{J} & 0 & 0 \\ 0 & -\dfrac{R_{\mathrm{o}}}{L_{\mathrm{so}}} & 0 \\ 0 & 0 & -\dfrac{R_{\mathrm{o}}}{L_{\mathrm{so}}} \end{bmatrix}$, $\beta'' = \begin{bmatrix} 0 & 0 & 0 \\ 0 & 0 & 1 \\ 0 & -1 & 0 \end{bmatrix}$.

In order to eliminate the influence of motor speed changes on fault diagnosis, the gain matrix of MIT-SMO can be designed as $\boldsymbol{P} = -\beta''$, $\omega_{\mathrm{e}} (\beta'' + \boldsymbol{P}) \boldsymbol{e} = 0$. Eq. (7.16) can be modified as

$$\frac{d\boldsymbol{s}_2}{dt} = \beta' \boldsymbol{e} + \hat{\boldsymbol{F}} - \boldsymbol{F} + \rho \operatorname{sgn}(\boldsymbol{e}) \tag{7.17}$$

According to Eqs. (7.16) and (7.17), the following sliding mode control function is designed to guarantee the convergence of the MIT-SMO.

$$\boldsymbol{Z} = -\beta' \boldsymbol{e} - \rho \operatorname{sgn}(\boldsymbol{e}) - \boldsymbol{k}_\alpha \boldsymbol{s}_2 - \boldsymbol{k}_\beta \operatorname{sgn}(\boldsymbol{s}_2) \tag{7.18}$$

where $\boldsymbol{k}_\alpha = k_l diag(1,1,1)$, $\boldsymbol{k}_\beta = k_v diag(1,1,1)$.

The following Lyapunov function is defined to prove the asymptotic stability of MIT-SMO. Combining it with Eq. (7.17) and Eq. (7.18), one yields

$$\begin{aligned}\frac{d\boldsymbol{V}_2}{dt} &= \boldsymbol{s}_2^T \frac{d\boldsymbol{s}_2}{dt} = \boldsymbol{s}_2^T (\beta'\boldsymbol{e} + \boldsymbol{Z} - \boldsymbol{F} + \rho\,\mathrm{sgn}(\boldsymbol{e}))\\ &= \boldsymbol{s}_2^T (-\boldsymbol{F} - \boldsymbol{k}_\alpha \boldsymbol{s}_2 - \boldsymbol{k}_\beta\,\mathrm{sgn}(\boldsymbol{s}_2))\\ &= -\boldsymbol{s}_2^T \boldsymbol{F} - k_l (s_{2\omega}^2 + s_{2d}^2 + s_{2q}^2) - \boldsymbol{s}_2^T \boldsymbol{k}_\beta\,\mathrm{sgn}(\boldsymbol{s}_2)\\ &\le -\boldsymbol{s}_2 \boldsymbol{F} - k_l (s_{2\omega}^2 + s_{2d}^2 + s_{2q}^2) - k_v \boldsymbol{s}_2\\ &= -\boldsymbol{s}_2 (\boldsymbol{F} + k_v) - k_l (s_{2\omega}^2 + s_{2d}^2 + s_{2q}^2)\end{aligned} \tag{7.19}$$

In engineering practice, the unknown disturbance $\boldsymbol{F}$ is bounded, i.e., $\|\boldsymbol{F}\| \le N_\delta$, where N_δ is the boundary value. When $k_v \ge -\|\boldsymbol{F}\|$, $k_l \ge 0$, then $\frac{d\boldsymbol{V}_2}{dt} \le 0$.

Combining Eq. (7.13) and Eq. (7.18), one yields

$$\begin{bmatrix}\frac{d\hat{\omega}_e}{dt}\\ \frac{d\hat{\psi}_d}{dt}\\ \frac{d\hat{\psi}_q}{dt}\end{bmatrix} = \begin{bmatrix}\frac{3n_p^2\psi_{ro}}{2J} & & \\ & 1 & \\ & & 1\end{bmatrix}\begin{bmatrix}i_q\\ u_d\\ u_q\end{bmatrix} - \begin{bmatrix}\frac{B_m}{J} & & \\ & \frac{R_o}{L_{so}} & \\ & & \frac{R_o}{L_{so}}\end{bmatrix}\begin{bmatrix}\hat{\omega}_e\\ \hat{\psi}_d\\ \hat{\psi}_q\end{bmatrix} + \omega_e\begin{bmatrix}0\\ \psi_q\\ -\psi_d\end{bmatrix} + \begin{bmatrix}\hat{F}_\omega\\ \hat{F}_d\\ \hat{F}_q\end{bmatrix} \tag{7.20}$$

The observed values of disturbance terms caused by parameter mismatch faults can be obtained as

$$\begin{bmatrix}\hat{F}_\omega\\ \hat{F}_d\\ \hat{F}_q\end{bmatrix} = \begin{bmatrix}\frac{B_m}{J} & & \\ & \frac{R_o}{L_{so}} & \\ & & \frac{R_o}{L_{so}}\end{bmatrix}\begin{bmatrix}e_\omega\\ e_d\\ e_q\end{bmatrix} - \rho\begin{bmatrix}\mathrm{sgn}(e_\omega)\\ \mathrm{sgn}(e_d)\\ \mathrm{sgn}(e_q)\end{bmatrix} - k_l\begin{bmatrix}s_{2\omega}\\ s_{2d}\\ s_{2q}\end{bmatrix} - k_v\begin{bmatrix}\mathrm{sgn}(s_{2\omega})\\ \mathrm{sgn}(s_{2d})\\ \mathrm{sgn}(s_{2q})\end{bmatrix} \tag{7.21}$$

7.5 Mismatch fault design of robust cascaded model-free fault-tolerant controller

According to Eq. (7.12), the voltage vector equation of the PMSM drive system under a parameter mismatch fault can be expressed as

$$\boldsymbol{u} = \alpha^{-1}\left[\frac{d\boldsymbol{x}}{dt} - \beta\boldsymbol{x} - \boldsymbol{F}\right] \tag{7.22}$$

where $\alpha^{-1} = diag(\alpha_\omega, \alpha_d, \alpha_q)$, $\beta = [\beta_\omega \quad \beta_d \quad \beta_q]^{\mathrm{T}}$.

Eq. (7.22) can be rewritten as

$$\begin{cases} i_q = \alpha_\omega\left(\dfrac{d\omega_e}{dt} + \dfrac{B_{\mathrm{m}}}{J}\omega_{\mathrm{e}} - F_\omega\right) \\ u_d = \alpha_d\left(\dfrac{d\psi_d}{dt} + \dfrac{R_{\mathrm{o}}}{L_{\mathrm{so}}}\psi_d - \omega_{\mathrm{e}}\psi_q - F_d\right) \\ u_q = \alpha_q\left(\dfrac{d\psi_q}{dt} + \dfrac{R_{\mathrm{o}}}{L_{\mathrm{so}}}\psi_q + \omega_{\mathrm{e}}\psi_d - F_q\right) \end{cases} \tag{7.23}$$

According to Eq. (7.23), the output vector reference value of the PMSM drive system under parameter mismatch fault can be expressed as

$$\begin{bmatrix} i_q^{ref} \\ u_d^{ref} \\ u_q^{ref} \end{bmatrix} = \begin{bmatrix} \alpha_\omega & & \\ & \alpha_d & \\ & & \alpha_q \end{bmatrix} \left\{ \begin{bmatrix} \dfrac{d\omega_e^{ref}}{dt} \\ \dfrac{du_d^{ref}}{dt} \\ \dfrac{du_q^{ref}}{dt} \end{bmatrix} - \begin{bmatrix} -\dfrac{B_m}{J} & & \\ & -\dfrac{R_o}{L_{so}} & \omega_e \\ & -\omega_e & -\dfrac{R_o}{L_{so}} \end{bmatrix} \begin{bmatrix} \omega_e \\ \psi_d \\ \psi_q \end{bmatrix} - \begin{bmatrix} F_\omega \\ F_d \\ F_q \end{bmatrix} + \begin{bmatrix} u_{c1} \\ u_{c2} \\ u_{c3} \end{bmatrix} \right\} \tag{7.24}$$

where u^{ref} represents the reference value of voltage; i^{ref} represents the reference value of current; $\boldsymbol{u}_c = [u_{c1} \quad u_{c2} \quad u_{c3}]^{\mathrm{T}}$ represents the input term to be designed for the model-free sliding mode control.

Expanding Eq. (7.24) yields

$$\begin{cases} i_q^{ref} = \alpha_\omega\left(\dfrac{d\omega_{\mathrm{e}}^{ref}}{dt} + \dfrac{B_{\mathrm{m}}}{J}\omega_{\mathrm{e}} - F_\omega + u_{c1}\right) \\ u_d^{ref} = \alpha_d\left(\dfrac{d\psi_d^{ref}}{dt} + \dfrac{R_{\mathrm{o}}}{L_{\mathrm{so}}}\psi_d - \omega_{\mathrm{e}}\psi_q - F_d + u_{c2}\right) \\ u_q^{ref} = \alpha_q\left(\dfrac{d\psi_q^{ref}}{dt} + \dfrac{R_{\mathrm{o}}}{L_{\mathrm{so}}}\psi_q + \omega_{\mathrm{e}}\psi_d - F_q + u_{c3}\right) \end{cases} \tag{7.25}$$

Combining with Eq. (7.23) and Eq. (7.25), one obtains

$$\begin{cases} \dfrac{d\omega_{e}^{ref}}{dt} - \dfrac{d\omega_{e}}{dt} + u_{c1} = 0 \\ \dfrac{d\psi_{d}^{ref}}{dt} - \dfrac{d\psi_{d}}{dt} + u_{c2} = 0 \\ \dfrac{d\psi_{q}^{ref}}{dt} - \dfrac{d\psi_{q}}{dt} + u_{c3} = 0 \end{cases} \tag{7.26}$$

Selecting speed and flux error as controller state variables, one yields

$$\begin{cases} \boldsymbol{x}_2 = \int \boldsymbol{x}_1 = \int \begin{bmatrix} \omega_{e}^{ref} - \omega_{e} \\ \psi_{d}^{ref} - \psi_{d} \\ \psi_{q}^{ref} - \psi_{q} \end{bmatrix} \\ \dfrac{d\boldsymbol{x}_1}{dt} = -\boldsymbol{u}_c = \begin{bmatrix} \dfrac{d\omega_{e}^{ref}}{dt} - \dfrac{d\omega_{e}}{dt} \\ \dfrac{d\psi_{d}^{ref}}{dt} - \dfrac{d\psi_{d}}{dt} \\ \dfrac{d\psi_{q}^{ref}}{dt} - \dfrac{d\psi_{q}}{dt} \end{bmatrix} \end{cases} \tag{7.27}$$

Selecting the following integral sliding surface

$$\boldsymbol{s}_1 = \boldsymbol{x}_1 + c\int \boldsymbol{x}_1 \tag{7.28}$$

where $c > 0$ is the constant to be determined.
The derivative of the sliding surface can be obtained as

$$\frac{d\boldsymbol{s}_1}{dt} = c\boldsymbol{x}_1 + \frac{d\boldsymbol{x}_1}{dt} = c\boldsymbol{x}_1 - \boldsymbol{u}_c \tag{7.29}$$

where $\boldsymbol{s}_1 = \begin{bmatrix} s_{1\omega} & s_{1d} & s_{1q} \end{bmatrix}^{\mathrm{T}}$.
The improved sliding mode reaching law is given as

$$\frac{d\boldsymbol{s}_1}{dt} = c\boldsymbol{x}_1 - \boldsymbol{u}_c = -\varepsilon \boldsymbol{D} sgn(\boldsymbol{s}_1) - kh(\boldsymbol{s}_1) \tag{7.30}$$

where $\boldsymbol{D} = diag\left(|s_{1\omega}|^{d}, |s_{1d}|^{d}, |s_{1q}|^{d}\right)$, $0 < d < 1$, ε and k are any positive real numbers; $h(s)$ is a nonlinear function.

$$h(\boldsymbol{s}_1)=\begin{cases}\boldsymbol{s}_1 & \boldsymbol{s}_1 \le \mu \\ \operatorname{sgn}(\boldsymbol{s}_1) & \boldsymbol{s}_1 > \mu\end{cases} \tag{7.31}$$

where μ is any positive real number.

To prove that the designed sliding mode controller is asymptotically stable, the following Lyapunov function $\boldsymbol{V}_1$ is defined as

$$\boldsymbol{V}_1 = \frac{1}{2}\boldsymbol{s}_1{}^T\boldsymbol{s}_1 \tag{7.32}$$

The derivation of Eq. (7.32) yields

$$\begin{aligned}\frac{d}{dt}\boldsymbol{V}_1 &= \boldsymbol{s}_1{}^T\frac{d\boldsymbol{s}_1}{dt}\\ &= \boldsymbol{s}_1{}^T(-\varepsilon \boldsymbol{D} sgn(\boldsymbol{s}_1) - kh(\boldsymbol{s}_1))\\ &= -\varepsilon \boldsymbol{s}_1{}^{d+1} - k\boldsymbol{s}_1{}^T h(\boldsymbol{s}_1)\end{aligned} \tag{7.33}$$

According to Eq. (7.33), when $\| \boldsymbol{s}_1 \| \le \mu$, the Eq. (7.33) is rewritten as

$$\frac{d}{dt}\boldsymbol{V}_1 = -\varepsilon \| \boldsymbol{s}_1 \|^{d+1} - k\left(s_{1\omega}^2 + s_{1d}^2 + s_{1q}^2\right) \le 0 \tag{7.34}$$

Similarly, when $\| \boldsymbol{s}_1 \| > \mu$, the Eq. (7.33) is expressed as

$$\frac{d}{dt}\boldsymbol{V}_1 = -\varepsilon \| \boldsymbol{s}_1 \|^{d+1} - k \| \boldsymbol{s}_1 \| \le 0 \tag{7.35}$$

According to Eqs. (7.34) and (7.35), it can be concluded that the robust cascaded model-free fault-tolerant controller is asymptotically stable.

The input function of the robust cascaded model-free fault-tolerant controller can be obtained as

$$\boldsymbol{u}_c = c\boldsymbol{x}_1 - \frac{d\boldsymbol{s}_1}{dt} = c\begin{bmatrix}\omega_e^{ref} - \omega_e \\ \psi_d^{ref} - \psi_d \\ \psi_q^{ref} - \psi_q\end{bmatrix} + \begin{bmatrix}\varepsilon |s_{1\omega}|^d sgn(s_{1\omega}) + kh(s_{1\omega}) \\ \varepsilon |s_{1d}|^d sgn(s_{1d}) + kh(s_{1d}) \\ \varepsilon |s_{1q}|^d sgn(s_{1q}) + kh(s_{1q})\end{bmatrix} \tag{7.36}$$

where c, d, ε, and k are positive constants to be designed.

According to Eq. (7.22) and Eq. (7.36), the robust cascaded model-free fault-tolerant controller under parameter mismatch faults can be designed as

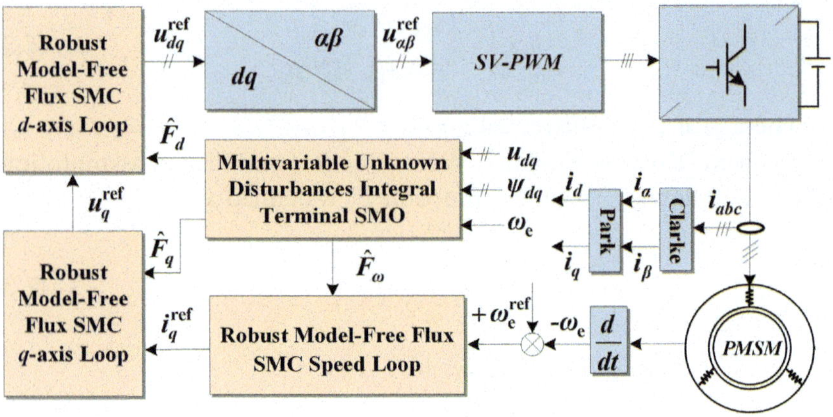

Fig. 7.2 The structural diagram of the RCM-FTC strategy with MIT-SMO under parameter mismatch fault.

$$\begin{bmatrix} i_q^{ref} \\ u_d^{ref} \\ u_q^{ref}\end{bmatrix} = \begin{bmatrix}\alpha_\omega & & \\ & \alpha_d & \\ & & \alpha_q\end{bmatrix}\left\{\begin{bmatrix}\frac{d\omega_e^{ref}}{dt}\\ \frac{du_d^{ref}}{dt} \\ \frac{du_q^{ref}}{dt}\end{bmatrix} - \begin{bmatrix} -\frac{B_m}{J} & & \\ & -\frac{R_o}{L_{so}} & \omega_e \\ & -\omega_e & -\frac{R_o}{L_{so}}\end{bmatrix}\begin{bmatrix}\omega_e\\ \psi_d \\ \psi_q\end{bmatrix} - \begin{bmatrix}F_\omega\\F_d\\F_q\end{bmatrix} + \begin{bmatrix}u_{c1}\\u_{c2}\\u_{c3}\end{bmatrix}\right\} \tag{7.37}$$

Fig. 7.2 shows the robust cascaded model-free fault-tolerant control block diagram based on the MIT-SMO for the PMSM drive system. The designed robust cascaded model-free fault-tolerant controller has two loops, including a speed control loop and a dq-axis stator flux control loop, both of which are implemented with the robust cascaded model-free control algorithm. The designed MIT-SMO can estimate unknown disturbances caused by parameter mismatch faults. The observed parameter mismatch fault disturbances are fed back to the robust cascaded model-free fault-tolerant controller. The robust cascaded model-free fault-tolerant controller will obtain the voltage vector required to control the PMSM drive system under parameter mismatch faults.

7.6 Simulation results

To validate the effectiveness of the RCM-FTC method with MIT-SMO, a three-phase PMSM drive system is built in MATLAB/Simulink. Simulations are conducted to compare the conventional MPC method and

Table 7.1 Main parameters and control parameters of PMSM.

Parameter	Value	Parameter	Value
Inductance (L_{so})	2.892/mH	Stator phase resistance (R_o)	0.02/Ω
Flux linkage of PM (Ψ_{ro})	0.782/Wb	Rotational inertia (J)	100/kg.m2
Rated torque (T_N)	2000/ N · m	Pole pairs (n_p)	4/-
Control parameter (ε)	50/-	Control parameter (k)	100/-
Control parameter (μ)	11/-	Control parameter (ρ)	5/-
Control parameter (k_1)	15,000/-	Control parameter (k_v)	1000/-

the proposed RCM-FTC method under parameter mismatch faults in the PMSM drive system. The motor parameters and control system parameters are provided in Table 7.1.

Simulation for the PMSM under flux linkage parameter mismatch ($\xi_1 = -50\%$, $\xi_2 = 0$)

In operating condition A, the PMSM operates with a load torque maintained at 2000 N · m and a rotational speed stabilized at 120 rad/s. At 0.7 s, a mismatch occurs in the flux linkage parameter of the PMSM, causing the flux linkage value to step down to 50% of its initial value, i.e., 0.391 Wb. Comparative simulation results of the conventional MPC strategy and the proposed RCM-FTC strategy under flux linkage parameter mismatch are shown in Figs. 7.3–7.6.

Fig. 7.3 presents the simulation comparison results of torque and current between the conventional MPC method and the proposed RCM-FTC method under a flux linkage parameter mismatch fault. From Fig. 7.3A, it can be observed that the torque ripple of the conventional MPC method is as high as 500 N · m under the flux linkage parameter mismatch fault. The reason for the torque ripple deterioration and stator current distortion is that the impact of the flux parameter mismatch on the conventional MFC strategy is more prominent. From Fig. 7.3B, it can be seen that the torque ripple of the PMSM drive system is reduced to 40 N · m by using the proposed RCM-FTC method under the flux linkage parameter mismatch fault. Compared with the conventional MFC strategy, the proposed RCM-FTC method can accurately estimate the

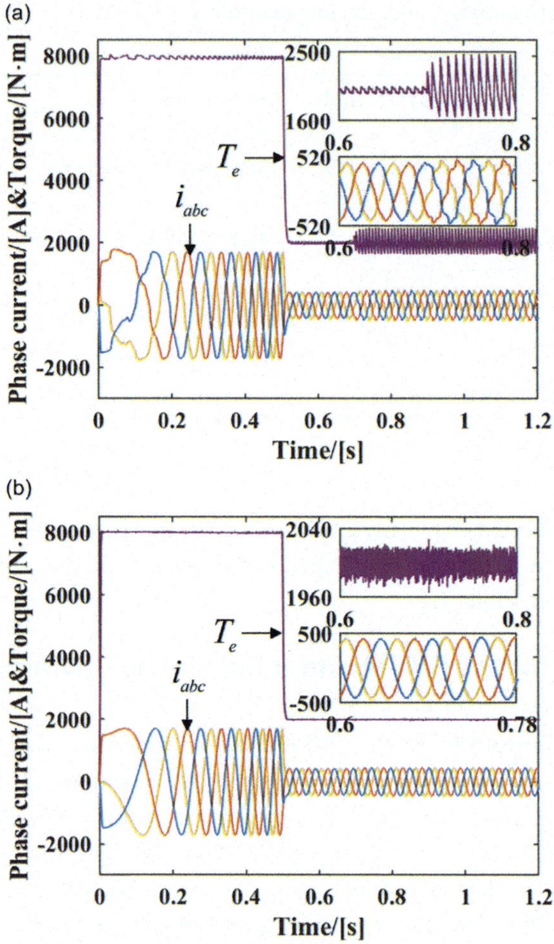

Fig. 7.3 Simulation comparison results between the conventional MPC method and the proposed RCM-FTC method under a flux linkage parameter mismatch fault. (a) Conventional MPC method. (b) Proposed RCM-FTC method.

unknown disturbance components of the motor, thereby significantly reducing torque ripple deterioration in the PMSM under flux linkage parameter mismatch, while also demonstrating notable improvement in stator current distortion. The result demonstrates that the PMSM drive system using the proposed RCM-FTC method has high robustness against flux linkage parameter mismatch fault.

Fig. 7.4 presents the observation results of speed and stator flux using the designed MIT-SMO under flux linkage parameter mismatch fault. As shown in Fig. 7.4A, under the condition of flux linkage parameter

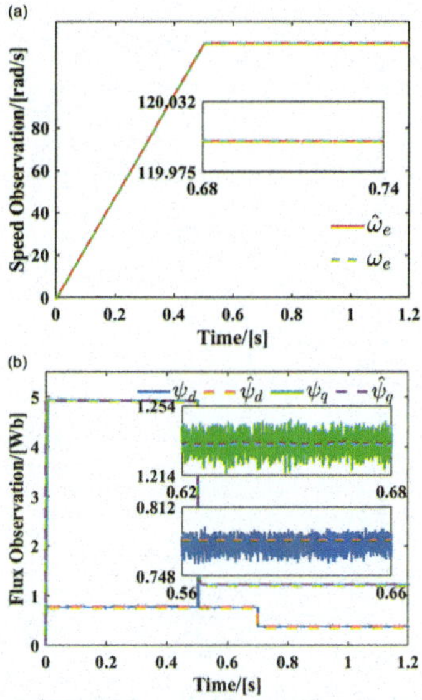

Fig. 7.4 Observation of speed and stator flux using the designed MIT-SMO under a flux linkage parameter mismatch fault. (a) Speed observation. (b) Flux speed observation.

mismatch, the designed MIT-SMO can accurately observe and track the actual speed response of the system, with a high degree of waveform consistency. Fig. 7.4B further demonstrates that the designed MIT-SMO is capable of accurately observing the dq-axis stator flux linkage response of the PMSM drive system under flux linkage parameter mismatch conditions without significant deviation. These results indicate that the designed MIT-SMO can simultaneously observe both speed and flux linkage disturbances under flux linkage parameter mismatch fault. This further confirms that the observer can accurately identify and diagnose parameter mismatch faults occurring during system operation, providing effective support for achieving high-performance control under flux linkage parameter mismatch fault.

Fig. 7.5 presents the stator flux simulation results of the conventional MPC method and the proposed RCM-FTC method under a flux linkage parameter mismatch fault. During the 0–0.7 s time interval, the proposed RCM-FTC method performs similarly to the conventional MPC method, with both achieving excellent stator flux control performance under normal conditions. As shown in Fig. 7.5A, when the conventional MPC method is employed, the αβ-axis stator flux waveform exhibits significant errors, and its response fails to track the reference value. Under the flux linkage parameter mismatch fault, the stator flux response of the PMSM

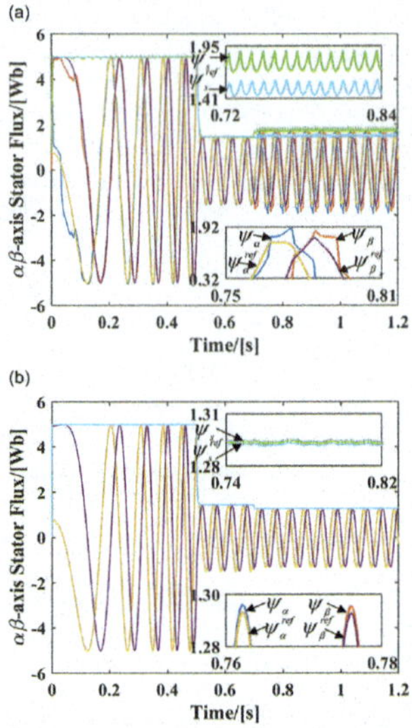

Fig. 7.5 Stator flux simulation results of the conventional MPC method and the proposed RCM-FTC method under a flux linkage parameter mismatch fault. (a) Conventional MPC method. (b) Proposed RCM-FTC method.

drive system reaches 1.81 Wb, with a stator flux error as high as 0.25 Wb. In contrast, as illustrated in Fig. 7.5B, the proposed RCM-FTC method enables the stator flux response to accurately track its reference value, reducing the stator flux error under the same fault condition to 0.002 Wb, corresponding to a reduction rate of 99.2%.

Fig. 7.6 presents the simulation results of three-dimensional rotor flux trajectories under a flux linkage parameter mismatch fault. Fig. 7.6A shows that the three-dimensional rotor flux linkage trajectory of the conventional MPC method exhibits large flux linkage oscillation and steady-state error under a flux linkage parameter mismatch fault. Fig. 7.6B demonstrates that the proposed RCM-FTC method can effectively suppress the flux linkage oscillation and steady-state error by utilizing the MIT-SMO. In addition, Fig. 7.6B illustrates that the proposed RCM-FTC method can effectively mitigate the flux linkage oscillation and steady-state error by diagnosing and compensating for the parameter mismatch fault.

Simulation for the PMSM under inductance parameter mismatch ($\xi_1 = 0$, $\xi_2 = -50\%$)

In operating condition A, the PMSM operates with a load torque maintained at 2000 N · m and a rotational speed stabilized at 120 rad/s. At 0.7 s,

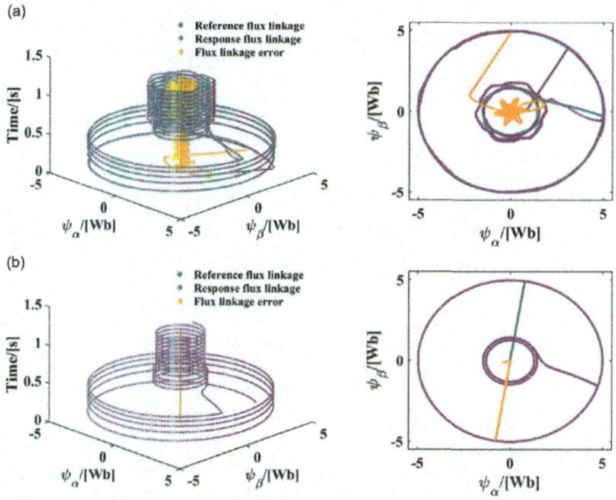

Fig. 7.6 Simulation results of three-dimensional rotor flux trajectories under a flux linkage parameter mismatch fault. (a) Conventional MPC method. (b) Proposed RCM-FTC method.

a mismatch occurs in the inductance parameter of the PMSM, causing the inductance value to step down to 50% of its initial value, i.e., 1.446 mH. Comparative simulation results of the conventional MPC strategy and the proposed RCM-FTC strategy under inductance parameter mismatch are shown in Figs. 7.7–7.10.

Fig. 7.7 presents the simulation comparison results of torque and current between the conventional MPC method and the proposed RCM-FTC method under an inductance parameter mismatch fault. As shown in Fig. 7.7A, when the conventional MPC method is employed, the PMSM drive system exhibits noticeable torque oscillation and stator current distortion due to the inductance parameter mismatch fault, with a torque ripple of approximately 100 N · m. In contrast, by adopting the proposed RCM-FTC method, which enables accurate estimation of the motor's unknown disturbance components, the torque ripple of the system is reduced to 40 N · m, as illustrated in Fig. 7.7B. Therefore, while the inductance parameter mismatch fault can induce significant torque ripple and stator current distortion in the PMSM drive system, the proposed RCM-FTC method effectively suppresses both the torque oscillation and current distortion caused by the fault, thereby enhancing the motor control performance.

Fig. 7.8 presents the observation results of speed and stator flux using the designed MIT-SMO under an inductance parameter mismatch fault. As shown in Fig. 7.8A, when an inductance parameter mismatch occurs

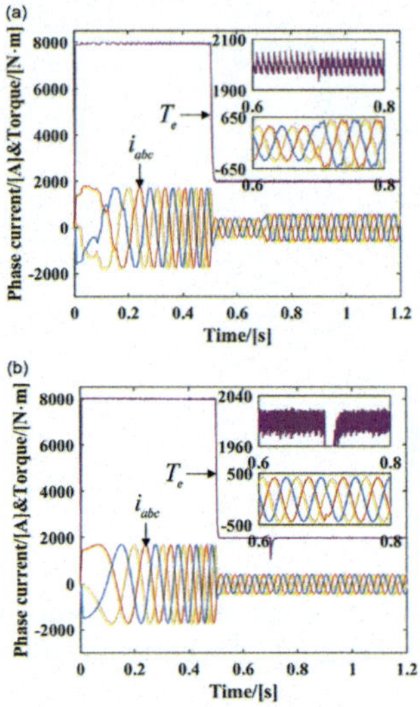

Fig. 7.7 Simulation comparison results between the conventional MPC method and the proposed RCM-FTC method under an inductance parameter mismatch fault. (a) Conventional MPC method. (b) Proposed RCM-FTC method.

in the system, the estimated speed curve output by the designed MIT-SMO closely matches the actual speed response, with the tracking error maintained at a low level, demonstrating good real-time performance and accuracy. Fig. 7.8B further indicates that the observed waveform is largely consistent with the actual system response, and the observer can accurately monitor the operating state of the stator flux in the PMSM drive system under the inductance parameter mismatch fault. It can be concluded that the designed MIT-SMO is capable of accurately observing the speed and stator flux disturbances caused by the inductance parameter mismatch fault, providing reliable observational information to support subsequent high-performance control and diagnostics.

Fig. 7.9 presents the stator flux simulation results of the conventional MPC method and the proposed RCM-FTC method under an inductance parameter mismatch fault. As shown in Fig. 7.9, both the conventional MPC method and the proposed RCM-FTC method demonstrate similarly excellent performance when the PMSM drive system operates under normal conditions. However, as illustrated in Fig. 7.9A, the αβ-axis stator flux response of the PMSM drive system exhibits significant fluctuations when using the conventional control method, with deviations reaching up to 0.8 Wb under the inductance parameter mismatch fault, resulting in substantial tracking errors. In contrast, Fig. 7.9B shows that the proposed

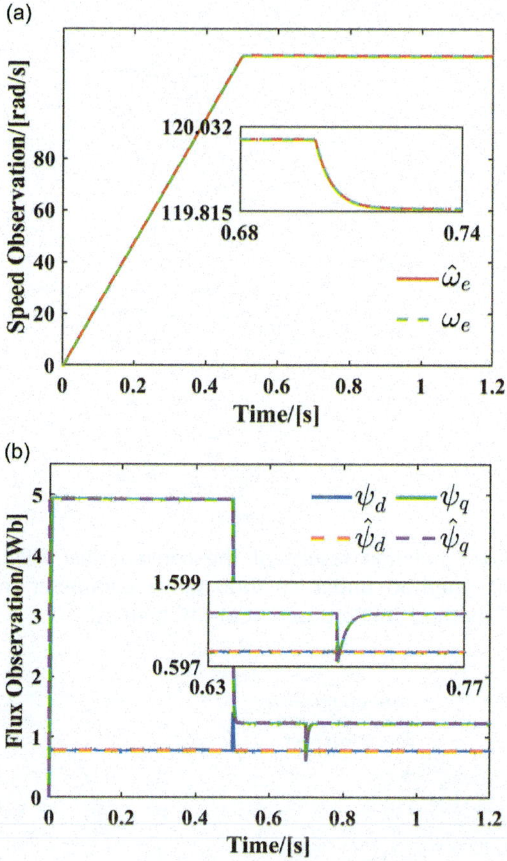

Fig. 7.8 Observation of speed and stator flux using the designed MIT-SMO under an inductance parameter mismatch fault. (a) Speed observation. (b) Flux speed observation.

RCM-FTC method achieves noticeably better stator flux tracking performance compared to the conventional MPC method, reducing the αβ-axis stator flux error to 0.002 Wb, which corresponds to a reduction of 99.75%. Therefore the proposed MF-FSMC strategy effectively suppresses the αβ-axis stator flux oscillations induced by the inductance parameter mismatch. Fig. 7.10 presents the simulation results of three-dimensional rotor flux trajectories under an inductance parameter mismatch fault. Fig. 7.10A shows that the three-dimensional rotor flux linkage trajectory of the PMSM drive system exhibits large flux linkage oscillations and steady-state errors when using the conventional MPC method. Fig. 7.10B shows that the proposed RCM-FTC method can effectively eliminate the influence of

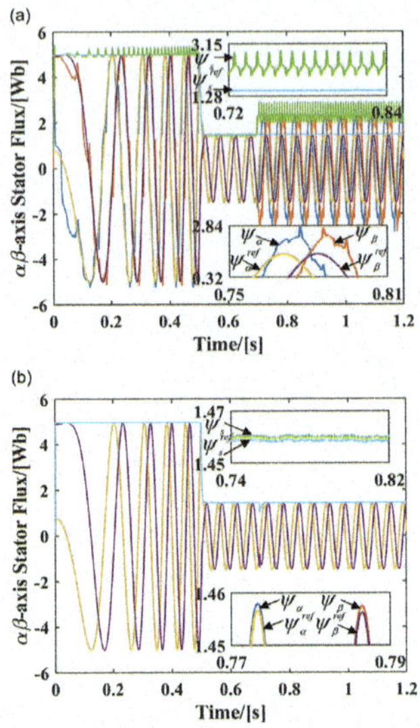

Fig. 7.9 Stator flux simulation results of the conventional MPC method and the proposed RCM-FTC method under an inductance parameter mismatch fault. (a) Conventional MPC method. (b) Proposed RCM-FTC method.

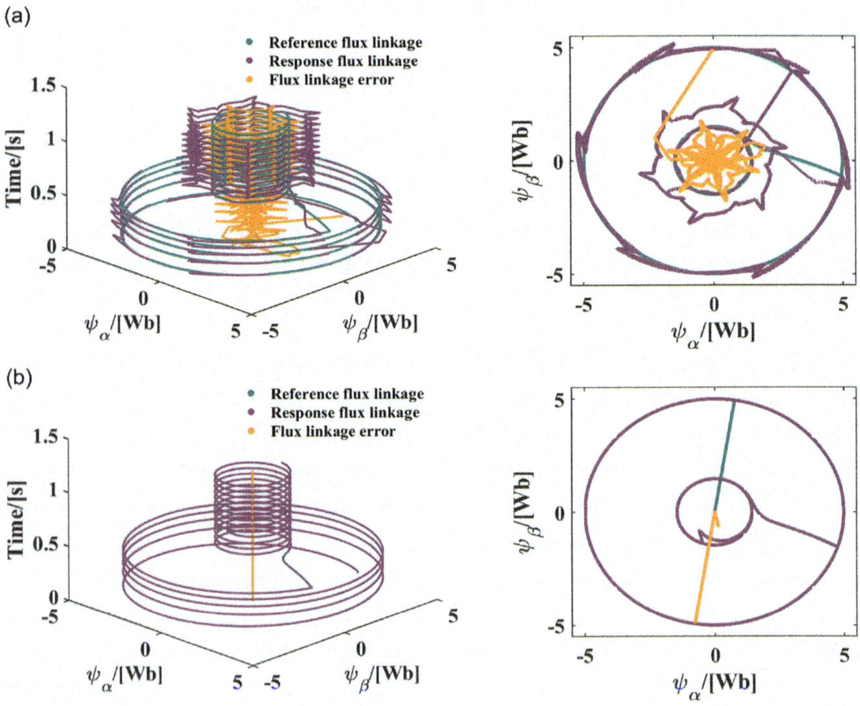

Fig. 7.10 Simulation results of three-dimensional rotor flux trajectories under an inductance parameter mismatch fault. (a) Conventional MPC method. (b) Proposed RCM-FTC method.

the inductance parameter mismatch fault, thereby suppressing flux linkage oscillations and static errors.

7.7 Conclusion

This chapter proposes a novel RCM-FTC method for parameter mismatch fault diagnosis in PMSM drive systems, which can improve the robustness of control systems under parameter mismatch faults. First, the parameter mismatch fault sensitivity of the PMSM drive system is analyzed, and a novel ultralocal model is established in the case of parameter mismatch faults. The robust cascaded model-free tolerant controller is then proposed for the speed loop and flux loop by using an improved reaching law. Furthermore, the MIT-SMO is designed to estimate unknown disturbances caused by parameter mismatch faults in the ultralocal model. The stability of the robust cascaded model-free tolerant controller and the MIT-SMO is proven using Lyapunov stability theory. Simulation results demonstrate the effectiveness of the proposed RCM-FTC method in significantly improving the dynamic and steady-state performance of the PMSM under parameter mismatch faults. Compared with the conventional MPC method, the proposed RCM-FTC method eliminates the dependence on the motor model and reduces stator flux/torque ripple while improving the fault tolerance of the PMSM drive system under parameter mismatch faults.

References

[1] S. Du, J. Hu, et al., A Hall Sensor-Based Position Measurement With On-Line Model Parameters Computation for Permanent Magnet Synchronous Linear Motor, IEEE Sens. J. 18 (13) (2018) 5245–5255.

[2] M.S. Khajueezadeh, S. FeizHoseini, et al., Analysis of Torsional Vibrations on the Resolver Under Eccentricity in PMSM Drive System, IEEE Sens. J. 22 (22) (2022) 21592–21599.

[3] M.S. KhajueeZadeh, M. Emadaleslami, et al., Comprehensive Investigation of the Resolver's Eccentricity Effect on the Field-Oriented Control of PMSM, IEEE Sens. J. 23 (17) (2023) 19145–19152.

[4] S. Huang, G. Wu, et al., Novel Predictive Stator Flux Control Techniques for PMSM Drives, IEEE Trans. Power Electron. 34 (09) (2019) 8916–8929.

[5] C. Zhang, Robust Fault-Tolerant Predictive Current Control for Permanent Magnet Synchronous Motors Considering Demagnetization Fault, IEEE Trans. Ind. Electron. 65 (07) (2018) 5324–5334.

[6] X. Zhang, B. Hou, et al., Deadbeat Predictive Current Control of Permanent-Magnet Synchronous Motors with Stator Current and Disturbance Observer, IEEE Trans. Power Electron. 32 (05) (2017) 3818–3834.

[7] X. An, G. Liu, et al., Adjustable Model Predictive Control for IPMSM Drives Based on Online Stator Inductance Identification, IEEE Trans. Ind. Electron. 69 (04) (2022) 3368–3381.

[8] H. Li, R. Zhang, et al., State Parameters Estimation for Distributed Drive Electric Vehicle Based on PMSMs' Sensorless Control, IEEE Sens. J. 24 (15) (2024) 24054–24069.
[9] L. Yaming, Z. Rongyun, et al., Distributed Electric Vehicle State Parameter Estimation Based on the ASO-SRGHCKF Algorithm, IEEE Sens. J. 22 (19) (2022) 18780–18792.
[10] Y. Zhou, S. Zhang, et al., Current Prediction Error Based Parameter Identification Method for SPMSM With Deadbeat Predictive Current Control, IEEE Trans. Energy Convers. 36 (03) (2021) 1700–1710.
[11] C. Xie, S. Zhang, et al., Parameter Identification for SPMSM With Deadbeat Predictive Current Control Using Online PSO, IEEE Trans. Transp. Electrific. 10 (02) (2024) 4055–4064.
[12] D.Q. Dang, M.S. Rafaq, et al., Online Parameter Estimation Technique for Adaptive Control Applications of Interior PM Synchronous Motor Drives, IEEE Trans. Ind. Electron. 63 (03) (2016) 1438–1449.
[13] M.N. Uddin, M.M.I. Chy, Online Parameter-Estimation-Based Speed Control of PM AC Motor Drive in Flux-Weakening Region, IEEE Trans. Ind. Appl. 44 (05) (2008) 1486–1494.
[14] Y. Wang, et al., A Robust DPCC for IPMSM Based on a Full Parameter Identification Method, IEEE Trans. Ind. Electron. 70 (08) (2023) 7695–7705.
[15] Y. Han, S. Chen, et al., Accurate SM Disturbance Observer-Based Demagnetization Fault Diagnosis With Parameter Mismatch Impacts Eliminated for IPM Motors, IEEE Trans. Power Electron. 38 (05) (2023) 5706–5710.
[16] B. Chen, W. Yao, et al., Parameter Sensitivity in Sensorless Induction Motor Drives With the Adaptive Full-Order Observer, IEEE Trans. Ind. Electron. 62 (07) (2015) 4307–4318.
[17] C. Bai, Z. Yin, et al., Robust Predictive Control for Linear Permanent Magnet Synchronous Motor Drives Based on an Augmented Internal Model Disturbance Observer, IEEE Trans. Ind. Electron. 69 (10) (2022) 9771–9782.
[18] X. Yuan, Y. Zuo, et al., Model-Free Predictive Current Control of SPMSM Drives Using Extended State Observer, IEEE Trans. Ind. Electron. 69 (07) (2022) 6540–6550.
[19] X. Sun, Z. Su, et al., Adaptive Model-Free Predictive Current Control for SPMSM Drives With Optimal Virtual Vector Modulation, IEEE/ASME Trans. Mechatron. 29 (04) (2024) 2569–2578.
[20] Y. Luo, K. Yang, et al., Luenberger Observer-Based Model Predictive Control for Six-Phase PMSM Motor With Localization Error Compensation, IEEE Trans. Ind. Electron. 70 (11) (2023) 10800–10810.
[21] O. Babayomi, Z. Zhang, Model-Free Predictive Control of Power Converters With Cascade-Parallel Extended State Observers, IEEE Trans. Ind. Electron. 70 (10) (2023) 10215–10226.
[22] S. Gao, Y. Wei, et al., Model-Free Hybrid Parallel Predictive Speed Control Based On Ultralocal Model of PMSM for Electric Vehicles, IEEE Trans. Ind. Electron. 69 (10) (2022) 9739–9748.

CHAPTER 8

Model predictive control of modular multiphase PMSM under normal conditions

8.1 Introduction

As a key and core component of complex electromechanical systems such as new energy urban rail vehicles, ship propulsion, high-speed elevators, wind power generation, aerospace, etc., the loss or stoppage of motor system functions can have fatal effects, and in severe cases, it can lead to catastrophic accidents [1–3]. The research on a new type of multiphase Permanent Magnet Synchronous Machine (PMSM) with strong fault tolerance, high reliability, and a simple structure has important theoretical and practical significance. Multiphase PMSM has two significant advantages: firstly, it has a large number of motor phases, which increases the frequency of electromagnetic torque pulsation, reduces its pulsation amplitude, and minimizes motor operating noise and vibration. Secondly, various short-circuit and open-circuit faults in multiphase motor drive systems can be converted into motor phase loss faults through fault isolation methods. Multiphase motors can operate smoothly under phase loss faults, with strong fault tolerance and reliability [4,5]. Therefore, in some special application fields that require higher and stricter reliability, compared with traditional three-phase PMSM systems, the multiphase PMSM systems with strong fault tolerance, high reliability, and a simple structure have more advantages.

In recent years, researchers have conducted extensive studies on multiphase PMSM and achieved numerous original results. References [6–8] have carried out in-depth research on key issues such as stator structure modularization, magnetic gap isolation, auxiliary fault-tolerant teeth, and fault-tolerant control strategies, effectively broadening the application of multiphase PMSM in harsh environments and high-reliability fields. Reference [9] proposed a unique modulation scheme to actively suppress harmonic currents in five-phase PMSM by synthesizing control voltages

Robust Model Predictive Control of Permanent Magnet Synchronous Motor Under Parameter Mismatch
https://doi.org/10.1016/B978-0-443-44209-4.00015-4

in the xy-space of the harmonic subspace. Reference [10] introduced a novel five-compatible fault-tolerant module for wheeled PMSM and analyzed its fault tolerance under various fault conditions, including single-phase open circuit, two-phase adjacent-phase open circuit, two-phase nonadjacent-phase open circuit, and single-phase short circuit. Reference [11] presented a unified open-phase fault diagnostic method for five-phase PMSM systems in multiple operating modes, which can be implemented uniformly through spatial current vectors and reduces the complexity of fault diagnosis. Reference [12] proposed a generalized fault-tolerant model predictive current control method for a five-phase PMSM drive, capable of addressing both single open-switch faults and arbitrary two open-switch faults. Reference [13] introduced a low-complexity model predictive current control strategy based on the deadbeat principle to address the issues of variable switching frequency and high computational burden in model predictive control of five-phase PMSM. Reference [14] proposed a fault-tolerant model predictive current control method to enhance the fault tolerance of five-phase PMSM drives under single and multiple open-circuit fault operations. Reference [15] developed a fault-tolerant control method for adjacent and nonadjacent two-phase open-circuit faults in a five-phase surface-mounted PMSM. This method not only considers the current reconstruction of two-phase open-circuit faults but also introduces a voltage reconstruction method. Reference [16] designed a novel multiphase consequent pole PMSM, which offers a relatively wide constant power speed range compared to traditional PMSM. References [17,18] designed a modular fault-tolerant multiphase PMSM, whose stator core is composed of stator laminations from multiple PMSM units with repeated characteristics. The motor units exhibit electrical, magnetic, and thermal isolation properties. The motor control algorithm is simple and easy to implement, not constrained by the number of power supply phases in the power system. In addition, due to the parallel control of each motor unit, the fault tolerance performance of the motor is significantly improved. Even if one or two motor units fail, the overall performance is minimally impacted.

In this chapter, a new modular multiphase PMSM with strong fault tolerance, high reliability, and a simple structure has been designed. The finite element simulation model of the modular multiphase PMSM is established, and the output characteristics of the modular multiphase PMSM are obtained through finite element calculations. Aiming at the special winding structure and characteristics of the new modular multiphase PMSM, the direct power predictive control strategy and cascade deadbeat predictive control strategy are proposed. The simulation verified the effectiveness of the direct power predictive control strategy and the cascade deadbeat predictive control strategy in modular multiphase PMSM systems. This chapter is organized as follows: the characteristic analysis of the modular PMSM is presented in Section 8.2, the direct power predictive control strategy is proposed in Section 8.3, the cascade deadbeat predictive control strategy is developed in Section 8.4, simulation results of the cascade deadbeat predictive control strategy are provided in Sections 8.5, and 8.6 concludes this chapter.

8.2 Characteristic analysis of modular multiphase PMSM

The modular multiphase PMSM is composed of multiple three-phase PMSM units with consistent characteristics. Each three-phase PMSM unit can be analogized to a traditional three-phase PMSM. The rotor of a modular multiphase PMSM is similar to that of a traditional three-phase PMSM. The stator core is made up of stator laminations of the three-phase PMSM units, and each motor unit has the characteristics of electrical, magnetic, and thermal isolation. Due to the unique body structure design of the modular multiphase PMSM, each motor unit has an independent set of three-phase stator windings. When a fault occurs in a certain motor unit, the entire motor can still continue to operate, greatly improving the system's fault tolerance and reliability. This chapter takes a modular multiphase PMSM composed of six motor units with consistent characteristics as an example. A finite element simulation model of the motor is established using finite element software, and the output characteristics of the motor are obtained through finite element calculations. Figs. 8.1 and 8.2 respectively show the stator schematic and structural schematic of a modular multiphase PMSM. Fig. 8.3 shows the output torque curve of a modular multiphase PMSM. Fig. 8.4 shows the no-load back electromotive force of a modular multiphase PMSM. By analyzing the characteristics of the modular multiphase PMSM, the following conclusions can be obtained.

1. As shown in Fig. 8.3, the no-load back electromotive forces output by the six three-phase PMSM units completely overlap and are all sinusoidal waveforms. For modular multiphase PMSMs, this indicates that as the phase number of the motor increases, the harmonic content of the air-gap magnetic field can be effectively suppressed.

2. From Fig. 8.4, it can be seen that the electromagnetic torques output by motor units 1 and 2, as well as motor units 1 and 3, are equal to and twice

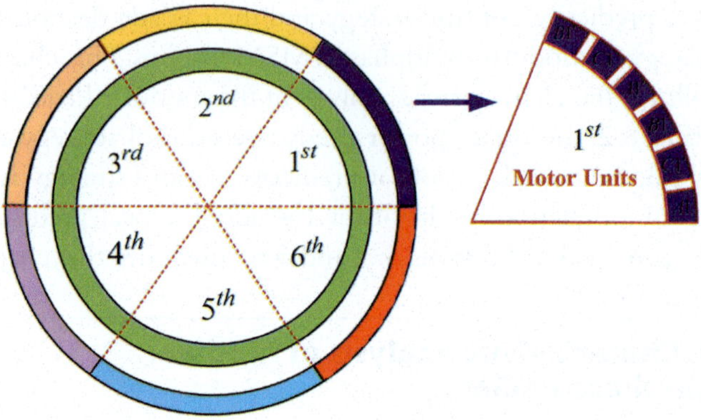

Fig. 8.1 The stator schematic of a modular multiphase PMSM.

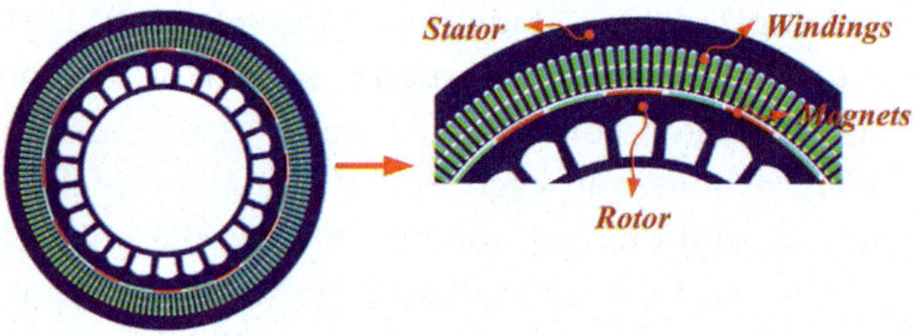

Fig. 8.2 The structural schematic of the modular multiphase PMSM.

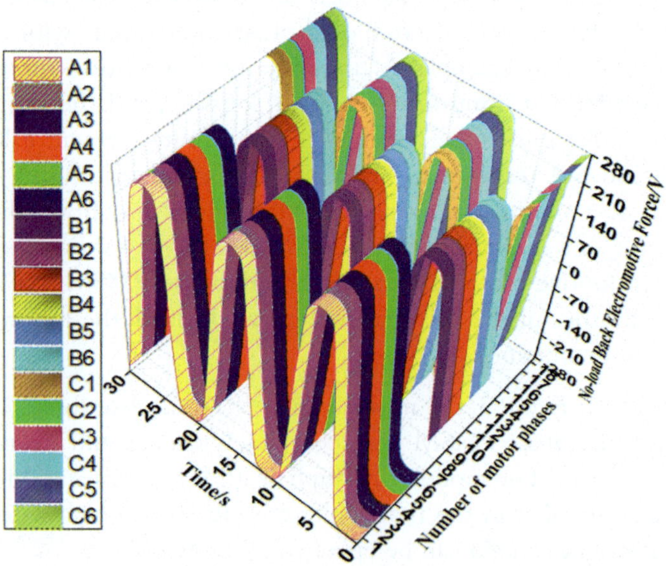

Fig. 8.3 Empty-load magnetic density cloud map and magnetic field line distribution map.

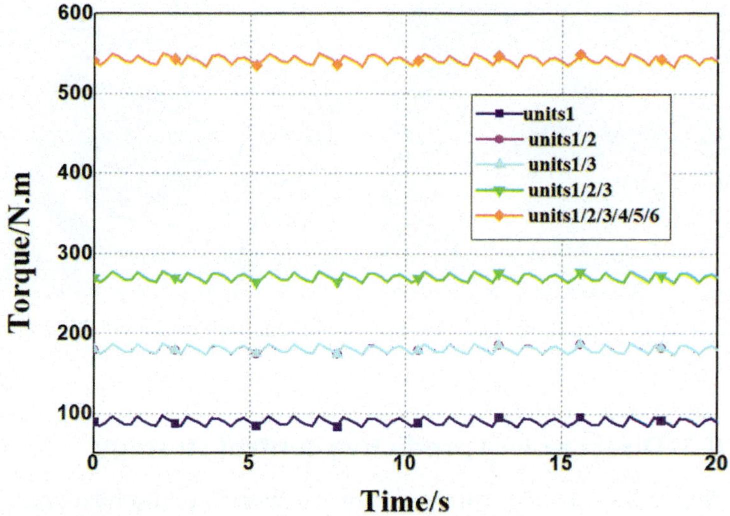

Fig. 8.4 The output torque curve of a modular multiphase PMSM.

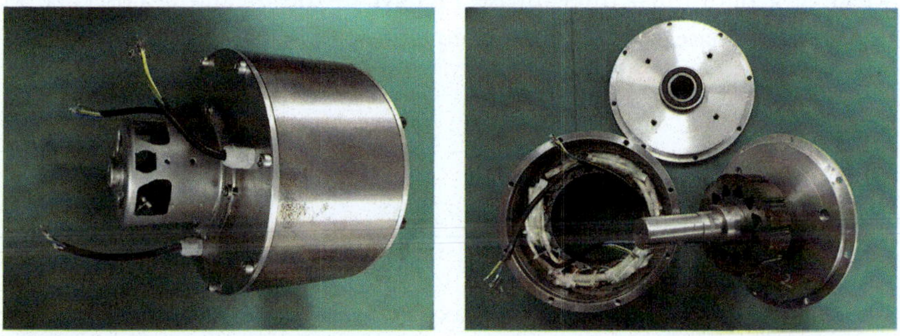

Fig. 8.5 The physical images of the nine-phase modular multiphase PMSM.

the electromagnetic torque output by independent motor units. The electromagnetic torque output by motor units 1, 2, and 3 is three times that of the independent motor units. Therefore it can be observed that the electromagnetic torque output by the modular multiphase PMSM is the sum of the electromagnetic torques output by each three-phase PMSM unit, which effectively satisfies the superposition characteristic of the motor output torque.

Based on the finite element analysis of the modular multiphase PMSM, we have designed a nine-phase modular multiphase PMSM with three motor units. The physical images of the nine-phase modular multiphase PMSM, as well as the rotor and stator, are shown in Figs. 8.5 and 8.6.

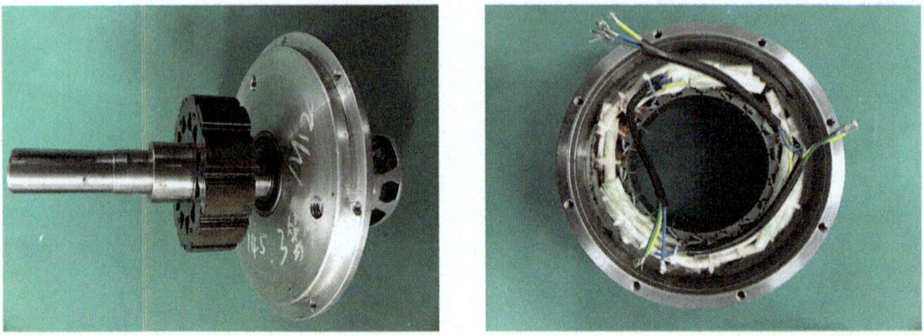

Fig. 8.6 The rotor and stator of a nine-phase modular multiphase PMSM.

8.3 Direct power predictive control strategy

According to the characteristic analysis of modular multiphase PMSM, it can be concluded that the mathematical model of modular multiphase PMSM in the *d*- and *q*-axis can be equivalent to the mathematical model of a traditional three-phase PMSM in the *d*- and *q*-axis [19]. The voltage coordinate transformation equation for each three-phase PMSM unit can be given as

$$\begin{bmatrix} u_{dj} \\ u_{qj} \end{bmatrix} = \frac{2}{3} \begin{bmatrix} \cos\theta & \cos\left(\theta - \frac{2\pi}{3}\right) & \cos\left(\theta + \frac{2\pi}{3}\right) \\ \sin\theta & \sin\left(\theta - \frac{2\pi}{3}\right) & \sin\left(\theta + \frac{2\pi}{3}\right) \end{bmatrix} \begin{bmatrix} u_{aj} \\ u_{bj} \\ u_{cj} \end{bmatrix} \tag{8.1}$$

Where u_{dj} and u_{qj} are the *d*- and *q*-axis stator voltages of the *j* three-phase PMSM unit, respectively; u_{aj}, u_{bj}, u_{cj} are the three-phase stator voltages of the *j* three-phase PMSM unit; θ is the rotor position angle.

By selecting the *d*- and *q*-axis currents as the system state variables, the state equation of the modular multiphase PMSM in the *d*- and *q*-axis can be obtained as

$$\frac{d\boldsymbol{x}_j}{dt} = \boldsymbol{A}_j\boldsymbol{x}_j + \boldsymbol{B}_j\boldsymbol{u}_j + \boldsymbol{D}_j \tag{8.2}$$

Where $\boldsymbol{x}_j = \begin{bmatrix} i_{dj} & i_{qj} \end{bmatrix}^{\mathrm{T}}$ and $\boldsymbol{u}_j = \begin{bmatrix} u_{dj} & u_{qj} \end{bmatrix}^{\mathrm{T}}$ are the state variables and system inputs of the *j* three-phase PMSM unit, respectively. The coefficient matrix is defined as $A_j = \begin{bmatrix} -\dfrac{R_j}{L_{dj}} & \omega_e \dfrac{L_{qj}}{L_{dj}} \\ -\omega_e \dfrac{L_{dj}}{L_{qj}} & -\dfrac{R_j}{L_{qj}} \end{bmatrix}$, $B_j = \begin{bmatrix} \dfrac{1}{L_{dj}} & 0 \\ 0 & \dfrac{1}{L_{qj}} \end{bmatrix}$, $D_j = \begin{bmatrix} 0 \\ -\dfrac{\omega_e}{L_{qj}} \psi_{ro} \end{bmatrix}$.

Where i_{dj} and i_{qj} are the *d*- and *q*-axis stator currents of the *j* three-phase PMSM unit, respectively; ω_e is the electrical angular velocity of the motor; ψ_{ro} is the magnetic flux of the rotor permanent magnet; R_j, L_{dj} and L_{qj} are the resistance and *dq*-axis inductance of the *j* three-phase PMSM unit, respectively.

The modular multiphase PMSM is composed of three-phase PMSM units with consistent characteristics. The resistance, *d*- and *q*-axis inductance of each three-phase PMSM unit are equal, i.e., $R_j = R_o$, $L_{dj} = L_{do}$ and $L_{qj} = L_{qo}$. In addition, the electromagnetic torque generated by each three-phase PMSM unit is equal in magnitude. Therefore the electromagnetic torque output by the modular multiphase PMSM is the sum of the electromagnetic torque generated by each three-phase PMSM unit. The electromagnetic torque equation of modular multiphase PMSM can be obtained as

$$T_e = \frac{3n_p}{2} \sum_{j=1}^{N} (\psi_{ro} i_{qj} + (L_{dj} - L_{qj}) i_{dj} i_{qj}) \tag{8.3}$$

Where T_e is the electromagnetic torque of modular multiphase PMSM.

The motion equation of a modular multiphase PMSM can be given as

$$T_e - T_L = \frac{J}{p} \frac{d\omega_e}{dt} \tag{8.4}$$

Where T_L is the load torque; J is the moment of inertia.

According to Eq. (8.2) and Eq. (8.4), the voltage equation of the surface-mounted modular multiphase PMSM in the *d*- and *q*-axis can be obtained as

$$\begin{cases} \dfrac{di_{dj}}{dt} = -\dfrac{R_o}{L_o} i_{dj} + \omega_e i_{qj} + \dfrac{1}{L_o} u_{dj} \\ \dfrac{di_{qj}}{dt} = -\dfrac{R_o}{L_o} i_{qj} - \omega_e i_{dj} - \dfrac{\psi_{ro}}{L_o} \omega_e + \dfrac{1}{L_o} u_{qj} \end{cases} \tag{8.5}$$

The electromagnetic torque equation of the surface-mounted modular multiphase PMSM is

$$T_e = \frac{3n_p}{2}\sum_{j=1}^{N}(\psi_{ro} i_{qj}) \tag{8.6}$$

Discretize Eq. (8.5), and the discrete voltage equation for the surface-mounted modular multiphase PMSM can be obtained as

$$\begin{cases} u_{dj}(k) = \dfrac{L_o}{T_s} i_{dj}(k+1) + \left(R_o - \dfrac{L_o}{T_s}\right) i_{dj}(k) - L_o \omega_e(k) i_{qj}(k) \\ u_{qj}(k) = \dfrac{L_o}{T_s} i_{qj}(k+1) + \left(R_o - \dfrac{L_o}{T_s}\right) i_{qj}(k) + L_o \omega_e(k) i_{dj}(k) + \psi_{ro}\omega_e(k) \end{cases} \tag{8.7}$$

The discrete power equation for the surface-mounted modular multiphase PMSM is

$$P(k) = T_e(k)\omega_r(k) \tag{8.8}$$

Where $\omega_r(k)$ is the mechanical speed.

In order to minimize the absolute deviation between the predicted values of stator flux and torque and their command values, the following cost function is constructed as

$$g = k_\psi \left|\psi_s^{ref} - \psi_s(k+1)\right| + \left|T_e^{ref} - T_e(k+1)\right| \tag{8.9}$$

Where k_ψ is the weight factor; T_e^{ref} and ψ_s^{ref} are the electromagnetic torque command value and stator flux command value, respectively; $T_e(k+1)$ and $\psi_s(k+1)$ are the electromagnetic torque prediction value and stator flux prediction value, respectively.

When the sampling period is short enough, the discrete state equation of the modular multiphase PMSM can be expanded using the first-order Taylor series, and the electromagnetic torque and stator flux prediction equations of the modular multiphase PMSM are derived accordingly.

$$\begin{cases} \psi_s(k+1) = \sqrt{\psi_{dj}^2(k+1) + \psi_{qj}^2(k+1)} \\ T_e(k+1) = \dfrac{3n_p}{2}\sum_{j=1}^{N}\left[\psi_{ro} i_{qj}(k+1)\right] \end{cases} \tag{8.10}$$

Where $\psi_{dj}(k+1)$ and $\psi_{qj}(k+1)$ are the d- and q-axis stator flux, respectively, and $i_{qj}(k+1)$ represent the q-axis current.

According to Eqs. (8.10) and (8.7), the torque prediction equation for the modular multiphase PMSM can be obtained as

$$\begin{cases} T_e(k+1) = \dfrac{3Nn_p}{2}\left[\psi_{ro} i_{qj}(k+1)\right] \\ i_{qj}(k+1) = \dfrac{1}{L_o} T_s u_{qj}(k) - \left(\dfrac{R_o}{L_o} T_s - 1\right) i_{qj}(k) - T_s \omega_e(k) i_{dj}(k) - \dfrac{1}{L_o} T_s \psi_{ro} \omega_e(k) \end{cases} \tag{8.11}$$

According to Eq. (8.7), the stator flux prediction equation for a modular multiphase PMSM can be obtained as

$$\begin{cases} \psi_{dj}(k+1) = T_s u_{dj}(k) + \left(1 - \dfrac{R_o}{L_o} T_s\right) \psi_{dj}(k) + T_s \omega_e(k) \psi_{qj}(k) + \dfrac{R_o}{L_o} \psi_{ro} T_s \\ \psi_{qj}(k+1) = T_s u_{qj}(k) + \left(1 - \dfrac{R_o}{L_o} T_s\right) \psi_{qj}(k) - T_s \omega_e(k) \psi_{dj}(k) \end{cases} \tag{8.12}$$

According to Eq. (8.8), the electromagnetic torque command value of the modular multiphase PMSM can be obtained as

$$T_e^{ref} = \frac{P^{ref}}{\omega_r(k)} \tag{8.13}$$

When using the maximum torque per ampere (MTPA) strategy, the mathematical relationship between the command values of stator flux and torque is

$$\psi_s^{ref} = \sqrt{\psi_d^2 + \psi_q^2} = \sqrt{(\psi_{ro})^2 + \left(L_o \cdot \dfrac{T_e^{ref}}{\frac{3Nn_p}{2}\psi_{ro}}\right)^2} \tag{8.14}$$

Fig. 8.7 shows the structural diagram of the modular multiphase PMSM system using the direct power prediction control method. From the analysis of the characteristics of the modular multiphase PMSM, it can be observed that each set of stator windings in the modular multiphase PMSM can be independently excited by a three-phase symmetrical voltage generated by a voltage source inverter, and the voltage-second characteristics of each winding inverter are consistent. The voltage source inverter control of the modular multiphase PMSM can be analogized to the control of traditional inverters. Since the modular multiphase PMSM consists of six motor units with consistent characteristics, controlling one motor unit can achieve control of the entire motor. As shown in Fig. 8.7, the implementation steps

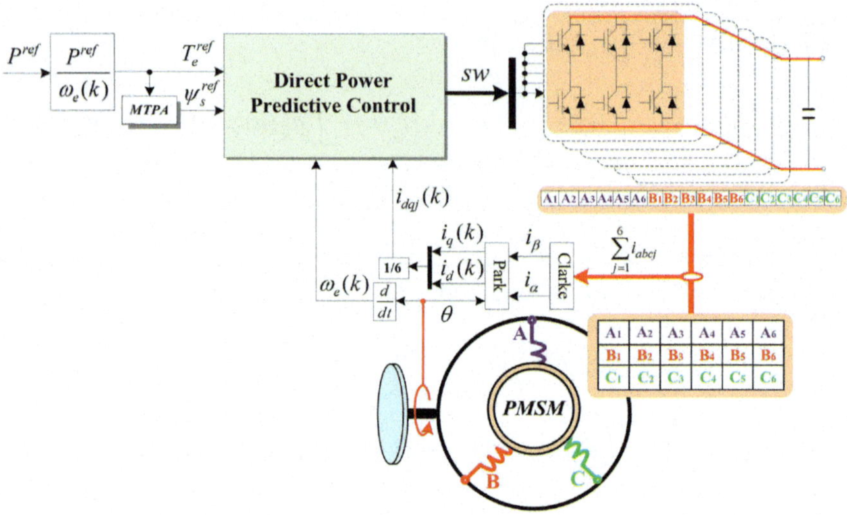

Fig. 8.7 Schematic diagram of a modular multiphase PMSM system using the direct power prediction control method.

of the direct power prediction control method are as follows: first, calculate the stator magnetic flux prediction value and the electromagnetic torque prediction value of the modular multiphase PMSM. Then, input the stator magnetic flux prediction value and the electromagnetic torque prediction value into the cost function to achieve the goal of controlling the modular multiphase PMSM. By adopting the proposed direct power prediction control method, the modular multiphase PMSM can achieve good electromagnetic torque and stator flux control performance.

To verify the effectiveness and feasibility of the direct power prediction control method for modular multiphase PMSM, the modular multiphase PMSM consisting of six motor units was built in Mat-lab. The stator resistance parameter used in the simulation is 0.026 Ω, the inductance parameter is 5.572 mH, and the permanent magnet flux parameter is 0.992 Wb. The power command value is set to

$$\begin{cases} P^{ref} = 2\times 10^5 t + 1\times 10^5 & 0 \le t \le 0.5s \\ P^{ref} = 2\times 10^5 & 0.5s \le t \le 1s \end{cases} \tag{8.15}$$

Fig. 8.8 shows the simulated waveforms of electromagnetic torque and stator flux when using the direct power predictive control method to control a modular multiphase PMSM. As shown in Fig. 8.8A, when the modular multiwinding permanent magnet motor reaches a steady state, the torque

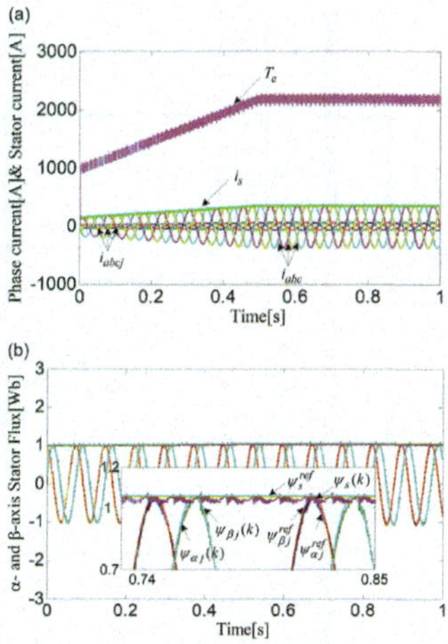

Fig. 8.8 Simulation results of the modular multiphase PMSM system using the direct power prediction control method. (A) Electromagnetic torque and phase current. (B) The α- and β-axis stator flux.

ripple of the motor is relatively small. Due to the modular multiphase PMSM being composed of six three-phase PMSM units, the stator current is six times the motor unit current, as shown in Fig. 8.8A. As shown in Fig. 8.8B, after adopting the direct power predictive control method, the predicted values of the α- and β-axis stator flux of the modular multiphase PMSM can accurately track their command values. It can be concluded that the proposed direct power predictive control method has achieved excellent torque and stator flux control performance. Moreover, the proposed direct power predictive control method can effectively eliminate steady-state errors in stator flux tracking and achieve lower stator flux and torque ripple.

8.4 Cascade deadbeat predictive control strategy

The deadbeat predictive control of modular multiphase PMSM consists of a speed outer-loop predictive controller and a current inner-loop predictive controller. This section will focus on the design of the speed outer-loop predictive controller and the current inner-loop predictive controller. The modular multiphase PMSM can be divided into six motor units, and all characteristics of each motor unit are mutually repetitive. The control strategy applicable to conventional three-phase PMSM can achieve control of the modular multiphase PMSM. Therefore this chapter proposes

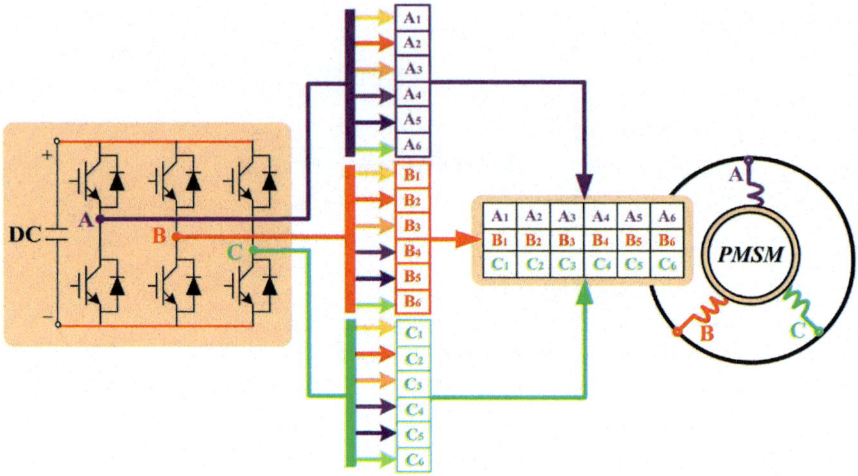

Fig. 8.9 Schematic diagram of the AC drive system for a modular multiphase PMSM.

the use of a three-phase voltage source inverter to control the entire modular multiphase PMSM, which can solve the problem of being limited by the number of power supply phases in the power system. Fig. 8.9 shows a schematic diagram of the AC drive system for a modular multiphase PMSM.

Design of current inner loop predictive controller

The general solution of the state Eq. (8.2) for a modular multiphase PMSM in the *d-q* axis system can be expressed as

$$\boldsymbol{x}_j(t) = e^{\boldsymbol{A}_j(t-t_o)}\boldsymbol{x}_j(t_o) + \int_{t_o}^{t} e^{\boldsymbol{A}_j(t-\tau)}(\boldsymbol{B}_j\boldsymbol{u}_j(\tau) + \boldsymbol{D}_j)d\tau \tag{8.16}$$

Let $t_o = kT_s$, $t_o = (k+1)T_s$, and T_s is the control period. When the control period is short enough, we can obtain

$$\begin{cases} e^{-\frac{R_j T_s}{L_{dj}}} \approx 1 - \dfrac{R_j T_s}{L_{dj}} \\ e^{-\frac{R_j T_s}{L_{qj}}} \approx 1 - \dfrac{R_j T_s}{L_{qj}} \\ \cos(\omega_e T_s) \approx 1 \\ \sin(\omega_e T_s) \approx \omega T_s \end{cases} \tag{8.17}$$

Discretize the state Eq. (8.2) to obtain the discrete state equation of the modular multiphase PMSM, as shown in Eq. (8.18).

$$\boldsymbol{u}_j(k) = \boldsymbol{F}_j^{-1}[\boldsymbol{i}_j(k+1) - \boldsymbol{E}_j(k)\boldsymbol{i}_j(k) - \boldsymbol{P}_j(k)] \tag{8.18}$$

where $\boldsymbol{i}_j(k) = \begin{bmatrix} i_{dj}(k) & i_{qj}(k) \end{bmatrix}^{\mathrm{T}}$, $\boldsymbol{u}_j(k) = \begin{bmatrix} u_{dj}(k) & u_{qj}(k) \end{bmatrix}^{T}$,

$$\boldsymbol{E}_j(k) = \begin{bmatrix} 1 - \dfrac{R_j}{L_{dj}} T_s & \dfrac{L_{qj}}{L_{dj}} T_s \omega_e(k) \\ -\dfrac{L_{dj}}{L_{qj}} T_s \omega_e(k) & 1 - \dfrac{R_j}{L_{qj}} T_s \end{bmatrix}, \quad \boldsymbol{F}_j = \begin{bmatrix} \dfrac{T_s}{L_{dj}} & 0 \\ 0 & \dfrac{T_s}{L_{qj}} \end{bmatrix},$$

$$\boldsymbol{P}_j(k) = \begin{bmatrix} 0 \\ -\dfrac{\psi_{ro}}{L_{qj}} T_s \omega_e(k) \end{bmatrix}.$$

In practical digital control systems, there is an irreversible one beat delay between the actual output voltage of the controller and the command voltage, resulting in the voltage vector $\boldsymbol{u}_j(k)$ calculated at time kT_s not being applied until time $(k+1)T_s$. In order to eliminate the negative impact of one beat delay on the dynamic and static performance of the system, this chapter uses a Smith predictor at kT_s to predict the sampled current value $\boldsymbol{i}_j(k+1)$ at $(k+1)T_s$. Then, using the current value $\boldsymbol{i}_j(k+1)$ predicted by Smith predictor, the voltage vector $\boldsymbol{u}_j(k+1)$ for the next control cycle is calculated to eliminate the impact of one beat delay. The logic diagram of delay compensation based on Smith prediction is shown in Fig. 8.10.

Due to the fact that the mechanical time constant of the motor is much greater than the electrical time constant, the speed will remain approximately constant within one control cycle, i.e. $\omega_e(k) \approx \omega_e(k+1)$. t$(k+1)T_s$, the voltage vector equation output by the deadbeat current controller can be obtained from Eq. (8.18) as

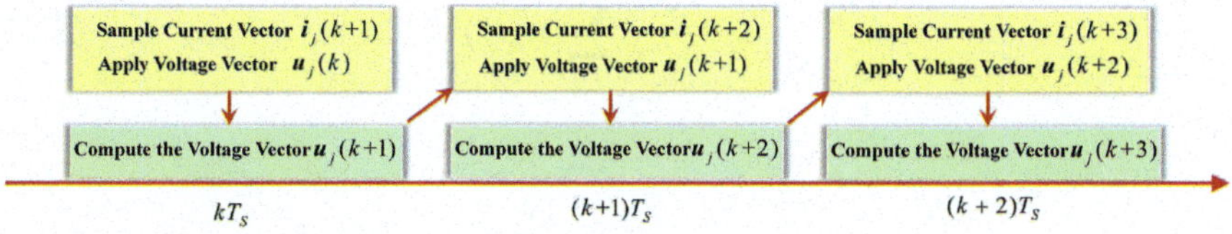

Fig. 8.10 Logic diagram of delay compensation based on Smith prediction.

$$\boldsymbol{u}_j^{ref}(k+1) = \boldsymbol{F}_j^{-1}[\boldsymbol{i}_j^{ref}(k+2) - \boldsymbol{E}_j(k)\boldsymbol{i}_j(k+1) - \boldsymbol{P}_j(k)] \tag{8.19}$$

Where $\boldsymbol{u}_j^{ref}(k+1) = \left[u_{dj}^{ref}(k+1) \quad u_{qj}^{ref}(k+1)\right]^{\mathrm{T}}$ is the voltage vector output by the deadbeat current controller; $\boldsymbol{i}_j^{ref}(k+2) = \left[i_{dj}^{ref}(k+2) \quad i_{qj}^{ref}(k+2)\right]^{\mathrm{T}}$ is the given current vector; $\boldsymbol{i}_j(k+1) = \left[i_{dj}(k+1) \quad i_{qj}(k+1)\right]^{\mathrm{T}}$ is the response current vector of the current motor.

According to Eq. (8.19), the voltage vector output by the deadbeat current controller depends on the acquisition of the sampled current value $\boldsymbol{i}_j(k+1)$ and the current command value $\boldsymbol{i}_j^{ref}(k+2)$. The Smith predictor is used to predict the sampled current value $\boldsymbol{i}_j(k+1)$.

$$\begin{cases} i_{dj}(k+1) = \left(1 - \dfrac{R_j}{L_{dj}}T_s\right)i_{dj}(k) + \dfrac{L_{qj}}{L_{dj}}T_s\omega_e(k)i_{qj}(k) + \dfrac{T_s}{L_{dj}}u_{dj}(k) \\ i_{qj}(k+1) = \left(1 - \dfrac{R_j}{L_{qj}}T_s\right)i_{qj}(k) - \dfrac{L_{dj}}{L_{qj}}T_s\omega_e(k)i_{dj}(k) - \dfrac{\psi_{ro}}{L_{qj}}T_s\omega_e(k) + \dfrac{T_s}{L_{qj}}u_{qj}(k) \end{cases} \tag{8.20}$$

The goal of the speed outer loop predictive controller is to enable the response speed to track the commanded speed in the shortest possible time, while obtaining accurate q-axis current command values. Due to the limitations of the system's moment of inertia and maximum torque driving capability, the sampling time T_{ds} of the speed outer loop will be longer than the sampling time T_s of the inner loop of the current. Although the traditional first-order Euler discretization method is simple and easy to implement, high bandwidth predictive speed control may experience oscillations, and the accuracy of the obtained shaft current command values is poor [20]. Therefore without significantly increasing computational complexity, this chapter adopts the second-order Euler discretization method to design a speed outer loop predictive controller.

When using the $i_d = 0$ control strategy, the electromagnetic torque of the modular multiphase PMSM obtained from Eq. (8.3) is

$$T_e = \frac{3n_p}{2}\sum_{j=1}^{N}(\psi_{ro}i_{qj}) \tag{8.21}$$

Combining Eq. (8.4) and Eq. (8.21) yields

$$\frac{d\omega_e}{dt} = \frac{n_p}{J}\left(\frac{3n_p}{2}\sum_{j=1}^{N}(\psi_{ro}i_{qj}) - T_{\mathrm{L}}\right) \tag{8.22}$$

Since $dT_L / dt = 0$, taking the derivative of Eq. (8.22) yields

$$\frac{d^2\omega_e}{dt^2} = \frac{3n_p^2}{2J}\frac{d\sum_{j=1}^{N}(\psi_{ro}i_{qj})}{dt} \tag{8.23}$$

The second-order extension of the rotational speed can be given as

$$\omega_e(k+1) = \omega_e(k) + \frac{d\omega_e}{dt}\cdot T_{\mathrm{ds}} + \frac{d^2\omega_e}{dt^2}\cdot\frac{T_{\mathrm{ds}}^2}{2} \tag{8.24}$$

Substitute Eq. (8.22) and Eq. (8.23) into Eq. (8.24). The q-axis current command value output by the predicted speed controller can be obtained as

$$\sum_{j=1}^{N}(\psi_{ro}i_{qj}(k+1)) = \frac{\omega_e(k+1) - \omega_e(k) + \dfrac{n_pT_L}{J}T_{ds} - \dfrac{57n_p^2}{40J}T_{ds}\sum_{j=1}^{N}(\psi_{ro}i_{qj}(k))}{\dfrac{3n_p^2}{40J}T_{ds}} \tag{8.25}$$

Due to the fact that each independent stator winding of the modular multiphase PMSM is identical, the stator windings of each three-phase PMSM unit are connected in parallel, i.e., $\sum_{j=1}^{N}(\psi_{ro}i_{qj}(k)) = N\psi_{ro}i_q(k)$. Let $i_q(k) = i_q^{ref}(k)$ and $\omega_e(k) = \omega_e^{ref}$, then Eq. (8.25) can be rewritten as

$$i_q^{ref}(k) = \frac{\omega_e^{ref} - \omega_e(k-1) + \dfrac{n_pT_L}{J}T_{ds} - \dfrac{57Nn_p^2}{40J}T_{ds}\psi_{ro}i_q(k-1)}{\dfrac{3Nn_p^2}{40J}\psi_{ro}T_{ds}} \tag{8.26}$$

According to Lagrange, the d- and q-axis current command values at $(k+2)T_s$ can be calculated as

$$\begin{cases} i_d^{ref}(k+2) = i_d^{ref}(k+1) = i_d^{ref}(k) = 0 \\ i_q^{ref}(k+2) = 3i_q^{ref}(k) - 3i_q^{ref}(k-1) + i_q^{ref}(k-2) \end{cases} \tag{8.27}$$

Design of load torque observer

According to Eq. (8.26), the key to achieving speed control of a modular multiphase PMSM lies in the observation of load torque. Combining Eq. (8.21) and Eq. (8.4) yields

$$\frac{d\omega_e}{dt}=\frac{3n_p^2}{2J}\sum_{j=1}^{N}(\psi_{ro}i_{qj})-\frac{n_p}{J}T_L \tag{8.28}$$

Design the following sliding mode observer, which yields

$$\frac{d\hat{\omega}_e}{dt}=\frac{3n_p^2}{2J}\sum_{j=1}^{N}(\psi_{ro}i_{qj})+h\,\mathrm{sgn}(e) \tag{8.29}$$

where $\hat{\omega}_e$ is the observed value of ω_e; $e=\omega_e-\hat{\omega}_e$ is the sliding surface; sgn(.) is the sign function; h is the matrix to be designed.

Combining Eq. (8.28) and Eq. (8.29), the observer error equation can be obtained as

$$\frac{de}{dt}=-\frac{n_p}{J}T_L-h\,\mathrm{sgn}(e) \tag{8.30}$$

When selecting an appropriate h, the observer error equation determined by Eq. (8.30) will converge to zero. Select the following positive definite function as the Lyapunov function.

$$V=\frac{1}{2}e^2 \tag{8.31}$$

Derive Eq. (8.31) and substitute Eq. (8.30) to obtain.

$$\frac{dV}{dt}=e\left[-\frac{n_p}{J}T_L-h\,\mathrm{sgn}(e)\right]\leq -e\frac{n_pT_L}{J}-h\,\|e\| \tag{8.32}$$

From an engineering perspective, the load torque should be bounded, meaning there exists a normal value T_N such that $T_L\leq T_N$. When designing $h\geq\frac{pT_N}{J}\geq\frac{pT_L}{J}$, then $dV/dt\leq 0$. According to the Lyapunov stability theory, the designed observer Eq. (8.29) is asymptotically stable.

Due to the fact that the predictive control algorithm is performed under discretization, Eq. (8.29) is discretized to obtain as

$$\hat{\omega}_e(k+1)=\hat{\omega}_e(k)+\frac{3n_p^2T_s}{2J}\sum_{j=1}^{N}(\psi_{ro}i_{qj}(k))+hT_s\,\mathrm{sgn}(e(k)) \tag{8.33}$$

When the system state reaches the sliding surface, according to the principle of sliding mode equivalent control, that is, $de/dt = e = 0$. It can be obtained from Eq. (8.30).

$$\hat{T}_L = -\frac{J}{n_p} h \operatorname{sgn}(e(k)) \tag{8.34}$$

To eliminate the chattering phenomenon of the sign function in the sliding mode observer, a smooth hyperbolic tangent function is used instead of the sign function. The hyperbolic tangent function is defined as

$$\operatorname{sgn}(v) = \tanh(v) = \frac{e^{2v} - 1}{e^{2v} + 1} \tag{8.35}$$

Fig. 8.11 shows the schematic diagram of the modular multiphase PMSM by using cascade deadbeat predictive control strategy. Input the d- and q-axis current command values $i_d^{ref}(k+2)$ and $i_q^{ref}(k+2)$, as well as the Smith predicted output $\boldsymbol{i}_j(k+1)$, into the current inner loop predictive controller

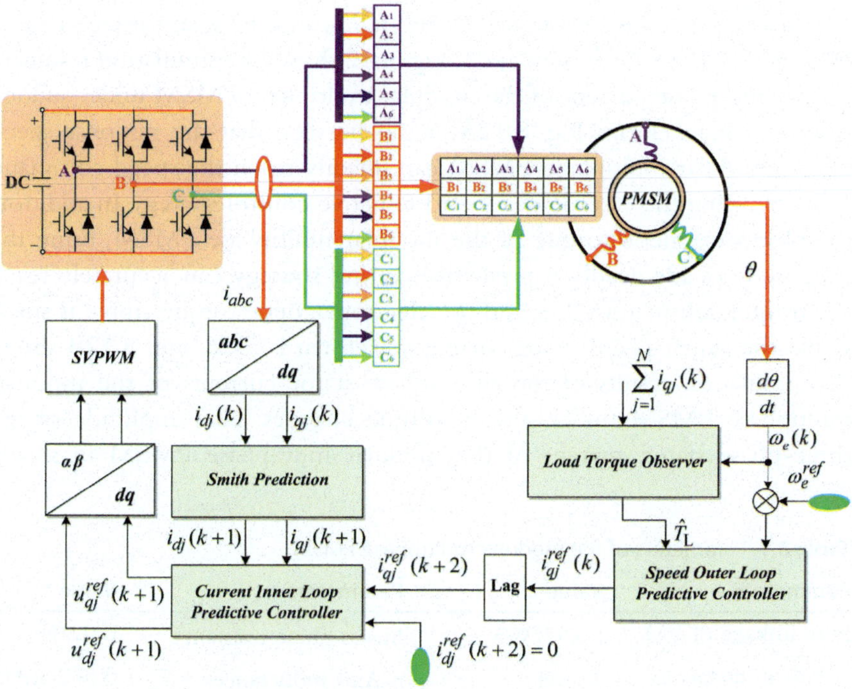

Fig. 8.11 Schematic diagram of the modular multiphase PMSM using a cascade deadbeat predictive control strategy.

Eq. (8.19) to calculate the voltage vector required for accurate control of the modular multiphase PMSM. By modulating the voltage vector with SVPWM, the required switch signal is generated and applied to the inverter to achieve control of the modular multiphase PMSM.

8.5 Simulation results

To verify the effectiveness and feasibility of the modular multiphase PMSM by using cascade deadbeat predictive control strategy in this chapter, a modular multiphase PMSM consisting of six motor units was built in MAT-Lab, and the motor parameters used for simulation are shown in Table 8.1. According to the Lyapunov stability theory and practical debugging of simulation programs, it is known that the designed sliding mode observer has the best observation effect when the parameter h of the sliding mode observer is $h = 1000$.

Case analysis 1: Sudden changes in torque

The operating condition of the modular multiphase PMSM in case analysis 1 is that the speed is maintained at 300 rad/s, and the load torque changes from 500 N·m to 1000 N·m at 2 s. Fig. 8.12 shows the simulation results of torque and phase current of the modular multiphase PMSM under sudden torque changes. From Fig. 8.12A, it can be seen that the response speed of the modular multiphase PMSM can steadily reach the speed command value by using the cascade deadbeat predictive control strategy. In addition, the electromagnetic torque of the modular multiphase PMSM, using the proposed cascade deadbeat predictive control strategy, can accurately reach the torque command value, and the electromagnetic torque ripple is small, while the single-phase stator current waveform is ideal. Fig. 8.12B shows the simulation results of the three-phase stator currents of the modular multiphase PMSM under sudden torque changes. The amplitude of the three-phase stator current of the modular multiphase PMSM in steady

Table 8.1 Parameters of the modular multiphase PMSM.

Parameters	Value	Parameters	Value
Flux linkage of PM	0.892/Wb	Stator phase resistance	0.02/Ω
d-Axis inductances	1/mH	q-Axis inductances	3.572/mH
Rated speed	3000/(r/min)	Rated torque	1000/(N.m)

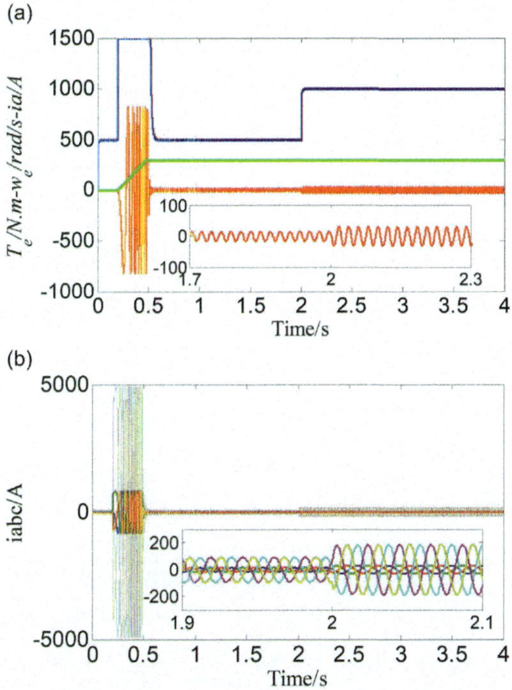

Fig. 8.12 Simulation results of torque and phase current under sudden torque changes. (A) Torque and single-phase current. (B) Three-phase stator current.

state is six times that of the stator current of a unit motor. The reason for this phenomenon is that modular multiphase PMSMs are composed of six motor units with consistent characteristics. In summary, the proposed cascade deadbeat predictive control strategy can achieve excellent electromagnetic torque and speed performance in modular multiphase PMSMs.

Fig. 8.13 shows the simulation results of load torque observation and current tracking for the modular multiphase PMSM under sudden torque changes. The load torque of the modular multiphase PMSM suddenly changes from 500 N·m to 1000 N·m at 2 s. The designed load torque observer can accurately observe the sudden change in load torque, as clearly shown in Fig. 8.13A. Fig. 8.13B presents the simulation results of *d*- and *q*-axis current tracking for the modular multiphase PMSM by adopting the proposed cascade deadbeat predictive control strategy. From Fig. 8.13B, it can be observed that the response value of the *d*-axis current accurately tracks the command value using the proposed cascade deadbeat predictive control strategy. When the modular multiphase PMSM adopts the proposed

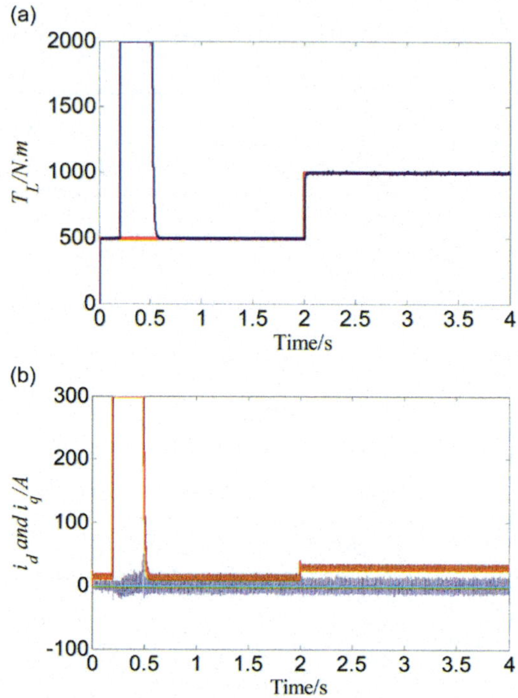

Fig. 8.13 Simulation results of load torque observation and current tracking under sudden torque changes. (A) Load torque observation. (B) *d*- and *q*-axis current tracking.

cascade deadbeat predictive control strategy, the *q*-axis response current of the modular multiphase PMSM can also accurately track the command value under sudden load torque changes, as clearly shown in Fig. 8.13B.

Fig. 8.14 shows the simulation results of the stator flux 3-D trajectory of the unit motor under a sudden torque change. From Fig. 8.14, it can be seen that the stator flux waveform of the unit motor changes relatively little during the torque mutation process because the stator flux variation of the unit motor is 1/6 of that of the modular multiphase PMSM. Fig. 8.15 shows the simulation results of the stator flux 3-D trajectory of the modular multiphase PMSM by adopting the proposed cascaded deadbeat predictive control strategy. By comparing the stator flux 3-D trajectory diagrams of the modular multiphase PMSM and the unit motor, it can be concluded that the stator flux amplitude of the modular multiphase PMSM is 6 times that of the unit motor, and the stator flux 3-D trajectory of the modular multiphase PMSM is similar to that of the unit motor. In summary, the proposed cascaded deadbeat predictive control strategy can achieve excellent α- and β-axis stator flux control performance in the modular multiphase PMSM.

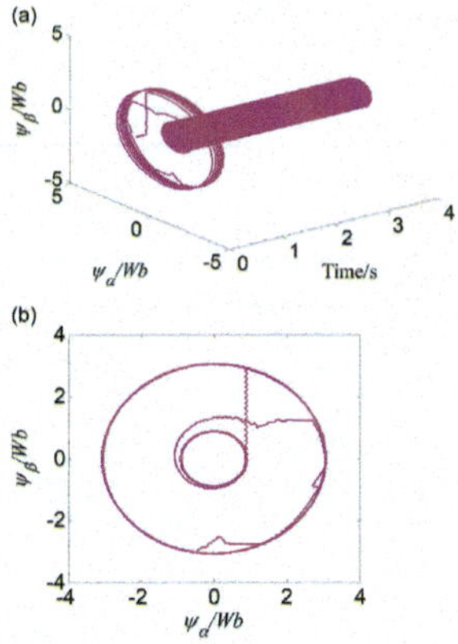

Fig. 8.14 Simulation results of the stator flux 3-D trajectory of a unit motor under a sudden torque change. (A) Stator flux 3-D trajectory. (B) α-β stator flux linkage.

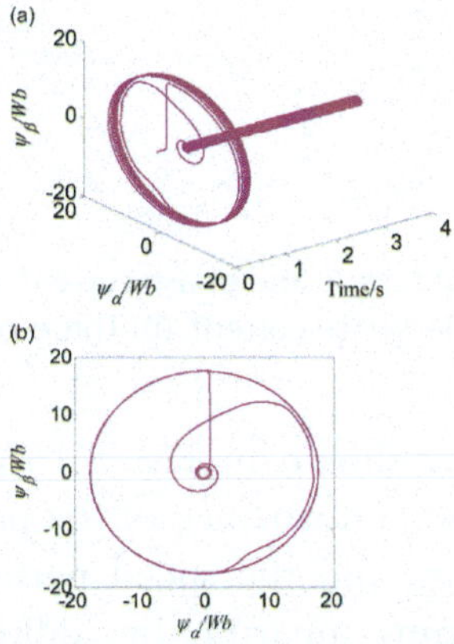

Fig. 8.15 Simulation results of the stator flux 3-D trajectory of a modular multiphase PMSM under a sudden torque change. (A) Stator flux 3-D trajectory. (B) α-β stator flux linkage.

Case analysis 2: Multiple sudden changes in torque

The operating condition of the modular multiphase PMSM in case analysis 2 is to maintain a speed of 300 rad/s, with a sudden increase in load torque from 0 N·m to 1000 N·m at 1.5 s and a sudden decrease to 0 N·m at 3 s. Fig. 8.16 shows the simulation results of torque and phase current of the modular multiphase PMSM by adopting the proposed cascade deadbeat predictive control strategy. When using the cascaded deadbeat predictive control strategy, the response speed of the modular multiphase PMSM can quickly reach the speed command value in the case of sudden

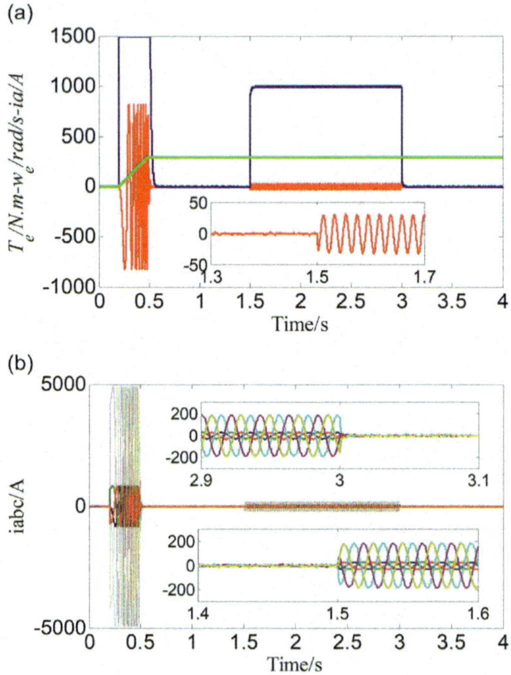

Fig. 8.16 Simulation results of torque and phase current under multiple sudden torque changes. (A) Torque and single-phase current. (B) Three-phase stator current.

torque changes, as clearly shown in Fig. 8.16A. The sudden change in torque does not affect the performance of the proposed cascade deadbeat predictive control strategy, and the speed performance of the modular multiphase PMSM remains superior. In addition, the electromagnetic torque of the modular multiphase PMSM achieves good control performance by using the proposed cascade deadbeat predictive control strategy. Fig. 8.16B shows the three-phase stator current relationship between the modular multiphase PMSM and the motor unit under multiple sudden torque changes. From Fig. 8.16B, it can be seen that when the electromagnetic torque is zero, the three-phase stator current of both the modular multiphase PMSM and the unit motor is zero. When the electromagnetic torque is 1000 N·m, the amplitude of the three-phase stator current of the modular multiphase PMSM is six times that of the unit motor, as clearly shown in Fig. 8.16B. In summary, the proposed cascade deadbeat predictive control strategy enables the modular multiphase PMSM to achieve lower torque ripple while maintaining good speed performance.

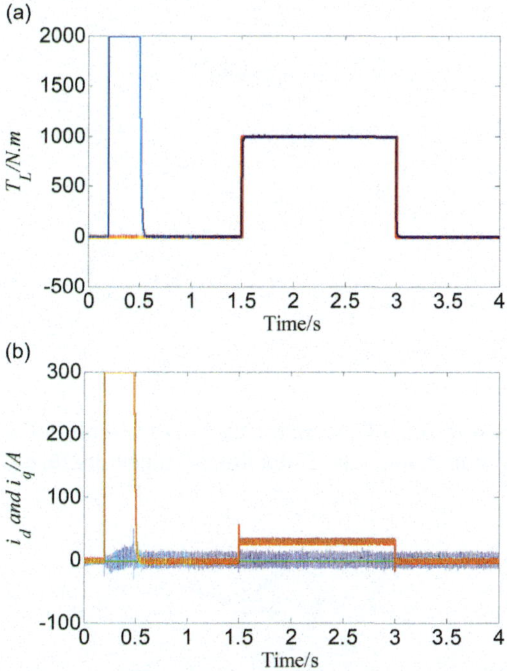

Fig. 8.17 Simulation results of load torque observation and current tracking under multiple sudden torque changes: (A) Load torque observation, (B) dq-axis current tracking.

Fig. 8.17 shows the simulation results of load torque observation and current tracking for the modular multiphase PMSM under multiple sudden torque changes. As shown in Fig. 8.17A, it can be observed that the load torque observer accurately detects changes in load torque, especially in cases of sudden variations in the load torque of the modular multiphase PMSM. Fig. 8.17B presents the simulation results of *d*- and *q*-axis current tracking for the modular multiphase PMSM using the proposed cascade deadbeat predictive control strategy. From Fig. 8.17B, it can be seen that the *d*- and *q*-axis response currents of the modular multiphase PMSM accurately track the commanded currents during multiple sudden torque changes. Therefore the modular multiphase PMSM achieves excellent electromagnetic torque and speed performance, as the use of the cascade deadbeat predictive control strategy ensures precise tracking of *d*- and *q*-axis currents.

Fig. 8.18 shows the simulation results of the stator flux 3-D trajectory of a unit motor using the proposed cascade deadbeat predictive control

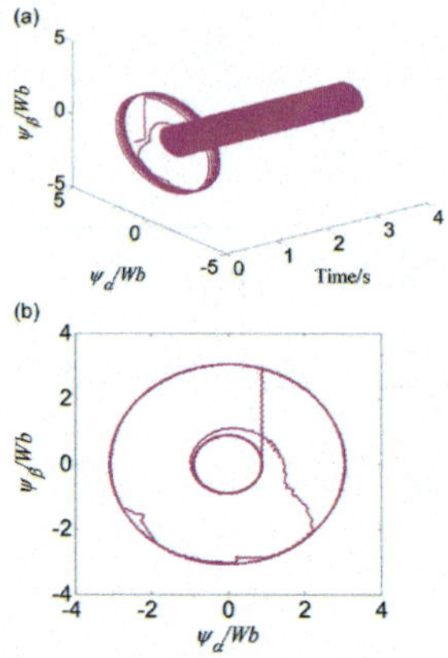

Fig. 8.18 Simulation results of the stator flux 3-D trajectory of a unit motor under multiple sudden torque changes. (A) Stator flux 3-D trajectory. (B) α-β stator flux linkage.

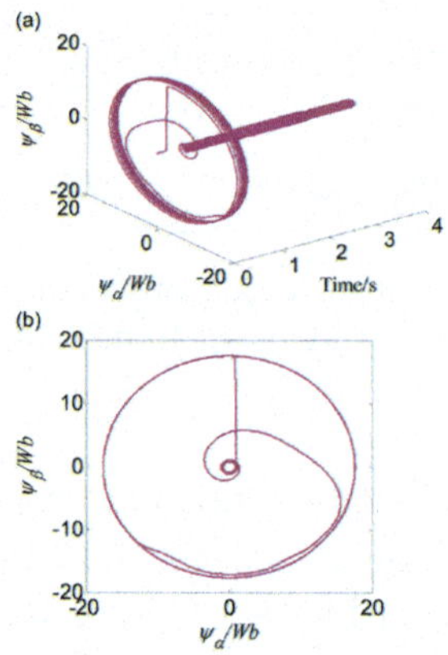

Fig. 8.19 Simulation results of the stator flux 3-D trajectory of a modular multiphase PMSM under multiple sudden torque changes. (A) Stator flux 3-D trajectory. (B) α-β stator flux linkage.

strategy. From Fig. 8.18, it can be seen that the proposed cascade deadbeat predictive control strategy achieves good stator flux control performance during the torque mutation process, as the *d*- and *q*-axis response currents of the modular multiphase PMSM accurately track the command current. Fig. 8.19 shows the simulation results of the stator flux 3-D trajectory of the modular multiphase PMSM under multiple sudden torque changes. By comparing the stator flux 3-D trajectories of the modular multiphase PMSM and the unit motor, it can be concluded that the stator flux 3-D trajectory of the modular multiphase PMSM is similar to that of the unit motor. Moreover, it can be observed that the α- and β-axis stator flux amplitudes of the modular multiphase PMSM are six times those of the unit motor, because the modular multiphase PMSM is composed of six motor units with consistent characteristics. In summary, the use of the proposed cascade deadbeat predictive control strategy achieves better stator flux control performance for the modular multiphase PMSM, as the *d*- and *q*-axis current tracking performance is improved.

8.6 Conclusion

Based on the comprehensive characteristics analysis of modular multiphase PMSM, this chapter proposes a direct power predictive control strategy and a cascade deadbeat predictive control strategy for modular multiphase PMSM, which can improve the control accuracy of torque/stator flux and achieve integrated predictive control of speed and current. The main conclusions drawn are as follows:

1. The direct power predictive control strategy proposed in this chapter for modular multiphase PMSM achieves good torque and stator flux control performance, effectively eliminating steady-state errors in stator flux tracking of modular multiphase PMSM. Meanwhile, this strategy improves the tracking accuracy of stator magnetic flux and reduces the pulsation of stator flux and torque.
2. The cascade deadbeat predictive control strategy based on load torque observation and Smith prediction was presented in this chapter, which can effectively eliminate the irreversible one-beat delay in digital control systems and suppress the oscillation caused by high-bandwidth predictive speed control. In addition, a sliding mode observer was designed to observe motor load disturbances, meeting the high-performance speed control system requirements of modular multiphase PMSM.
3. The modular multiphase PMSM studied in this chapter exhibits better dynamic and transient performance compared to a standard three-phase PMSM. The stator current of a modular multiphase PMSM is significantly smaller than that of a three-phase PMSM under the same load torque. The reduction in stator current indicates that the risk of demagnetization of permanent magnets can be greatly minimized.
4. The modular multiphase PMSM studied in this chapter has the advantages of strong fault tolerance, high reliability, simple structure, and easy modular assembly. In addition, the control algorithm for modular

multiphase PMSM is simple and easy to implement and is not limited by the number of power supply phases in the power system. It is beneficial for expanding the application of PMSM in harsh environments with high reliability requirements, providing reference value for practical engineering.

References

[1] E. Levi, Advances in converter control and innovative exploitation of additional degrees of freedom for multiphase machines, IEEE Trans. Ind. Electron. 63 (01) (2015) 433–448.

[2] A. Mohammadpour, L. Parsa, Global fault tolerant control technique for multiphase permanent magnet machines, IEEE Trans. Ind. Appl. 51 (01) (2015) 178–186.

[3] N. Nguyen, F. Meinguet, E. Semail, et al., Fault tolerant operation of an open-end winding five-phase PMSM drive with short-circuit inverter fault, IEEE Trans. Ind. Electron. 63 (01) (2016) 595–605.

[4] F.M. A, F. Betin, C.G. A, et al., Fuzzy logic and sliding-mode controls applied to six-phase induction machine with open phases, IEEE Trans. Ind. Electron. 57 (01) (2010) 354–364.

[5] Z. Sun, J. Wang, J.G. W, et al., Enhanced optimal torque control of fault-tolerant PM machine under flux-weakening operation, IEEE Trans. Ind. Electron. 57 (01) (2010) 344–353.

[6] F. Li, W. Hua, M. Tong, et al., Nine-phase flux-switching permanent magnet brushless machine for low-speed and high-torque applications, IEEE Trans. Magn. 51 (03) (2015) 1–4.

[7] X. Xue, W. Zhao, J. Zhu, et al., Design of five-phase modular flux-switching permanent-magnet machines for high reliability applications, IEEE Trans. Magn. 49 (07) (2013) 3941–3944.

[8] L. Feng, H. Wei, C. Ming, et al., Analysis of fault tolerant control for a nine-phase flux-switching permanent magnet machine, IEEE Trans. Magn. 50 (11) (2014) 1–4.

[9] W. Wusen, S. Zaixin, et al., Decoupled Modulation Scheme for Harmonic Current Suppression in Five-Phase PMSM, IEEE Trans. Power Electron. 37 (08) (2022) 8795–8799.

[10] Z. Ping, S. Yi, et al., Investigation of a Novel Five-Phase Modular Permanent-Magnet In-Wheel Motor, IEEE Trans. Magn. 47 (10) (2011) 4084–4087.

[11] X. Xiaolong, H. Jun, et al., Unified Open-Phase Fault Diagnosis of Five-Phase PMSM System in Normal Operation and Fault-Tolerant Operation Modes, IEEE Trans. Transp. Electrific. 11 (05) (2025) 12063–12075.

[12] H. Wentao, Z. Xiaofeng, et al., Generalized Fault-Tolerant Model Predictive Control of Five-Phase PMSM Drives Under Single/Two Open-Switch Faults, IEEE Trans. Ind. Electron. 70 (08) (2023) 7569–7579.

[13] M.S.R. Saeed, S. Wensheng, et al., Low-Complexity Deadbeat Model Predictive Current Control With Duty Ratio for Five-Phase PMSM Drives, IEEE Trans. Power Electron. 35 (11) (2020) 12085–12099.

[14] D. Jiachen, H. Wentao, et al., Torque Ripple Reduction of Model Predictive Controlled Five-Phase PMSM Drives With Open-Circuit Faults, IEEE Trans. Transp. Electrific. 9 (02) (2023) 2228–2237.

[15] W. Huanran, G. Chunyang, et al., A Double-Phase Open-Circuit Fault Tolerant Control Method for Five-Phase PMSM With Zero Sequence Voltage Compensation, IEEE Trans. Ind. Appl. 61 (04) (2025) 6434–6443.

[16] M. Onsal., Y. Demir, Aydin M.A New Nine-Phase permanent magnet synchronous motor with consequent pole rotor for high-power traction applications, IEEE Trans. Magn. 53 (11) (2017) 1–6.
[17] F. Luise, S. Pieri, M. Mezzarobba, et al., Regenerative testing of a concentrated-winding permanent-magnet synchronous machine for offshore wind generation—Part I: test concept and analysis[J], IEEE Trans. Ind. Appl. 48 (06) (2012) 1779–1790.
[18] F. Luise, S. Pieri, M. Mezzarobba, et al., Regenerative testing of a concentrated-winding permanent-magnet synchronous machine for offshore wind generation—Part II: test implementation and results[J], IEEE Trans. Ind. Appl. 48 (06) (2012) 1791–1796.
[19] M. Onsal., Y. Demir, M. Aydin, A new nine-phase permanent magnet synchronous motor with consequent pole rotor for high-power traction applications, IEEE Trans. Magn. 53 (11) (2017) 1–6.
[20] C. Garcia, J. Rodriguez, C. Silva, et al., Full predictive cascaded speed and current control of an induction machine, IEEE Trans. Energy Convers. 31 (03) (2016) 1059–1067.

CHAPTER 9

Cascade speed and torque predictive control of modular multiphase PMSM

9.1 Introduction

Modular multiphase permanent magnet synchronous motor (PMSM) drives are considered for many special high-power applications, including high-power traction vehicles, marine propulsion, and wind turbine generation, due to their excellent advantages over three-phase counterparts in terms of fault-tolerant capability, operating efficiency, and torque density [1–3]. Compared with conventional multiphase PMSMs, the novel modular multiphase PMSM with special mechanical degree winding placement has attracted increasing attention due to its outstanding advantages of higher torque quality and lower cogging. In addition, the converter control design of modular multiphase PMSMs is simple, as they can be driven by multiple independent three-phase converters [4,5]. Under steady-state conditions, the modular multiphase PMSM can achieve low torque ripple and a reduced stator current harmonic distortion rate by adopting conventional PI controllers [4–6]. However, since the parameters of the conventional PI controller need to be adjusted for different working conditions, it cannot guarantee excellent control performance of the modular multiphase PMSM across the entire operating range.

With the development of advanced microprocessor technology, predictive control (PC) has been widely used in the field of power electronics applications because of its simple implementation and high bandwidth. There are two main methods of predictive control in the field of power electronics applications: finite control set predictive control (FCSPC) [7] and continuous control set predictive control (CCSPC) [8]. The CCSPC method generates the switching state of the converter actuation by solving the solution of an optimization problem in the modulation stage. The FCSPC method uses the discreteness and load model of the control object to solve the optimization problem. The FCSPC method has been used

Robust Model Predictive Control of Permanent Magnet Synchronous Motor Under Parameter Mismatch
https://doi.org/10.1016/B978-0-443-44209-4.00002-6

in many specialized applications, such as multilevel converters [9], matrix converters [10], and multiphase applications [11]. Compared with the conventional PI controller, predictive torque control (PTC) has received extensive attention for multiphase PMSM control as it has the advantages of faster dynamic response and less torque ripple across the entire operating range [3,12].

The PTC method has been widely adopted in high-performance adjustable speed drives due to its advantages of faster dynamic response and smaller torque ripple. A finite control set model PTC with an adaptive observer was presented in [12], which can achieve excellent performance over a wide speed range. In [13], a modified double-vector model PTC was proposed by reasonably selecting an active vector instead of a zero vector, which can significantly reduce torque ripples while maintaining good dynamic performance. In [14], a newly developed parallel PTC strategy without a weighting factor was proposed, addressing the issue of optimal weighting factor selection in the conventional PTC method. In [15], a disturbance observer-based PTC approach for induction motor systems was presented, which can improve load disturbance rejection ability and robustness. A novel model PTC method combined with space vector modulation was proposed in [16], enhancing the fault-tolerant performance of a five-phase PMSM under an open-circuit fault. In [17], the deadbeat model PTC with discrete space vector modulation was presented for PMSM drives, which can reduce the computational burden and torque ripple. In [12–17], the internal torque loop of PMSM is controlled by the FCSPC strategy to achieve excellent performance, but the external speed loop is still controlled by a classical PI controller.

To improve the control performance of the internal loop and external loop, many predictive cascaded control methods have been studied. In [18,19], a nonlinear predictive control method using the cascaded structure was proposed for a PMSM drive, which can achieve precise speed trajectory tracking and regulate the current component. In [20], a novel full predictive cascaded speed and current control method was developed to avoid the use of the conventional speed PI controller, which can obtain fast speed responses and zero steady-state error. In [21], a cascade model predictive control scheme with an embedded disturbance model was proposed for a PMSM, which can reduce the effects of disturbances and nonlinearities. In [22], a predictive cascaded stator flux control algorithm for PMSM drives was proposed, which can eliminate the influence of rotor position error and flux linkage parameter perturbation.

In this chapter, the cascade speed and torque predictive control of a modular multiphase PMSM is revisited in the internal and external control loops. The cascade speed and torque predictive control method, considering unknown load torque disturbance, is proposed, which can improve the dynamic and static response as well as the tracking performance of the modular multiphase PMSM. The predictive cascaded control structure has two loops: an external predictive speed loop and an internal predictive torque loop. The external predictive speed loop, based on model predictive control deadbeat, is designed to eliminate the influence of load torque disturbance by using an unknown load torque disturbance observer. The internal predictive torque loop utilizes the finite control set model predictive control to realize the control of torque and stator flux. This chapter is organized as follows. The mathematical model and drive system structure of the modular multiphase PMSM are developed in Section 9.2. The cascade speed and torque predictive control with an unknown load torque disturbance observer is presented in Section 9.3. The simulation results are provided in Section 9.4. Section 9.5 concludes this chapter.

9.2 Mathematical model and drive system structure of modular multiphase PMSM

Mathematical model of the proposed modular multiphase PMSM

The structural diagrams of the modular multiphase PMSM with the unconventional stator winding placement are illustrated in Fig. 9.1. The proposed modular multiphase PMSM is composed of three sets of segregated motor units, which share a set of permanent magnet rotors. Each motor unit has isolation characteristics in terms of magnetic, electrical, and thermal properties. The stator winding distribution of each motor unit is similar to that of the conventional three-phase PMSM. The winding arrangement of the proposed three-modular three-phase PMSM is synthesized by using multiple winding groups in phase with each other. Under no-load conditions, the flux line distribution and magnetic density of the proposed three-modular three-phase PMSM are shown in Fig. 9.1. From Fig. 9.1, it can be observed that the distribution of magnetic lines is repeated three times along the circumference of the stator. There is no saturation phenomenon, and the stator teeth are the parts with the maximum magnetic density (i.e., 1.42 T), which is suitable for most operating conditions.

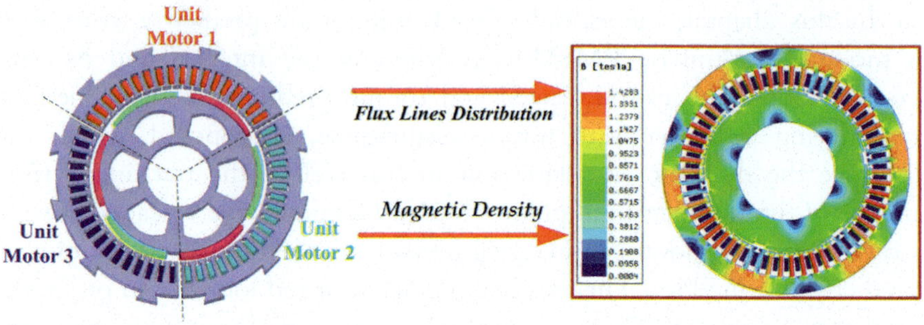

Fig. 9.1 The structural diagrams and magnetic density of a three-modular, three-phase PMSM.

According to the characteristic analysis of the proposed modular multiphase PMSM, it is evident that the modular multiphase PMSM can be modeled as three isolated three-phase PMSMs, which greatly reduces the complexity of the established mathematical model. The *d-q* axis model of the modular multiphase PMSM in the synchronous rotor coordinate system is similar to that of the conventional three-phase PMSM [4,6]. Therefore the *d-q* axis voltage transformation can be expressed as

$$\begin{bmatrix} u_{dj} \\ u_{qj} \end{bmatrix} = \frac{2}{3}\begin{bmatrix} \cos\theta & \cos\left(\theta - \frac{2\pi}{3}\right) & \cos\left(\theta + \frac{2\pi}{3}\right) \\ \sin\theta & \sin\left(\theta - \frac{2\pi}{3}\right) & \sin\left(\theta + \frac{2\pi}{3}\right) \end{bmatrix}\begin{bmatrix} u_{aj} \\ u_{bj} \\ u_{cj} \end{bmatrix} \tag{9.1}$$

Where u_{aj}, u_{bj} and u_{cj} represent the A-, B-, and C-axis voltages of the *j*th motor unit, respectively; u_{dj} and u_{qj} represent the *d*- and *q*-axis voltages of the *j*th motor unit, respectively; θ is the rotor position angle.

According to Eq. (9.1), the *d-q* axis voltage equation of a modular multiphase PMSM can be obtained as:

$$\begin{cases} \dfrac{di_{dj}}{dt} = -\dfrac{R_o}{L_o} i_{dj} + \omega_e i_{qj} + \dfrac{1}{L_o} u_{dj} \\ \dfrac{di_{qj}}{dt} = -\dfrac{R_o}{L_o} i_{qj} - \omega_e i_{dj} - \dfrac{\psi_{ro}}{L_o}\omega_e + \dfrac{1}{L_o} u_{qj} \end{cases} \tag{9.2}$$

where i_{dj} and i_{qj} are the *d*- and *q*-axis currents of the *j*th motor unit, respectively; R_o and L_o are the stator resistance and stator inductances, respectively; ψ_{ro} is the flux linkage; ω_e is the electrical rotor speed.

According to the torque characteristics, the electromagnetic torque equation of a modular multiphase PMSM, considering the resistance torque between motor units, can be expressed as:

$$T_e = \frac{3n_p}{2}\sum_{j=1}^{N}(\psi_{ro} i_{qj}) \tag{9.3}$$

The mechanical equation of a modular multiphase PMSM can be described as:

$$T_e - T_L = \frac{J}{n_p}\frac{d\omega_e}{dt} \tag{9.4}$$

where n_p is the number of pole pairs; J is the moment of inertia; T_e and T_L are the electromagnetic torque and load torque of modular multiphase PMSM, respectively.

Drive system structure of the proposed modular multiphase PMSM

According to the characteristics of modular multiphase PMSM, it is known that the three-phase winding of the motor unit does not share any slot with other motor units, and the mutual inductance between the motor units can be ignored. Therefore the modular multiphase PMSM can be driven by three drive units independently to obtain the required torque. In addition, each set of three-phase windings of the modular multiphase PMSM is isolated from each other. Thus each independent three-phase winding group can be excited independently using conventional two-level voltage source inverters (VSIs). The three-segment stator winding of the modular multiphase PMSM is respectively connected with three VSIs in parallel, as shown in Fig. 9.2. Because the mathematical model and characteristics of

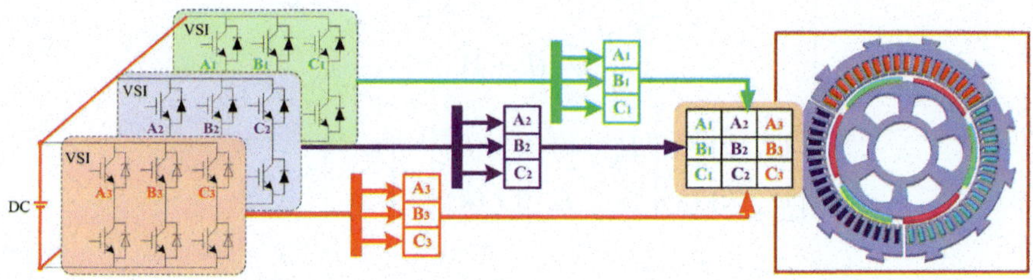

Fig. 9.2 Drive system structure diagram of a modular multiphase PMSM.

each motor unit are the same as those of a conventional PMSM, the controller design of the VSIs can be similar to that of a conventional inverter, which simplifies the control algorithm of the modular multiphase PMSM.

9.3 Cascade speed and torque predictive control with unknown load torque disturbance observer

Predictive torque control of modular multiphase PMSM

The model predictive torque control method is a discrete advanced control method depend strongly on the mathematical model of modular multiphase PMSM. To realize the model predictive torque control method, the state equation of modular multiphase PMSM should be discretized by using the first-order Taylor expansion. If the M-PTC sampling period T_s is short enough, according to Eq. (9.2), the discrete state equation of modular multiphase PMSM can be expressed as

$$\begin{cases} u_{dj}(k) = \dfrac{L_o}{T_s} i_{dj}(k+1) + \left(R_o - \dfrac{L_o}{T_s} \right) i_{dj}(k) - L_o \omega_e(k) i_{qj}(k) \\ u_{qj}(k) = \dfrac{L_o}{T_s} i_{qj}(k+1) + \left(R_o - \dfrac{L_o}{T_s} \right) i_{qj}(k) + L_o \omega_e(k) i_{dj}(k) + \psi_{ro} \omega_e(k) \end{cases} \tag{9.5}$$

According to Eq. (9.5), the predictive stator flux state equation of the modular multiphase PMSM can be obtained as:

$$\begin{cases} \begin{bmatrix} u_{dj}(k) \\ u_{qj}(k) \end{bmatrix} = \begin{bmatrix} \left(\dfrac{R_o}{L_o} - \dfrac{1}{T_s} \right) & -\omega(k) \\ \omega(k) & \left(\dfrac{R_o}{L_o} - \dfrac{1}{T_s} \right) \end{bmatrix} \begin{bmatrix} \psi_{dj}(k) \\ \psi_{qj}(k) \end{bmatrix} \\ \qquad + \begin{bmatrix} \dfrac{1}{T_s} \psi_{dj}(k+1) \\ \dfrac{1}{T_s} \psi_{qj}(k+1) \end{bmatrix} + \begin{bmatrix} -\dfrac{R_o}{L_o} \psi_{ro} \\ 0 \end{bmatrix} \\ \begin{bmatrix} \psi_{dj}(k) \\ \psi_{qj}(k) \end{bmatrix} = \begin{bmatrix} L_o & 0 \\ 0 & L_o \end{bmatrix} \begin{bmatrix} i_d(k) \\ i_q(k) \end{bmatrix} + \begin{bmatrix} \psi_{ro} \\ 0 \end{bmatrix} \end{cases} \tag{9.6}$$

According to Eq. (9.6), the d- and q-axis voltage equations of the modular multiphase PMSM can be rewritten as:

$$\begin{cases} \begin{bmatrix} \psi_{dj}(k+1) \\ \psi_{qj}(k+1) \end{bmatrix} = \begin{bmatrix} T_s u_{dj}(k) \\ T_s u_{qj}(k) \end{bmatrix} + \begin{bmatrix} \left(1 - \dfrac{R_o}{L_o} T_s\right) & T_s \omega_e(k) \\ -T_s \omega_e(k) & \left(1 - \dfrac{R_o}{L_o} T_s\right) \end{bmatrix} \begin{bmatrix} \psi_{dj}(k) \\ \psi_{qj}(k) \end{bmatrix} \\ \qquad + \begin{bmatrix} \dfrac{R_o}{L_o} \psi_{ro} \\ 0 \end{bmatrix} \\ \begin{bmatrix} \psi_{dj}(k) \\ \psi_{qj}(k) \end{bmatrix} = \begin{bmatrix} L_o & 0 \\ 0 & L_o \end{bmatrix} \begin{bmatrix} i_d(k) \\ i_q(k) \end{bmatrix} + \begin{bmatrix} \psi_{ro} \\ 0 \end{bmatrix} \end{cases} \tag{9.7}$$

At the $(k+1)T_s$ moment, the total stator flux prediction equation of modular multiphase PMSM can be obtained as follow according to Eq. (9.7).

$$\psi_{sj}(k+1) = \sqrt{\psi_{dj}^2(k+1) + \psi_{qj}^2(k+1)} \tag{9.8}$$

where $\psi_s(k+1)$ are the total stator flux predictive value.

According to Eq. (9.5), the q-axis current prediction equation of a modular multiphase PMSM can be calculated as follows:

$$i_{qj}(k+1) = \frac{1}{L_o} T_s u_{qj}(k) - \left(\frac{R_o}{L_o} T_s - 1\right) i_{qj}(k) - T_s \omega_e(k) i_{dj}(k) - \frac{1}{L_o} T_s \psi_{ro} \omega_e(k) \tag{9.9}$$

At the $(k+1)T_s$ moment, the torque prediction equation of modular multiphase PMSM can be expressed as follow according to Eqs. (9.9) and (9.3).

$$T_e = \frac{3 n_p N \psi_{ro}}{2} \cdot \begin{Bmatrix} \dfrac{1}{L_o} T_s u_{qj}(k) - \left(\dfrac{R_o}{L_o} T_s - 1\right) i_{qj}(k) \\ -T_s \omega_e(k) i_{dj}(k) - \dfrac{1}{L_o} T_s \psi_{ro} \omega_e(k) \end{Bmatrix} \tag{9.10}$$

To better control the torque and stator flux of modular multiphase PMSM, the cost function of stator flux and torque is defined as

$$g = k_\psi \left|\psi_{sj}^{ref} - \psi_{sj}(k+1)\right| + \left|T_e^{ref} - T_e(k+1)\right| \tag{9.11}$$

where k_ψ is a weighting factor; T_e^{ref} and ψ_s^{ref} are the torque reference value and flux linkage reference value, respectively.

The relationship between the torque reference value and the stator flux reference value can be expressed using the maximum torque per ampere (MTPA) control strategy.

$$\psi_{sj}^{ref} = \sqrt{\psi_d^2 + \psi_q^2} = \sqrt{(\psi_{ro})^2 + \left(L_o \cdot \frac{T_e^{ref}}{\frac{3Nn_p}{2}\psi_{ro}} \right)^2} \tag{9.12}$$

The structural diagram of the modular multiphase PMSM with a cascade speed and torque predictive control strategy is illustrated in Fig. 9.3. The control of the VSI can be similar to that of a conventional inverter because the mathematical model and characteristics of each motor unit are the same as those of a regular PMSM. To simplify the control algorithm of the modular multiphase PMSM, a group of predictive control algorithms is used to send three groups of identical modulation signals simultaneously to control three parallel VSIs. The proposed cascade speed and torque predictive control strategy can enhance the dynamic and static response as well as the tracking accuracy of the modular multiphase PMSM.

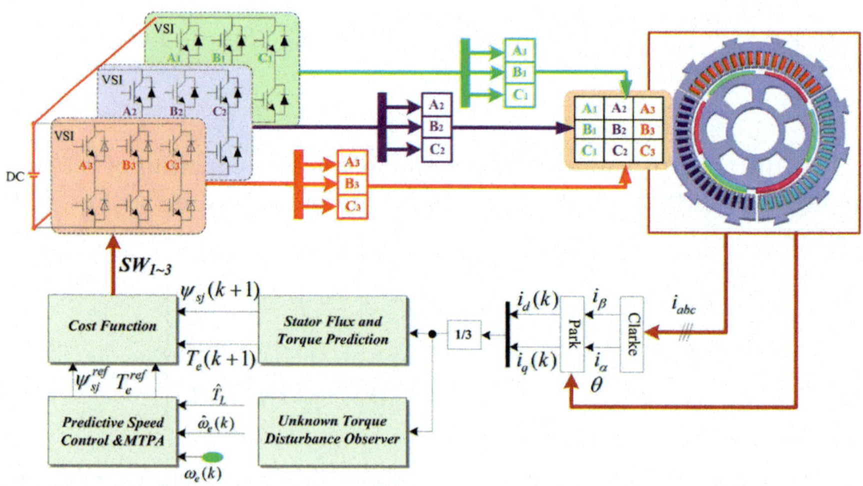

Fig. 9.3 Structural diagram of a modular multiphase PMSM with a speed and torque predictive control strategy.

Predictive speed control with an unknown load torque disturbance observer

In this section, the unknown load torque disturbance observer-based speed prediction controller is designed to obtain the torque reference value. Substituting Eq. (9.3) into Eq. (9.4) yields:

$$\frac{d\omega_e}{dt} = \frac{n_p}{J}\left(\frac{3n_p}{2}\sum_{j=1}^{N}(\psi_{ro} i_{qj}) - T_L\right) \tag{9.13}$$

The load torque is basically considered constant in the sampling time, therefore ${dT_L}/{dt} = 0$. The derivation of equation Eq. (9.13) can be obtained as

$$\frac{d^2\omega_e}{dt^2} = \frac{3n_p^{\,2}}{2J}\frac{d\sum_{j=1}^{N}(\psi_{ro} i_{qj})}{dt} \tag{9.14}$$

The second-order expansion for the rotor speed can be expressed as

$$\omega_e(k+1) = \omega_e(k) + \frac{d\omega_e}{dt}\cdot T_{ds} + \frac{d^2\omega_e}{dt^2}\cdot\frac{T_{ds}^{\,2}}{2} \tag{9.15}$$

Substituting Eqs. (9.13) and (9.14) into Eq. (9.15), one yields

$$\begin{aligned}\omega_e(k+1) = \omega_e(k) &+ \frac{57Nn_p^{\,2}\psi_{ro}}{40J} i_{qj}(k)T_{ds} - \frac{n_p}{J}T_L T_{ds} \\ &+ \frac{3Nn_p^{\,2}\psi_{ro}}{40J} i_{qj}(k+1)T_{ds}\end{aligned} \tag{9.16}$$

According to Eq. (9.3), the relationship between the torque predictive value and the q-axis current predictive value can be obtained as

$$T_e(k+1) = \frac{3Nn_p\psi_{ro}}{2} i_{qj}(k+1) \tag{9.17}$$

To improve the robustness of the predictive controller, the speed response value is replaced by the observed value. Taking ω_e^{ref} and T_e^{ref} as the references of $\omega_e(k+1)$ and $T_e(k+1)$, respectively, the torque reference value can be expressed as according to Eq. (9.16).

$$T_e^{ref} = \frac{20J \cdot \left[\omega_e^{ref} - \hat{\omega}_e(k) - \frac{57Nn_p{}^2\psi_{ro}}{40J} i_{qj}(k)T_{ds} + \frac{n_p}{J}\hat{T}_L T_{ds} \right]}{n_p T_{ds}} \tag{9.18}$$

where ω_e^{ref} and T_e^{ref} are the speed reference value and torque reference value, respectively; $\hat{\omega}_e(k)$ and $\hat{T}_L$ are the observed values of the speed response value and unknown load torque, respectively.

The integral terminal sliding surface is defined as

$$s_\omega = (\hat{\omega}_e - \omega_e) + \lambda \int_0^t \mathrm{sgn}(\hat{\omega}_e - \omega_e)\,d\tau \tag{9.19}$$

Where λ is the designed parameters, and $\lambda > 0$; *sgn* () is sign function; $\hat{\omega}_e$ is the observed value of ω_e.

According to Eq. (9.13), the speed equation can be rewritten as

$$\frac{d\omega_e}{dt} = \frac{3Nn_p^2}{2J}(\psi_{ro} i_{qj}) - \frac{n_p}{J} T_L \tag{9.20}$$

According to Eq. (9.20), the unknown load torque disturbance observer, based on the integral terminal sliding surface, is designed as follows:

$$\frac{d\hat{\omega}_e}{dt} = \frac{3Nn_p^2}{2J}(\psi_{ro} i_{qj}) - U_\omega \tag{9.21}$$

where U_ω is the sliding mode control function.

Combining with Eq. (9.19) and Eq. (9.21) yields

$$\frac{ds_\omega}{dt} = \frac{n_p}{J} T_L - U_\omega + \lambda\,\mathrm{sgn}\left(\widehat{\omega}_e - \omega_e\right) \tag{9.22}$$

with

$$U_\omega = \lambda\,\mathrm{sgn}(e_\omega) + \eta s_\omega + \delta\,\mathrm{sgn}(s_\omega) \tag{9.23}$$

To prove the asymptotic stability of the unknown load torque disturbance observer, the following Lyapunov function is considered as

$$\begin{aligned}\frac{dV}{dt} &= s_{\omega}\cdot\frac{ds_{\omega}}{dt}=s_{\omega}\left(\frac{n_p}{J}T_L-U_{\omega}+\lambda\,\text{sgn}\left(\widehat{\omega}_e-\omega_e\right)\right)\\ &= s_{\omega}\left(\frac{n_p}{J}T_L-\eta s_{\omega}-\delta\,\text{sgn}\left(s_{\omega}\right)\right)\\ &= -\eta s_{\omega}^2+\left(\frac{n_p}{J}s_{\omega}T_L-\delta\,\|s_{\omega}\|\right)\end{aligned} \tag{9.24}$$

According to an engineering point of view, the unknown load torque disturbance is bounded, that is $\|T_L\|\leq N_{\omega}$, where N_{ω} is the boundary values. When η and δ are set to $\eta\geq 0$ and $\delta\geq\frac{n_p}{J}N_{\omega}$, then $dV/dt\leq 0$. According to Eq. (9.22), the observed value of unknown load torque disturbance can be obtained as

$$\hat{T}_L=\frac{J}{n_p}U_{\omega}(k) \tag{9.25}$$

The discrete forms of the unknown load torque disturbance observer can be deduced according to Eq. (9.21).

$$\hat{\omega}_e(k+1)=\hat{\omega}_e(k)+\frac{3Nn_p^2}{2J}T_s\left[\psi_{ro}i_{qj}(k)\right]-T_sU_{\omega}(k) \tag{9.26}$$

9.4 Simulations results

To validate the proposed cascade speed and torque predictive control strategy, the drive system structure of modular multiphase PMSM, as shown in Fig. 9.4, is established in the MATLAB/Simulink. The main parameters of modular multiphase PMSM used in Simulink are listed in Table 9.1. The weight factor of the proposed cascade speed and torque predictive control strategy is 5000, and the sampling frequency is set to 10 kHz. In addition, the gain parameters of the unknown torque disturbance observer are respectively set as λ=5, η=1000, and δ=200 to guaranteed the stability and convergence.

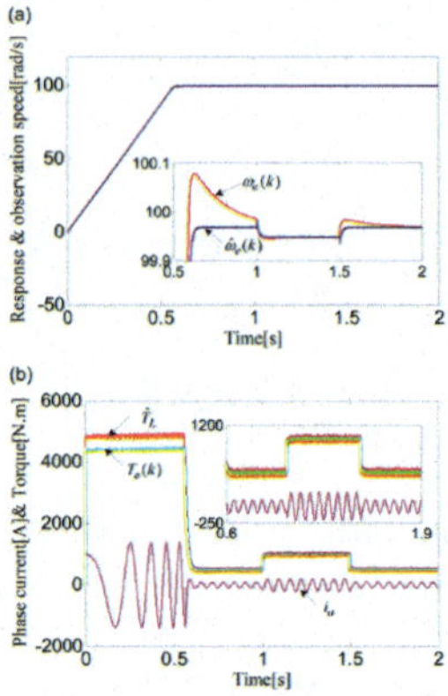

Fig. 9.4 Simulation results of response speed and electromagnetic torque under sudden torque changes. (A) Response speed observation. (B) Torque and single-phase current.

Table 9.1 Main parameters of the modular multiphase PMSM.

Parameter	Value	Parameter	Value
Inductance (L_o)	3.572/mH	Stator phase resistance (R_o)	0.02/Ω
Flux linkage of PM (ψ_f)	0.892/Wb	Rotational inertia (J)	100/kg m^2

Case analysis 1: Simulation results of modular multiphase PMSM under sudden torque changes

The operating conditions of the modular multiphase PMSM in case analysis 1 are to maintain a speed of 100 rad/s, with a sudden increase in load torque from 500 N·m to 1000 N·m at 1 s, and a sudden decrease from 1000 N·m to 500 N·m at 1.5 s. Fig. 9.4 shows the simulation results of response speed and electromagnetic torque under sudden torque changes. As shown in Fig. 9.4A, it can be observed that the proposed cascade speed and torque predictive control algorithm can accurately achieve speed prediction control of the modular multiphase PMSM under sudden torque changes. In addition, the designed unknown load torque disturbance observer can accurately observe the response speed of the modular multiphase PMSM in the case of sudden torque changes, as clearly shown in Fig. 9.4A. Fig. 9.4B presents the simulation results of electromagnetic torque and load torque observation of the modular multiphase PMSM using the proposed cascade speed and torque predictive control method. As shown in Fig. 9.4B, the designed unknown load torque disturbance observer can accurately observe the electromagnetic torque changes of the modular multiphase PMSM. Moreover, it can be observed that the proposed cascade speed and torque predictive control method can achieve precise control of electromagnetic torque in the modular multiphase PMSM under sudden load torque changes. In summary, the proposed cascade speed and torque predictive control method can accurately achieve electromagnetic torque and speed control of the modular multiphase PMSM under sudden torque changes.

Fig. 9.5 shows the simulation results of the stator current and stator flux of a modular multiphase PMSM using the proposed cascade speed and torque predictive control method. The modular multiphase PMSM is composed of three motor units with consistent characteristics, so the stator current amplitude of the modular multiphase PMSM is three times that of a single motor unit. As clearly shown in Fig. 9.5A, the stator current of the proposed modular multiphase PMSM is significantly lower than that of the conventional three-phase PMSM under the same load torque. The advantage of a smaller stator current is that it can reduce the risk of permanent magnet demagnetization faults. Fig. 9.5B illustrates the simulation results of the α-β stator flux linkage of the modular multiphase PMSM under sudden torque changes. Note that there is a small tracking error between the α-β stator flux linkage response value and the reference value of the modular multiphase PMSM when adopting the proposed cascade speed and torque predictive control method. However, the α-β stator flux linkage response value of the modular multiphase PMSM can accurately track the

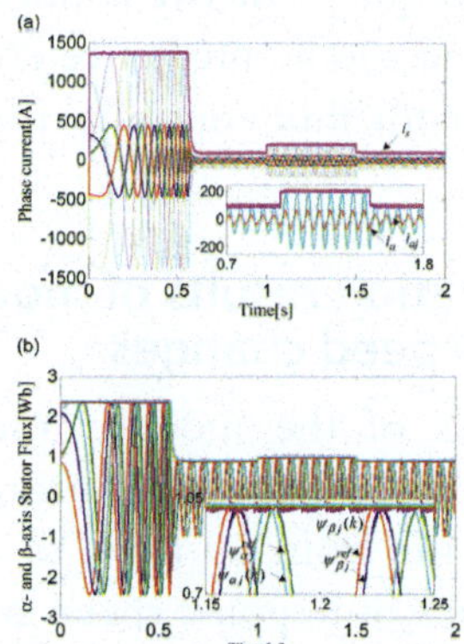

Fig. 9.5 Simulation results of the stator current and stator flux under sudden torque changes. (A) Three-phase stator current. (B) The α- and β-axis stator flux.

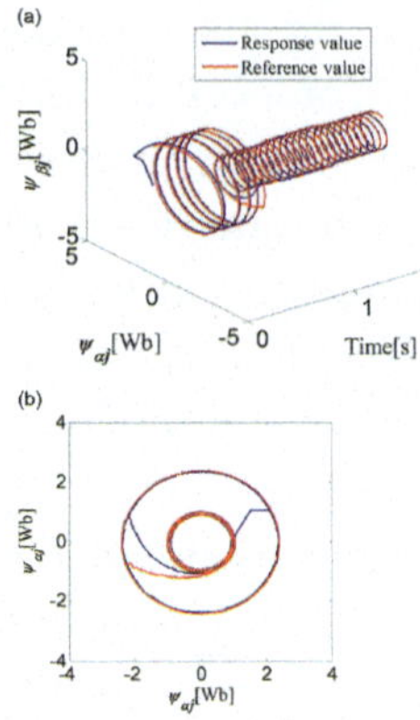

Fig. 9.6 Simulation results of the stator flux 3-D trajectory under sudden torque changes. (A) Stator flux 3-D trajectory. (B) α-β stator flux linkage.

reference value by adopting the proposed cascade speed and torque predictive control method, as shown clearly in Fig. 9.5B.

Fig. 9.6 presents the simulation results of the stator flux 3-D trajectory by adopting the proposed cascade speed and torque predictive control method. The stator flux 3-D trajectory of the modular multiphase PMSM shows little change under load torque variation, as clearly illustrated in Fig. 9.6A. Compared with the proposed modular multiphase PMSM, the stator flux 3-D trajectory of the conventional PMSM exhibits significant changes under load torque variation. The reason for the large changes in the three-dimensional rotor flux trajectories is that the high stator current of the conventional PMSM affects the control performance of the proposed cascade speed and torque predictive control method. In summary, the proposed cascade speed and torque predictive control method can achieve excellent α- and β-axis stator flux control performance in modular multiphase PMSM.

Case analysis 2: Simulation results of modular multiphase PMSM under sudden speed changes

The operating conditions of the modular multiphase PMSM in case analysis 2 are such that the load torque is maintained at 1000 N·m, and the speed suddenly increases from 50 rad/s to 100 rad/s at 1 s. Fig. 9.7 shows the simulation results of response speed and electromagnetic torque under sudden speed changes. As shown in Fig. 9.7A, the speed of the modular multiphase PMSM can be accurately controlled using the proposed cascade speed and torque predictive control algorithm under sudden

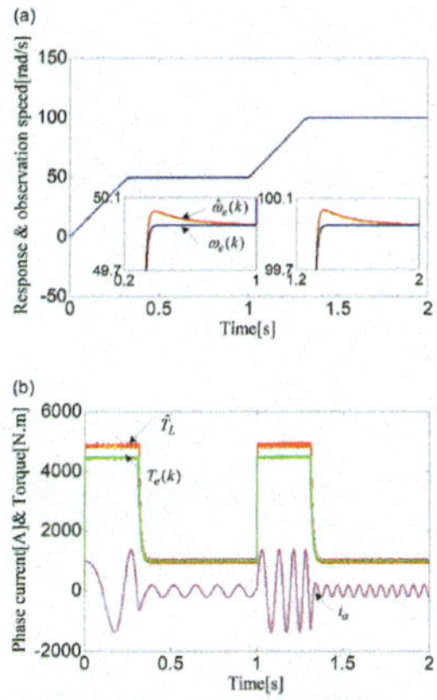

Fig. 9.7 Simulation results of response speed and electromagnetic torque under sudden speed changes: (A) Response speed observation; (B) Torque and single-phase current.

speed changes. Moreover, the designed unknown load torque disturbance observer can accurately observe the response speed changes of the modular multiphase PMSM, as clearly shown in Fig. 9.7A. From Fig. 9.7A, it can be seen that the deviation between the observed speed values and the actual speed response values is very small when using the designed unknown load torque disturbance observer. Fig. 9.7B presents the simulation results of electromagnetic torque and load torque observation of the modular multiphase PMSM under sudden speed changes. As shown in Fig. 9.7B, the designed unknown load torque disturbance observer can accurately observe the load torque of the modular multiphase PMSM under sudden speed changes. Moreover, the electromagnetic torque of the modular multiphase PMSM can be accurately controlled using the proposed cascade speed and torque predictive control algorithm. From Fig. 9.7B, it can be seen that the electromagnetic torque of the modular multiphase PMSM quickly reaches a stable state in the case of sudden speed changes. In summary, the proposed cascade speed and torque predictive control method can accurately achieve electromagnetic torque and speed control of the modular multiphase PMSM under sudden speed changes.

Fig. 9.8 shows the simulation results of the stator current and stator flux of a modular multiphase PMSM under sudden speed changes. From Fig. 9.8A, it can be seen that the stator current amplitude of the modular multiphase PMSM is three times that of a single motor unit. The reason for this phenomenon is that the modular multiphase PMSM is composed

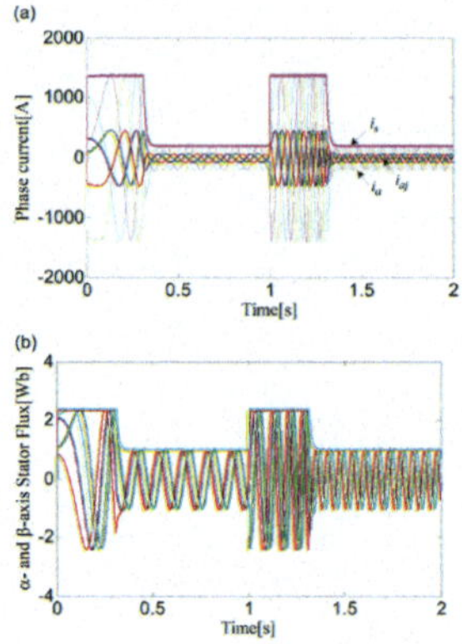

Fig. 9.8 Simulation results of stator flux and stator current under sudden speed changes: (A) Three-phase stator current, (B) The α- and β-axis stator flux.

of three motor units with consistent characteristics. The response speed of the modular multiphase PMSM quickly reaches stability, and the three-phase stator current remains unchanged. The response speed of the modular multiphase PMSM changes at 1 s, and the stator current also changes but quickly stabilizes, as clearly shown in Fig. 9.8A. Fig. 9.8B presents the simulation results of the α-β stator flux linkage of the modular multiphase PMSM using the proposed cascade speed and torque predictive control algorithm. From Fig. 9.8B, it can be observed that the α-β stator flux linkage response value of the modular multiphase PMSM can accurately track the reference value under sudden speed changes. Note that there is a significant change in the α-β stator flux linkage of the modular multiphase PMSM at 1 s, but the response value of the α-β stator flux linkage still maintains accurate tracking with the command value, as clearly shown in Fig. 9.8B. The reason for this phenomenon is that the speed of the modular multiphase PMSM suddenly changes at 1 s. In summary, the proposed cascade speed and torque predictive control method can achieve precise control of the three-phase stator current and stator flux linkage of the modular multiphase PMSM under sudden speed changes.

Fig. 9.9 presents the simulation results of the stator flux 3-D trajectory of a modular multiphase PMSM under sudden speed changes. Under the condition of sudden speed changes, the stator flux 3-D trajectory of the modular multiphase PMSM motor undergoes significant changes when using the proposed cascade speed and torque predictive control method, as clearly shown in Fig. 9.9A. The reason for this phenomenon is that sudden speed changes can cause an increase in the three-phase stator current of

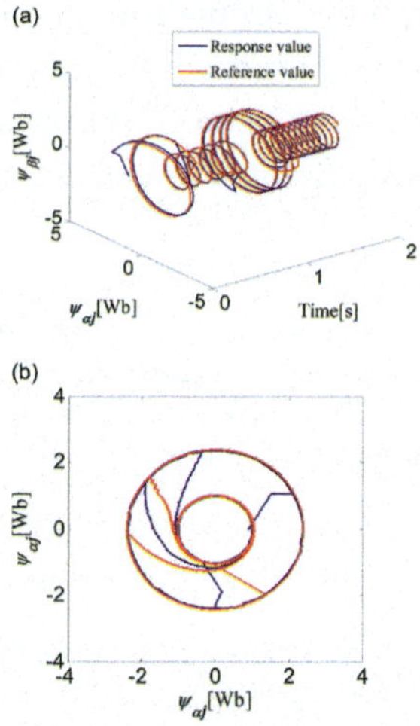

Fig. 9.9 Simulation results of the stator flux 3-D trajectory under sudden speed changes. (A) Stator flux 3-D trajectory. (B) α-β stator flux linkage.

the modular multiphase PMSM. When the response speed of the modular multiphase PMSM reaches a stable state, the 3-D trajectory of its stator flux also stabilizes. In summary, the proposed cascade speed and torque predictive control method can achieve excellent α- and β-axis stator flux control performance under sudden speed changes.

9.5 Conclusion

In this chapter, the cascade speed and torque predictive control method, considering the unknown load torque disturbance, has been proposed to control the speed loop and torque loop of modular multiphase PMSM. The proposed cascade speed and torque predictive control strategy includes an external speed predictive control loop and an internal torque predictive control loop. The external speed predictive control loop, based on model predictive control deadbeat, is designed to overcome the issues of response speed fluctuation caused by unknown load torque disturbance. The internal torque predictive control loop utilizes the finite control set model predictive control to achieve the control of torque and stator flux. Moreover, the unknown load torque disturbance observer, based on an integral terminal sliding surface, is designed to estimate changes in load torque. Simulation results performed on the modular multiphase PMSM confirm the validity and feasibility of the proposed cascade speed and torque predictive control method implementation. The simulation results demonstrate that the modular multiphase PMSM exhibits advantages such as fast speed response, accurate stator flux tracking, and reduced stator current when adopting the cascade speed and torque predictive control method.

References

[1] P.V. I, W. Jiabin, et al., Demagnetization assessment of fractional-slot and distributed wound 6-phase permanent magnet machines, IEEE Trans. Magn. 51 (06) (2015) 1–11.

[2] J. Karttunen, S. Kallio, et al., Decoupled vector control scheme for dual three-phase permanent magnet synchronous machines, IEEE Trans. Ind. Electron. 61 (05) (2014) 2185–2196.

[3] W. Gongping, H. Sheng, et al., Robust Predictive Torque Control of N*3-phase PMSM for High Power Traction Application, IEEE Trans. Power Electron. 35 (10) (2020) 10799–10809.

[4] M. Onsal, Y. Demir, et al., A new nine-phase permanent magnet synchronous motor with consequent pole rotor for high-power traction applications, IEEE Trans. Magn. 53 (11) (2017) 1–6.

[5] H. Xun, J. Dong, et al., Two-segment three-phase PMSM drive with carrier phase-shift PWM, IEEE APEC (2018) 848–854.

[6] H. Xun, J. Dong, et al., Two-Segment Three-Phase PMSM Drive With Carrier Phase-Shift PWM for Torque Ripple and Vibration Reduction, IEEE Trans. Power Electron. 34 (01) (2018) 588–599.

[7] P. Cortes, M. Kazmierkowski, et al., Predictive control in power electronics and drives, IEEE Trans. Ind. Electron. 55 (12) (2008) 4312–4324.

[8] S. Mariethoz, A. Domahidi, et al., High-Bandwidth explicit model predictive control of electrical drives, IEEE Trans. Ind. Appl. 48 (06) (2012) 1980–1992.

[9] C.D. K, L.K. B, Dynamic performance improvement of AC/DC converter using model predictive direct power control with finite control set, IEEE Trans. Ind. Electron. 62 (02) (2015) 757–767.

[10] R. Vargas, J. Rodriguez, et al., Predictive control of an induction machine fed by a matrix converter with increased efficiency and reduced common-mode voltage, IEEE Trans. Energy Convers. 29 (02) (2014) 473–485.

[11] H. Guzman, D.M. J, et al., Speed control of five-phase induction motors with integrated open-phase fault operation using model-based predictive current control techniques, IEEE Trans. Ind. Electron. 61 (09) (2014) 4474–4484.

[12] W. Fengxiang, D.S. A, et al., Finite control set model predictive torque control of induction machine with a robust adaptive observer, IEEE Trans. Ind. Electron. 64 (04) (2017) 2631–2641.

[13] C. Wei, Z. Sike, et al., A modified double vectors model predictive torque control of permanent magnet synchronous motor, IEEE Trans. Power Electron. 34 (11) (2019) 11419–11428.

[14] W. Fengxiang, X. Haotian, et al., Parallel predictive torque control for induction machines without weighting factors, IEEE Trans. Power Electron. 35 (02) (2019) 1779–1788.

[15] W. Junxiao, W. Fengxiang, et al., Design and implementation of disturbance compensation-based enhanced robust finite control set predictive torque control for induction motor systems, IEEE Trans. Ind. Inf. 13 (05) (2017) 2645–2656.

[16] H. Wentao, H. Wei, et al., Model predictive torque control with SVM for five-phase PMSM under open-circuit fault condition, IEEE Trans. Power Electron. 35 (05) (2019) 5531–5540.

[17] W.Yuanlin, W. Xiaocan, et al., Deadbeat model-predictive torque control with discrete space-vector modulation for PMSM drives, IEEE Trans. Ind. Electron. 64 (05) (2017) 3537–3547.
[18] R. Errouissi, M. Ouhrouche, et al., Robust cascaded nonlinear predictive control of a permanent magnet synchronous motor with antiwindup compensator, IEEE Trans. Ind. Electron. 59 (08) (2011) 3078–3088.
[19] R. Errouissi, M. Ouhrouche, et al., Robust Nonlinear Generalized Predictive Control of a Permanent Magnet Synchronous Motor with an Anti-windup Compensator, IEEE ISIE, 2010, pp. 3184–3189.
[20] C. Garcia, J. Rodriguez, et al., Full predictive cascaded speed and current control of an induction machine, IEEE Trans. Energy Convers. 31 (03) (2016) 1059–1067.
[21] C. Shan, W. Liuping, et al., A cascade MPC control structure for a PMSM with speed ripple minimization, IEEE Trans. Ind. Electron. 60 (08) (2012) 2978–2987.
[22] H. Shoudao, W. Gongping, et al., Novel predictive stator flux control techniques for PMSM drives, IEEE Trans. Power Electron. 34 (09) (2018) 8916–8929.

CHAPTER 10

Robust predictive torque control of modular multiphase PMSM considering parameter mismatch

10.1 Introduction

Permanent magnet synchronous motors (PMSMs) have become the key and core components of complex electromechanical systems, such as urban rail vehicles, ship propulsion, high-speed elevators, and wind power generation, due to their advantages of fast dynamic torque response, high overload capability, and wide speed range [1–4]. Although three-phase PMSMs are quite common in various industrial fields, multiphase PMSMs have unique advantages in special applications requiring redundancy and high power [5,6]. Compared with traditional three-phase PMSMs, multiphase PMSMs can provide higher torque and power density under the same voltage vector [7,8], making them especially suitable for high-power traction applications. With the improvement of power levels, multiphase PMSMs have become inevitable in the field of high-power traction [9]. Compared with conventional multiphase PMSMs, the proposed modular multiphase PMSM structure offers significantly higher torque quality and lower cogging. Furthermore, the control strategy of modular multiphase PMSMs is simpler because they can be driven by several independent three-phase voltage source inverters (VSIs) [9–11].

The efficient control strategy is an essential part of multi-three-phase PMSM drives. The improvement of the control strategy can greatly enhance the performance of the drive system and the overall application. In recent years, several methods have been investigated to control multi-three-phase PMSMs more effectively. In [9,12,13], a novel multi-three-phase PMSM was proposed for high-power traction applications. Compared with the conventional multiphase PMSM, the control algorithm is easy to implement and is similar to the traditional three-phase PMSM control algorithm. In [14,15], a two-segment three-phase PMSM system powered by

Robust Model Predictive Control of Permanent Magnet Synchronous Motor Under Parameter Mismatch
https://doi.org/10.1016/B978-0-443-44209-4.00008-7

two inverters with phase-shift PWM has been developed, which can fulfill the requirements of high power and torque ripple mitigation. In [16,17], a dual three-level T-type neutral-point clamped inverter-fed dual PMSM topology was proposed for high-power automotive applications, which can reduce current stress and capacitors' voltage deviation. In [12–17], the PI controller was adopted to control the multi-three-phase PMSM, which can achieve excellent steady-state control performance. However, in practice, the multi-three-phase PMSM is a time-varying control system with parameter variations and uncertain disturbances. It is challenging for the PI controller to achieve satisfactory dynamic control performance across the entire operating range of the time-varying system [18].

The quick dynamic response and torque smoothness are important criteria that determine the performance of the multi-three-phase PMSM system. Compared with classical PI control, predictive torque control (PTC) has attracted extensive attention in PMSM control due to its excellent dynamic and steady-state performance, as well as reduced torque ripple [19–27]. The stator flux and torque error-based cost function is usually utilized in the PTC method, which can accurately predict the behavior of stator flux and torque. Thus the weighting factor of the cost function is a critical factor in determining the control performance between stator flux and torque. However, obtaining a suitable weighting factor over the entire operating range of the motor is a very complex task [28]. In [19], a novel predictive stator flux and torque control technique without a weighting factor was developed, which avoids the need for tuning the weighting factor calculation for correct operation. In [20], an improved finite control set-model predictive control with an imposed optimized weighting factor was proposed, which achieves reduced torque ripple over a wide speed range. In [21,22], a predictive control strategy using a discrete-time state-space model was presented for PMSM, which eliminates the requirement for weighting factors in the PTC method. Although these methods have been proven to be very effective in eliminating the weighting factors of the cost function, the influence of parameter mismatch on the predicted torque and stator flux is not considered [19–22].

The parameters utilized in the predictive controller are not able to remain actual modular multiphase PMSM parameter values all the time due to identification errors or high-temperature operation [23]. Model parameter mismatch deteriorates the control performance of the PTC method and results in torque ripple increases if the parameter mismatch is not considered in the modified controller design. The design of a disturbance observer is the main approach to eliminate parameter mismatch. In [24–26], a robust fault-tolerant predictive control strategy based on a composite discrete observer was developed for PMSMs, which can reduce the impact of motor parameter perturbation. In [27], a flux immunity robust predictive control with an incremental model and extended state observer was presented for PMSM drives, which can improve robustness against inductance mismatch and overcome rotor flux effects. In [29], a robust sensor less PTC method with full-and reduced-order observers was proposed, which can improve the robustness of uncertain stator and rotor resistances. In [30], a novel encoder less finite control set PTC method with a robust adaptive observer was developed, which can achieve strong robustness for uncertain stator flux and rotor flux parameters. In [31], a disturbance observer-based PTC approach was investigated, which can improve the system's disturbance rejection ability against parameter uncertainties and time delays. Although the methods in [24–27,29–31] can achieve excellent control performance under parameter mismatch, the PTC method based on disturbance observer compensation relies heavily on the accurate measurement of voltage and current.

Instead of adopting observer compensation, a modified PTC method with a weighting factor was proposed in [32], which can reduce torque ripple in the presence of parametric uncertainty by improving prediction accuracy. In [33], an extension of the PTC method was proposed using two-degree-of-freedom control, which can make the PMSM drive system more robust in the presence of parameter mismatch. However, only the case of inductance parameter mismatch was considered in [32,33], and the flux linkage parameter mismatch can also seriously affect the performance of the PTC method. In this chapter, an R-PTC technique for the modular multiphase PMSM is proposed to enhance robustness against parameter mismatch and operate without the weighting factor of the cost function. A proportional controller is used to replace the disturbance observer compensation in this chapter to eliminate the influence of inductance parameter mismatch and flux linkage parameter mismatch. Compared to the conventional PTC method, the R-PTC technique with a proportional controller can effectively eliminate the steady-state errors of the stator flux and achieve lower torque ripple. Simulation and experimental results both validate the excellent performance of the R-PTC technique. This chapter is organized as follows: The mathematical model and drive system structure of the modular multiphase PMSM are presented in Section 10.2. The parameter sensitivity analysis of the conventional PTC method is discussed in Section 10.3. The R-PTC method for modular multiphase PMSM with parameter mismatch is proposed in Section 10.4. The simulation setup is described in Section 10.5, and Section 10.6 concludes this chapter.

10.2 Model and drive structure of modular multiphase PMSM

Using the concept of a unit motor, the modular multiphase PMSM is developed. The proposed modular multiphase PMSM consists of N motor units with repetitive characteristics. The stator core of the modular multiphase PMSM is composed of unit motor stator punches, and each unit motor is independently connected. Therefore, the modular multiphase PMSM can be modularized and assembled with a simple structure, and each unit motor has isolation characteristics in terms of electrical, magnetic, and thermal properties. Taking the modular multiphase PMSM as an example, the structural diagram of the modular multiphase PMSM is shown in Fig. 10.1. The stator winding arrangement is an important part of the modular multiphase PMSM. The stator winding arrangement of the unit motor can be similar to that of the traditional three-phase PMSM. The star of slots for the modular multiphase PMSM is shown in Fig. 10.1, with each phase sharing six identical slot vectors.

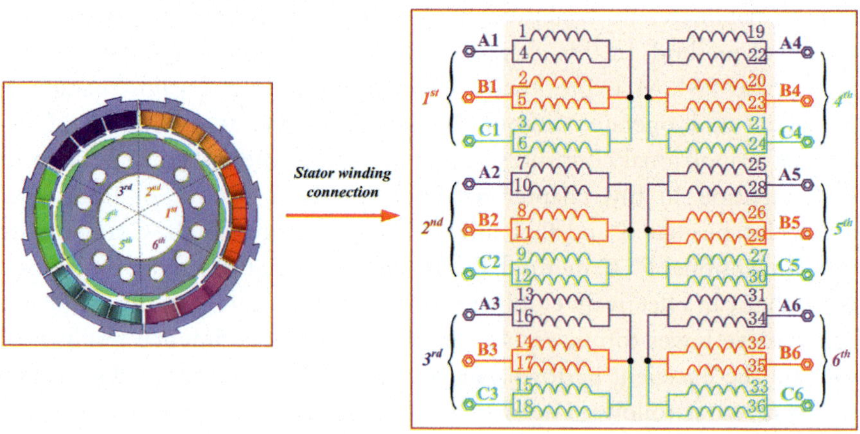

Fig. 10.1 Structure diagram and stator winding connection of a modular multiphase PMSM.

Mathematical model of modular multiphase PMSM

According to the characteristic analysis of the modular multiphase PMSM, the mathematical model of the modular multiphase PMSM can be equivalent to that of the traditional three-phase PMSM. According to [9,13], the voltage state-space equations of the modular multiphase PMSM can be expressed as the traditional voltage equations of the PMSM. Thus the *d*- and *q*-axis voltage transformation of one winding set can be described as follows.

$$\begin{cases} \begin{bmatrix} u_{dj} \\ u_{qj} \end{bmatrix} = \dfrac{2}{3} \begin{bmatrix} \cos\theta & \cos\left(\theta - \dfrac{2\pi}{3}\right) & \cos\left(\theta + \dfrac{2\pi}{3}\right) \\ \sin\theta & \sin\left(\theta - \dfrac{2\pi}{3}\right) & \sin\left(\theta + \dfrac{2\pi}{3}\right) \end{bmatrix} \begin{bmatrix} u_{aj} \\ u_{bj} \\ u_{cj} \end{bmatrix} \\ j=1,2,3,\cdots N \end{cases} \tag{10.1}$$

where j stands for any unit of the motor, u_{dj} and u_{qj} represent the *d*- and *q*-axis voltages of the j th winding, respectively; u_{aj} , u_{bj} , and u_{cj} represent the *a, b*-, and *c*-axis voltages of the j th winding, respectively; and θ is the rotor position angle.

The modular multiphase PMSM is composed of three-phase PMSM units with identical characteristics. Therefore, the stator resistance and stator inductance of each three-phase PMSM unit are equal. The *d*- and *q*-axis voltage equations of the modular multiphase PMSM unit are as [12–14]

$$\begin{cases} \dfrac{di_{dj}}{dt} = -\dfrac{R_o}{L_o} i_{dj} + \omega_e i_{qj} + \dfrac{1}{L_o} u_{dj} \\ \dfrac{di_{qj}}{dt} = -\dfrac{R_o}{L_o} i_{qj} - \omega_e i_{dj} - \dfrac{\psi_{ro}}{L_o} \omega_e + \dfrac{1}{L_o} u_{qj} \end{cases} \tag{10.2}$$

where i_{dj} and i_{qj} are the *d*- and *q*-axis currents, respectively; R_o and L_o are the stator resistance and stator inductance, respectively; ω_e is the electrical rotor speed, and ψ_{ro} is the flux linkage of permanent magnets.

According to the characteristics of modular multiphase PMSM, it is known that the output electromagnetic torque of a modular multiphase PMSM is the sum of the electromagnetic torque produced by each three-phase PMSM unit. The electromagnetic torque produced by the modular multiphase PMSM is as [12,13]

$$T_e = \frac{3n_p}{2} \sum_{j=1}^{N} (\psi_{ro} i_{qj}) = \frac{3Nn_p}{2} \psi_{ro} i_{qj} \tag{10.3}$$

The mechanical dynamic model of a modular multiphase PMSM can be described as:

$$T_e - T_L = \frac{J}{n_p}\frac{d\omega_e}{dt} \tag{10.4}$$

where n_p is the number of pole pairs; J is the moment of inertia; T_e and T_L are the electromagnetic torque and load torque of the modular multiphase PMSM, respectively.

Drive system structure of modular multiphase PMSM

Taking modular multiphase PMSM as an example, the modular multiphase PMSM is driven by six drive units to obtain the required torque. The six-segment stator winding of the modular multiphase PMSM is respectively connected with six VSIs in parallel, as shown in Fig. 10.2. Each group of stator windings is excited independently with three-phase symmetrical voltage generated by a VSI. The VSIs of *the* modular multiphase PMSM are electrically separated from each other. According to the characteristics of the modular multiphase PMSM, the mathematical model and characteristics of each motor unit are identical to those of a regular PMSM, and all the characteristics of each motor unit are repetitive. In addition, the voltage-seconds characteristic of any two VSIs should be the same [14–16]. The control of a VSI can be analogous to that of a conventional inverter [12,34]. In order to make the control algorithm simple and easy to implement, this paper uses a set of the PTC algorithm to send out six groups of the same modulation signals at the same time, as shown in Fig. 10.2. The six groups of the same modulation signals are input into six parallel inverters, respectively.

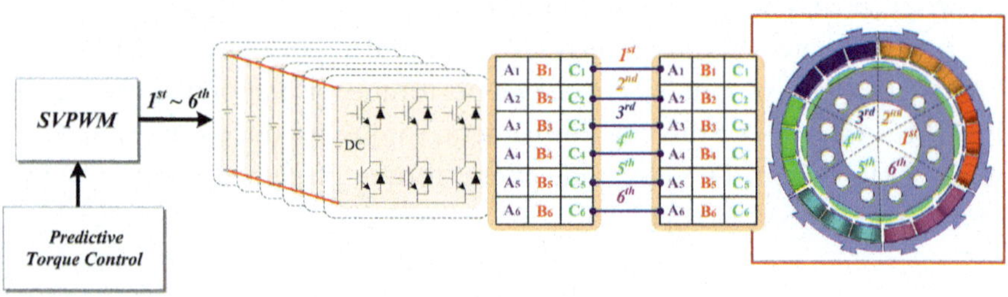

Fig. 10.2 Drive system structure diagram of a modular multiphase PMSM.

10.3 Parameter sensitivity analysis of conventional PTC

The purpose of the PTC strategy is to control the torque and stator flux of the motor. The structural diagram of the conventional PTC method is illustrated, as shown in Fig. 10.2. In order to minimize the absolute error between the predicted stator flux and torque values and their reference values, the cost function can be defined as [32].

$$g = k_{\psi} \left|\psi_s^{ref} - \psi_s(k+1)\right| + \left|T_e^{ref} - T_e(k+1)\right| \tag{10.5}$$

where k_{ψ} is a weighting factor; T_e^{ref} and ψ_s^{ref} are the torque reference value and flux linkage reference value, respectively; $T_e(k+1)$ and $\psi_s(k+1)$ are the predicted torque value and predicted flux linkage value, respectively (Fig. 10.3).

If the PTC sampling period T_s is short enough, the discrete form of modular multiphase PMSM can be modeled by the first-order Taylor expansion. According to Eq. (10.2) and Eq. (10.3), the torque and stator flux prediction equation of modular multiphase PMSM can be expressed as

$$\begin{cases} \psi_s(k+1) = \sqrt{\psi_{dj}^2(k+1) + \psi_{qj}^2(k+1)} \\ T_e(k+1) = \dfrac{3n_p}{2} \sum_{j=1}^{N} (\psi_{ro} i_{qj}(k+1)) \end{cases} \tag{10.6}$$

The discrete form of Eq. (10.2) can be obtained as

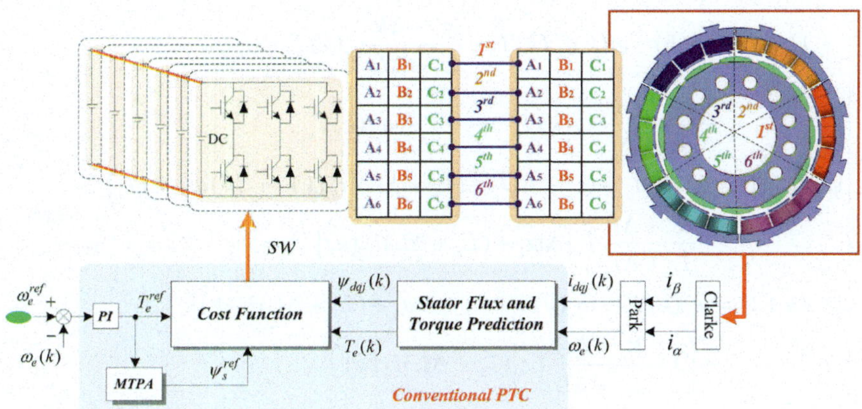

Fig. 10.3 Structural diagram of the conventional PTC method.

$$\begin{cases} i_{qj}(k+1)=i_{qj}(k)+\dfrac{T_s}{L_o}\left[-L_o\omega(k)i_{dj}(k)-R_o i_{qj}(k)-\omega(k)\psi_{ro}+u_{qj}(k)\right] \\ \psi_{dj}(k+1)=u_{dj}(k)T_s+\left(1-\dfrac{R_o}{L_o}T_s\right)\psi_{dj}(k)+\omega_e(k)T_s\psi_{qj}(k)+\dfrac{R_o}{L_o}\psi_{ro}T_s \\ \psi_{qj}(k+1)=u_{qj}(k)T_s-\omega_e(k)T_s\psi_{dj}(k)+\left(1-\dfrac{R_o}{L_o}T_s\right)\psi_{qj}(k) \end{cases} \tag{10.7}$$

where $\psi_{dj}(k)$ and $\psi_{qj}(k)$ are the *d*- and *q*-axis stator flux components, respectively.

The PTC strategy is a discrete control algorithm based on the precise mathematical model of modular multiphase PMSM. From Eq. (10.7), it can be found that the torque and stator flux prediction equation are absolutely dependent on the motor parameters (i.e., ψ_{ro}, R_o, and L_o), which means that these parameters mismatch would result in inaccurate prediction of modular multiphase PMSM drive system. Furthermore, the motor parameters mismatch leads to increasing torque and flux ripples and deteriorates the performance of the PTC algorithm.

In order to evaluate the sensitivity of the PTC algorithm response to a parameter, let us analyze the torque prediction model according to Eq. (10.7). Introducing $R=R_o+\Delta R$, $L=L_o+\Delta L$, and $\psi_r=\psi_{ro}+\Delta\psi_r$, where R_o, L_o, and ψ_{ro} are actual values and ΔR, ΔL, and $\Delta\psi_r$ are uncertain components of the corresponding parameter. The Eq. (10.7) is rewritten as follows when parameters mismatch are considered.

$$\begin{cases} i_{qj}(k+1)=i_{qj}(k)-\omega(k)i_{dj}(k)T_s-\dfrac{R_0+\Delta R}{L_0+\Delta L}i_{qj}(k)T_s \\ \qquad -\dfrac{\varphi_{ro}+\Delta\varphi_r}{L_0+\Delta L}\omega(k)T_s+\dfrac{1}{L_o+\Delta L}u_{qj}(k)T_s \\ \psi_{dj}(k+1)=u_{dj}(k)T_s+\psi_{ro}+\Delta\psi_r+(L_o+\Delta L)\left[i_{dj}(k)+i_{qj}(k)\omega_\varepsilon(k)T_s\right] \\ \qquad -\dfrac{R_o+\Delta R}{L_o+\Delta L}T_s\left[\Delta\psi_r+(L_o+\Delta L)i_{dj}(k)\right] \\ \psi_{qj}(k+1)=u_{qj}(k)T_s-\omega_e(k)T_s\left[\psi_{ro}+\Delta\psi_r+(L_o+\Delta L)i_{dj}(k)\right] \\ \qquad +\left(1-\dfrac{R_o+\Delta R}{L_o+\Delta L}T_s\right)(L_o+\Delta L)i_{qj}(k) \end{cases} \tag{10.8}$$

It should be noted that the value of stator resistance voltage is much smaller than the amplitude of the voltage source when the motor operates stably (e.g., $\left|\frac{R_o+\Delta R}{L_o+\Delta L} i_{dj,qj}(k)\right| << \left|\frac{1}{L_o+\Delta L} u_{dj,qj}(k)\right|$). Thus the term $\left|\frac{R_o+\Delta R}{L_o+\Delta L} i_{dj,qj}(k)\right|$ does not have any negative effect on torque and stator flux prediction. It can be inferred that the stator resistance parameter mismatch does not have any considerable effect on torque prediction as well as stator flux linkage prediction. However, the inductance parameter mismatch and flux linkage parameter mismatch will lead to nonoptimal switching state and increase the prediction bias. Torque and stator flux ripple of the modular multiphase PMSM drive system will also increase if the model parameters mismatch are not considered in the design of the robust predictive controller.

10.4 R-PTC method of modular multiphase PMSM with parameters mismatch

According to Eq. (10.2), the d- and q-axis voltage equations of a modular multiphase PMSM under inductance parameter mismatch and flux linkage parameter mismatch can be expressed as:

$$\begin{cases} \frac{di_{dj}}{dt} = -\frac{R_o}{L_o} i_{dj} + \omega_e i_{qj} + \frac{1}{L_o} u_{dj} - \frac{1}{L_o} f_{dj} \\ \frac{di_{qj}}{dt} = -\omega_e i_{dj} - \frac{R_o}{L_o} i_{qj} - \frac{\psi_{ro}}{L_o} \omega_e + \frac{1}{L_o} u_{qj} - \frac{1}{L_o} f_{qj} \end{cases} \tag{10.9}$$

with

$$\begin{cases} f_{dj} = \frac{\Delta L}{(L_o + \Delta L)} (u_{dj} - R_o i_{dj} + L_o \omega_e i_{qj} + \Delta L \omega_e i_{qj}) - \Delta L \omega_e i_{qj} \\ f_{qj} = \frac{\Delta L}{(L_o + \Delta L)} (u_{qj} - R_o i_{qj} - L_o \omega_e i_{dj} - \Delta L \omega_e i_{dj} - \omega_e \psi_{ro} - \omega_e \Delta \psi_r) + \Delta L \omega_e i_{dj} + \omega_e \Delta \psi_r \end{cases} \tag{10.10}$$

where f_{dj} and f_{qj} are the d- and q-axis uncertain disturbances caused by inductance parameter mismatch and flux linkage parameter mismatch, respectively.

From Eq. (10.9), the d- and q-axis flux linkage equations of the modular multiphase PMSM can be obtained as

$$\begin{cases} \begin{bmatrix} \frac{d\psi_{dj}}{dt} \\ \frac{d\psi_{qj}}{dt} \end{bmatrix} = \begin{bmatrix} u_{dj} \\ u_{qj} \end{bmatrix} + \begin{bmatrix} -\frac{R_o}{L_o} & \omega_e \\ -\omega_e & -\frac{R_o}{L_o} \end{bmatrix} \begin{bmatrix} \psi_{dj} \\ \psi_{qj} \end{bmatrix} + \begin{bmatrix} \frac{R_o}{L_o}\psi_{ro} \\ 0 \end{bmatrix} - \begin{bmatrix} f_{dj} \\ f_{qj} \end{bmatrix} \\ \begin{bmatrix} \psi_{dj} \\ \psi_{qj} \end{bmatrix} = \begin{bmatrix} L_o & 0 \\ 0 & L_o \end{bmatrix} \begin{bmatrix} i_{dj} \\ i_{qj} \end{bmatrix} + \begin{bmatrix} \psi_{ro} \\ 0 \end{bmatrix} \end{cases} \tag{10.11}$$

The discrete expression of Eq. (10.11) is

$$\begin{aligned} \begin{bmatrix} \psi_{dj}(k+1) \\ \psi_{qj}(k+1) \end{bmatrix} = \begin{bmatrix} u_{dj}(k)T_s \\ u_{qj}(k)T_s \end{bmatrix} + \begin{bmatrix} 1-\frac{R_o}{L_o}T_s & \omega_e(k)T_s \\ -\omega_e(k)T_s & 1-\frac{R_o}{L_o}T_s \end{bmatrix} \begin{bmatrix} \psi_{dj}(k) \\ \psi_{qj}(k) \end{bmatrix} \\ + \begin{bmatrix} \frac{R_o}{L_o}\psi_{ro}T_s \\ 0 \end{bmatrix} - \begin{bmatrix} T_s & 0 \\ 0 & T_s \end{bmatrix} \begin{bmatrix} f_{dj}(k) \\ f_{qj}(k) \end{bmatrix} \end{aligned} \tag{10.12}$$

In a feedback control system, the most effective means of eliminating static error is to add error feedback adjustment. Therefore we consider adding predictive error feedback in the flux linkage equations, Eq. Equation (10.12) is modified to:

$$\begin{aligned} \begin{bmatrix} \psi_{dj}(k+1) \\ \psi_{qj}(k+1) \end{bmatrix} = \begin{bmatrix} u_{dj}(k)T_s \\ u_{qj}(k)T_s \end{bmatrix} + \begin{bmatrix} 1-\frac{R_o}{L_o}T_s & \omega_e(k)T_s \\ -\omega_e(k)T_s & 1-\frac{R_o}{L_o}T_s \end{bmatrix} \begin{bmatrix} \psi_{dj}(k) \\ \psi_{qj}(k) \end{bmatrix} \\ + \begin{bmatrix} \frac{R_o}{L_o}\psi_{ro}T_s \\ 0 \end{bmatrix} - \begin{bmatrix} Q_1 & 0 \\ 0 & Q_2 \end{bmatrix} \begin{bmatrix} \Delta\psi_{dj}(k) \\ \Delta\psi_{qj}(k) \end{bmatrix} \end{aligned} \tag{10.13}$$

In order to eliminate the influence of parameter mismatches, the torque prediction equation is modified to

$$T_e(k) = \frac{3Nn_p}{2}\psi_{ro}\frac{\psi_{qj}(k)}{L_o} + Q_3\Delta T_e(k) \tag{10.14}$$

where $\Delta T_e(k)$ is the torque prediction error between the predicted value and reference value; $\Delta\psi_{dj}(k)$ and $\Delta\psi_{qj}(k)$ are the d- and q-axis flux linkage error between the predicted value and reference value, respectively; Q_1, Q_2 and Q_3 are the proportional controller parameter.

In steady state, the torque prediction error caused by parameter mismatch can be considered as a constant during one sampling period (i.e., $\Delta T_e(k+1)=\Delta T_e(k)$). According to Eq. (10.14), we can get

$$T_e(k+1)-T_e(k)=\frac{3Nn_p}{2}\psi_{ro}\left[\frac{\psi_{qj}(k+1)}{L_o}-\frac{\psi_{qj}(k)}{L_o}\right] \tag{10.15}$$

Substituting Eq. (10.13) into Eq. (10.15), the relationship between torque and q-axis voltage can be obtained as follows:

$$u_{qj}(k)T_s = G(k) \tag{10.16}$$

where

$$G(k)=\frac{2L_o}{3Nn_p\psi_{ro}}[T_e(k+1)-T_e(k)]+\omega_e(k)T_s\psi_{dj}(k)+\frac{R_o}{L_o}T_s\psi_{qj}(k)+Q_2\Delta\psi_{qj}(k)$$

The influence of resistance voltage can be neglected, therefore, $\frac{R_o}{L_o}T_s\approx 0$. From Eq. (10.13), the relationship between stator flux amplitude and voltage can be expressed as

$$\begin{aligned}[\psi_s(k+1)]^2 &= [\psi_{dj}(k+1)]^2+[\psi_{qj}(k+1)]^2\\ &=[u_{dj}(k)T_s+\psi_{dj}(k)+\omega_e(k)T_s\psi_{qj}(k)-Q_1\Delta\psi_{dj}(k)]^2\\ &\quad+[u_{qj}(k)T_s+\psi_{qj}(k)-\omega_e(k)T_s\psi_{dj}(k)-Q_2\Delta\psi_{qj}(k)]^2\end{aligned} \tag{10.17}$$

From Eq. (10.17), we can derive

$$u_{dj}(k)T_s=\sqrt{[\psi_s(k+1)]^2-F_1(k)}-F_2(k) \tag{10.18}$$

where $F_1(k)=[G(k)+\psi_{qj}(k)-\omega_e(k)T_s\psi_{dj}(k)-Q_2\Delta\psi_{qj}(k)]^2$ and $F_2(k)=[\psi_{dj}(k)+\omega_e(k)T_s\psi_{qj}(k)-Q_1\Delta\psi_{dj}(k)]$.

These values for ψ_s^{ref} and T_e^{ref} are used as the reference in Eqs. (10.16) and (10.18) by considering $\psi_s(k+1)=\psi_s^{ref}$ and $T_e(k+1)=T_e^{ref}$. Thus the d- and q-axis voltage equations are modified to

$$\begin{cases} u_{dj}(k)=\dfrac{\sqrt{[\psi_s^{ref}]^2-F_1(k)}-F_2(k)}{T_s}\\ u_{qj}(k)=\dfrac{G(k)}{T_s}\end{cases} \tag{10.19}$$

with

$$\begin{cases} G(k) = \dfrac{2L_o}{3Nn_p\psi_{ro}}[T_e^{ref} - T_e(k)] + \omega_e(k)T_s\psi_{dj}(k) + \dfrac{R_o}{L_o}T_s\psi_{qj}(k) + Q_2\Delta\psi_{qj}(k) \\ F_1(k) = [G(k) + \psi_{qj}(k) - \omega_e(k)T_s\psi_{dj}(k) - Q_2\Delta\psi_{qj}(k)]^2 \\ F_2(k) = [\psi_{dj}(k) + \omega_e(k)T_s\psi_{qj}(k) - Q_1\Delta\psi_{dj}(k)] \end{cases} \tag{10.20}$$

According to the maximum torque per ampere (MTPA), the relationship between the stator flux reference and the torque reference can be obtained as [21].

$$\psi_s^{ref} = \sqrt{\psi_{dj}^2 + \psi_{qj}^2} = \sqrt{\psi_{ro}^2 + (L_o \cdot \dfrac{T_e^{ref}}{\dfrac{3Nn_p}{2}\psi_{ro}})^2} \tag{10.21}$$

The structural diagram of the modular multiphase PMSM drive system with R-PTC strategy is shown in Fig. 10.4. The proposed R-PTC strategy is used to substitute conventional PTC to enhance robustness against

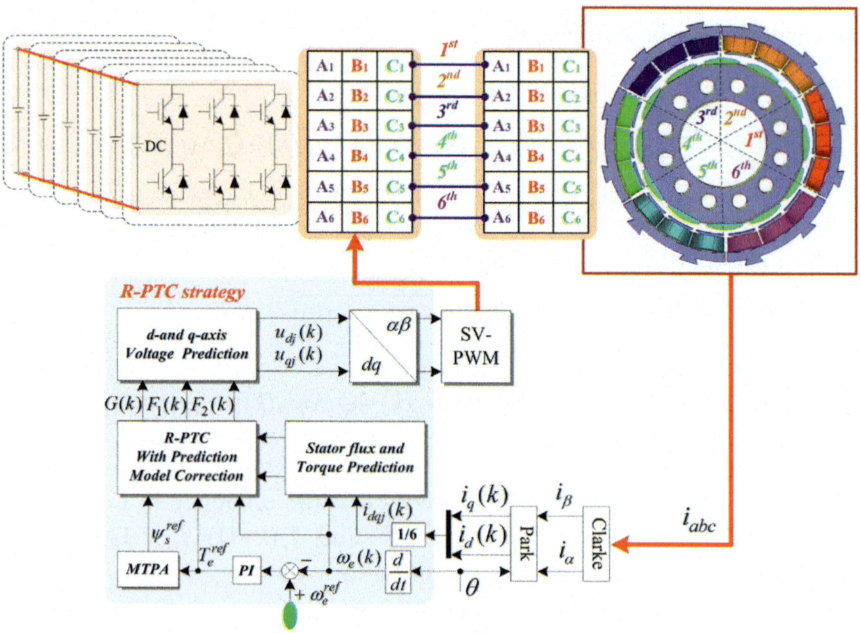

Fig. 10.4 Structural diagram of the modular multiphase PMSM drive system with R-PTC strategy.

parameters mismatch and eliminate the weighting factor of cost function. The predicted value can accurately track the reference value when there is no parameter mismatch in predictive controller (i.e., $Q_1=Q_2=Q_3=0$). When parameter mismatch occurs in predictive controller, the errors between predicted value and reference value can be eliminated by adjusting proportional parameters (i.e., Q_1 , Q_2 , Q_3).

10.5 Simulations results

In order to demonstrate the effectiveness of the proposed R-PTC strategy, the proposed R-PTC strategy in modular multiphase PMSM drive system are carried out in MATLAB/ Simulink. Some simulation results are performed in comparisons with conventional PTC. The main parameters of 6*3-phase SPMSM used in the simulation are given in Table 10.1. The weight factor of conventional PTC method is 5000. The proportion coefficient and integral coefficient in PI controller of speed control loop are 8000 and 0.5, respectively. In addition, the sampling frequency is 10 kHz; the proportional parameters of proposed R-PTC strategy are $Q_1=1$ and $Q_2=Q_3=50$.

Control performance comparison between conventional PTC and proposed R-PTC without parameter mismatches

The operating conditions of the modular multiphase PMSM in case analysis A are that the load torque remains at 2000 N·m, and the speed suddenly increases from 100 rad/s to 200 rad/s at 1 s. Comparative simulation results of the conventional PTC strategy and the proposed R-PTC strategy without parameter mismatch are depicted in Figs. 10.5–10.8. Fig. 10.5 presents the comparative simulation results of phase current and torque using the conventional PTC strategy and the proposed R-PTC strategy. The peak-to-peak torque ripple of the proposed R-PTC and the conventional PTC are basically the same, both of which are about ± 200 N·m. From Fig. 10.5, it can be seen that the proposed R-PTC strategy for the

Table 10.1 Main parameters of the modular multiphase PMSM.

Parameter	Value	Parameter	Value
Inductance (L_o)	3.572/ mH	Stator phase resistance (R_o)	0.02/Ω
Flux linkage of PM (Ψ_{ro})	0.892/Wb	Rotational inertia (J)	100/kg m^2

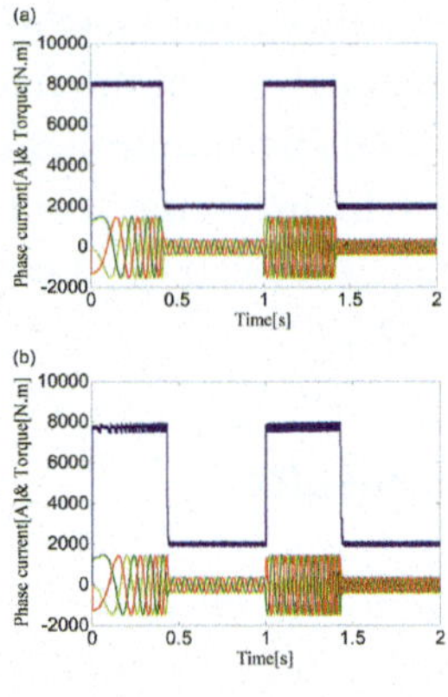

Fig. 10.5 Comparative simulation results of the phase current and torque without parameter mismatch. (A) Proposed R-PTC. (B) Conventional PTC.

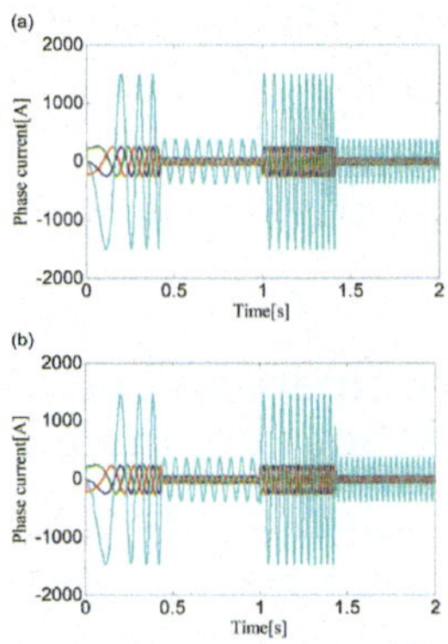

Fig. 10.6 Comparative simulation results of the three-phase stator current without parameter mismatch: (A) Proposed R-PTC, (B) Conventional PTC.

modular multiphase PMSM can achieve better control performance than the conventional PTC strategy without parameter mismatch. Fig. 10.6 presents the comparative simulation results of the three-phase stator current using the conventional PTC strategy and the proposed R-PTC strategy. From Fig. 10.6, it can be observed that the three-phase stator current of the modular multiphase PMSM is six times that of a single motor when using the proposed R-PTC strategy. The reason for this phenomenon is that the modular multiphase PMSM is composed of six motor units with consistent characteristics. Note that when the speed suddenly increases at 1 s, the stator current of the modular multiphase PMSM will also suddenly increase until it stabilizes.

Fig. 10.7 shows the comparative simulation results of the α-β stator flux linkage using the conventional PTC strategy and the proposed R-PTC strategy. The stator flux linkage ripples of the modular multiphase PMSM are respectively ± 0.02 Wb and ± 0.04 Wb for the proposed R-PTC and the conventional PTC, respectively, as presented in Figs. 10.7A and 10.6B. The error between the predicted α-β flux linkage value and its reference value is very small in the absence of parameter mismatch, as clearly shown in Fig. 10.7A,B. Fig. 10.8 presents the comparative simulation results of three-dimensional rotor flux trajectories without parameter mismatch.

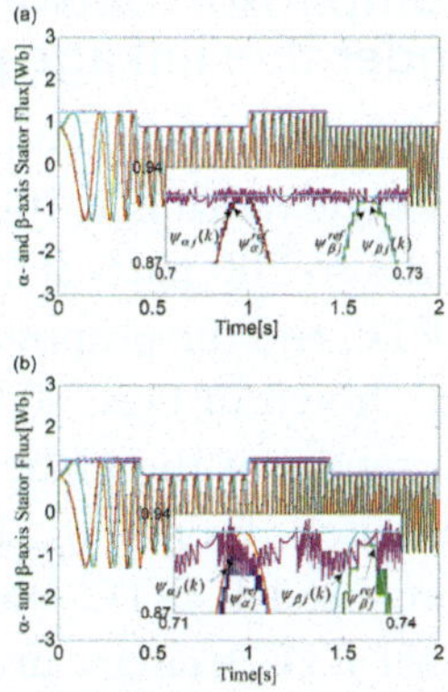

Fig. 10.7 Comparative simulation results of the α-β stator flux linkage without parameter mismatch: (A) Proposed R-PTC, (B) Conventional PTC.

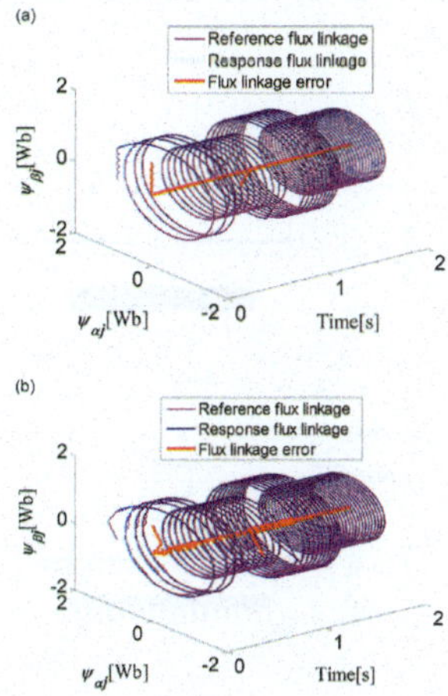

Fig. 10.8 Comparative simulation results of three-dimensional rotor flux trajectories without parameter mismatch: (A) Proposed R-PTC, (B) Conventional PTC.

From Fig. 10.8, it can be observed that the α-β stator flux linkage response value of the modular multiphase PMSM can accurately track the reference value using both the conventional PTC strategy and the proposed R-PTC strategy. It is worth noting that there is a significant overshoot phenomenon in the α-β stator flux linkage due to the sudden increase in speed at 1 s. From Figs. 10.7 and 10.8, it can be concluded that the α-β stator flux linkage of the modular multiphase PMSM can achieve good control performance without parameter mismatch using both the conventional PTC strategy and the proposed R-PTC strategy.

Control performance comparison between conventional PTC and proposed R-PTC under flux linkage parameter mismatch

The operating conditions of the modular multiphase PMSM in case analysis B are such that the load torque remains at 1000 N·m, and the speed suddenly increases from 100 rad/s to 200 rad/s at 1 s. Comparative simulation results of the conventional PTC and the proposed R-PTC under flux linkage parameter mismatch are shown in Figs. 10.9–10.12. Fig. 10.9 presents the comparative simulation results of the phase current and torque under flux linkage parameter mismatch. From Fig. 10.9A, it can be seen that the torque ripple of the proposed R-PTC strategy is significantly lower than that of the conventional PTC strategy under flux linkage parameter mismatch. The reason is that the proposed R-PTC strategy can effectively eliminate the influence of flux linkage parameter mismatch by adjusting proportional parameters. Therefore, the proposed R-PTC strategy can effectively suppress the torque ripple of the modular multiphase PMSM

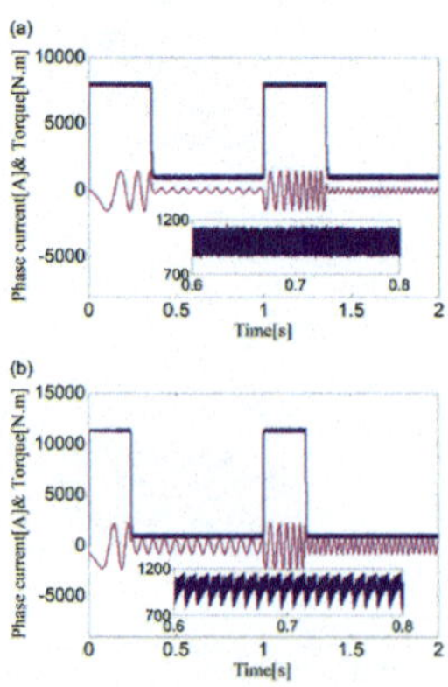

Fig. 10.9 Comparative simulation results of the phase current and torque under flux linkage parameter mismatch. (A) Proposed R-PTC. (B) Conventional PTC.

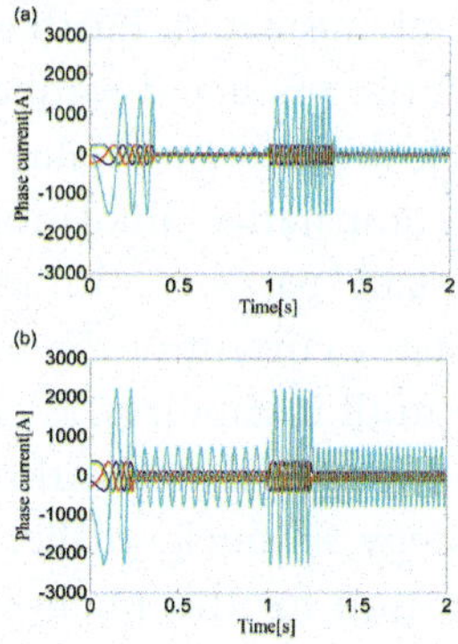

Fig. 10.10 Comparative simulation results of the three-phase stator current under flux linkage parameter mismatch: (A) Proposed R-PTC, (B) Conventional PTC.

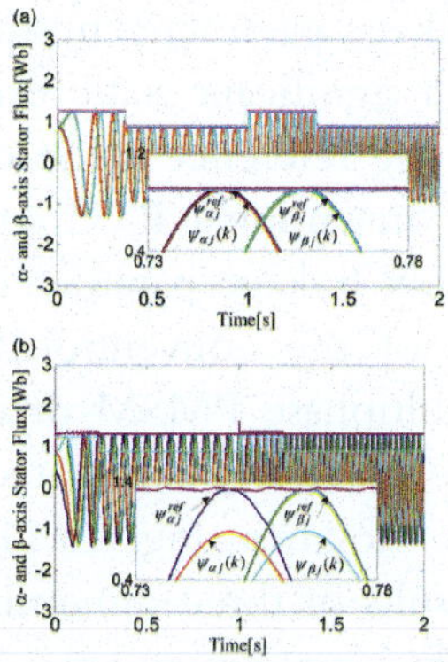

Fig. 10.11 Comparative simulation results of the α-β stator flux linkage under flux linkage parameter mismatch: (A) Proposed R-PTC, (B) Conventional PTC.

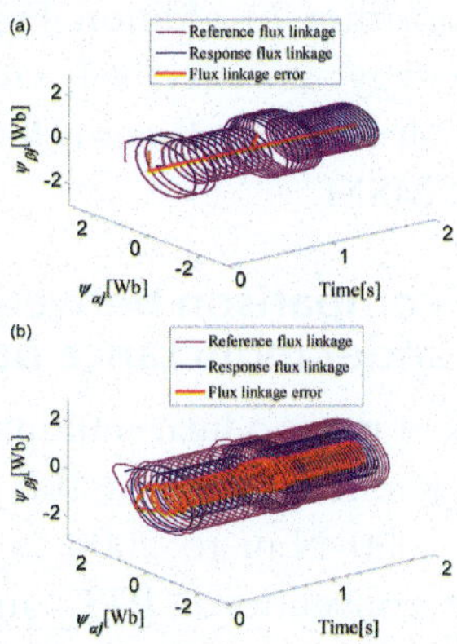

Fig. 10.12 Comparative simulation results of three-dimensional rotor flux trajectories under flux linkage parameter mismatch. (A) Proposed R-PTC. (B) Conventional PTC.

under flux linkage parameter mismatch. Furthermore, compared with the conventional PTC strategy, the predicted torque of the proposed R-PTC strategy can accurately track the reference value.

Fig. 10.10 presents the comparative simulation results of the three-phase stator current under flux linkage parameter mismatch. When the speed suddenly increases at 1 s, the stator current of the modular multiphase PMSM will also suddenly increase until it stabilizes, as clearly shown in Fig. 10.10. From Fig. 10.10A, it can be seen that the three-phase stator current of the proposed R-PTC strategy is much smaller than that of the conventional PTC strategy. The reason for this phenomenon is that the conventional PTC strategy requires a larger current to maintain electromagnetic torque due to the reduction of the flux linkage parameter.

Fig. 10.11 presents the comparative simulation results of the α-β stator flux linkage under flux linkage parameter mismatch. From Fig. 10.11B, it can be seen that there is a significant static error between the predicted α-β stator flux value and the reference value of the modular multiphase PMSM when using the conventional PTC strategy. The reason for this phenomenon is that the flux linkage parameter mismatch seriously deteriorates the performance of the conventional PTC strategy. However, the impact of modular multiphase PMSM drive systems on flux linkage parameter mismatch can be effectively eliminated by adopting the proposed R-PTC strategy, as clearly shown in Fig. 10.11A. Fig. 10.12 presents the comparative simulation results of three-dimensional rotor flux trajectories using the conventional PTC strategy and the proposed R-PTC strategy. From Fig. 10.12, it can be observed that the flux linkage parameter mismatch has a significant influence on stator flux prediction and control in the conventional PTC algorithm. In addition, Fig. 10.12 demonstrates that the proposed R-PTC strategy can effectively suppress the impact of flux linkage parameter mismatch on the α-β stator flux control performance of the modular multiphase PMSM.

Control performance comparison between conventional PTC and proposed R-PTC under inductance parameter mismatch

The operating conditions of the modular multiphase PMSM in case analysis C are that the speed reference remains at 100 rad/s, and the load torque suddenly increases from 1000 N·m to 2000 N·m at 1.2 s. Comparative simulation results of the conventional PTC and the proposed R-PTC under inductance parameter mismatch are shown in Figs. 10.13–10.16. Fig. 10.13 presents the comparative simulation results of the phase current

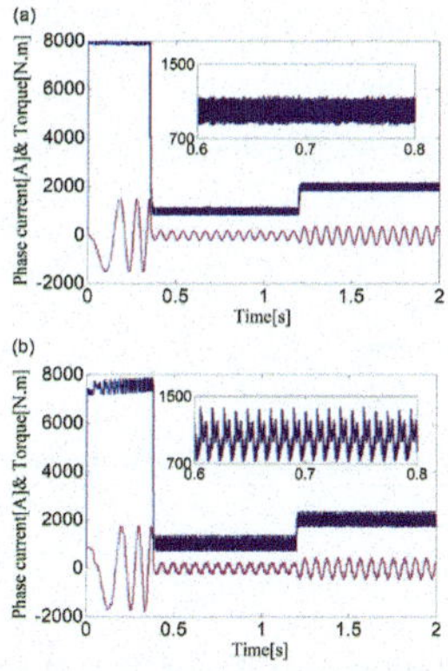

Fig. 10.13 Comparative simulation results of the phase current and torque under inductance parameter mismatch: (A) Proposed R-PTC and (B) Conventional PTC.

and torque using the conventional PTC strategy and the proposed R-PTC strategy. From Fig. 10.13B, it can be seen that the inductance parameter mismatch significantly increases the torque ripple of the modular multiphase PMSM when using the conventional PTC strategy. Compared with the conventional PTC strategy, the proposed R-PTC strategy can effectively suppress the impact of inductance parameter mismatch on the modular multiphase PMSM, as clearly shown in Fig. 10.13A. From Fig. 10.13, it can be observed that the peak-to-peak torque ripple of the modular multiphase PMSM decreases from $\pm$ 700 to $\pm$ 300 N·m when using the proposed R-PTC strategy. Therefore, the proposed R-PTC strategy can effectively suppress the torque ripple of the modular multiphase PMSM under inductance parameter mismatch. Fig. 10.14 presents the comparative simulation results of the three-phase stator current using the conventional PTC strategy and the proposed R-PTC strategy. From Fig. 10.14, it can be seen that the stator current of the modular multiphase PMSM significantly increases at 1.2 s due to the sudden increase in load torque from 1000 N·m to 2000 N·m at 1.2 s. From Fig. 10.14B, it can be observed that the stator current of the modular multiphase PMSM using the conventional PTC strategy significantly distorts in the case of inductance parameter mismatch. The proposed R-PTC strategy can significantly suppress the influence of inductance parameter mismatch on the stator current of the modular multiphase PMSM, as clearly shown in Fig. 10.14A.

Fig. 10.15 presents the comparative simulation results of the α-β stator flux linkage using the conventional PTC strategy and the proposed R-PTC strategy. From Fig. 10.15B, it can be seen that the α-β stator flux linkage of the modular multiphase PMSM exhibits significant pulsation

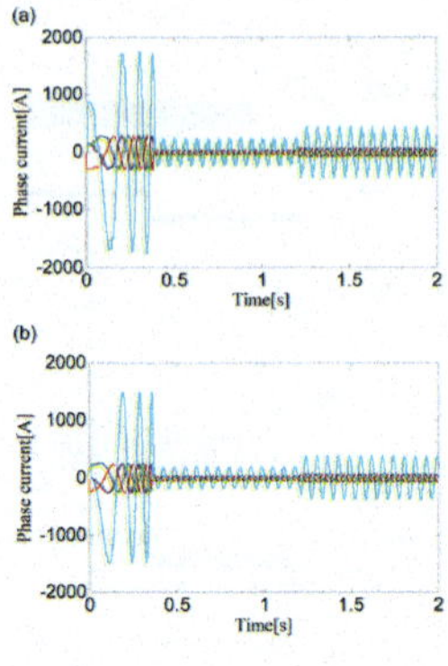

Fig. 10.14 Comparative simulation results of the three-phase stator current under inductance parameter mismatch: (A) Proposed R-PTC and (B) Conventional PTC.

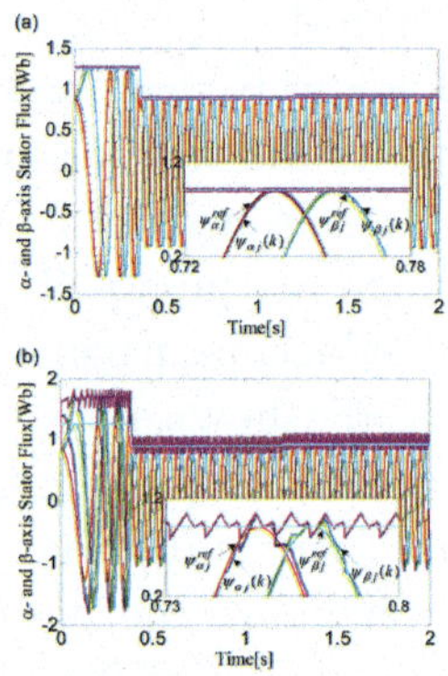

Fig. 10.15 Comparative simulation results of the α-β stator flux linkage under inductance parameter mismatch: (A) Proposed R-PTC, (B) Conventional PTC.

when using the conventional PTC strategy. The reason for this phenomenon is that the flux linkage parameter mismatch seriously deteriorates the performance of the conventional PTC strategy. In addition, the inductance parameter mismatch significantly impacts the α-β stator flux linkage control performance of the conventional PTC strategy. However, the proposed R-PTC strategy effectively suppresses the impact of inductance parameter mismatch on the α-β stator flux linkage control performance, as clearly shown in Fig. 10.15A. Fig. 10.16 presents the comparative simulation results of three-dimensional rotor flux trajectories using the conventional PTC strategy and the proposed R-PTC strategy. From Fig. 10.16B, it is evident that the inductance parameter mismatch has a significant impact on

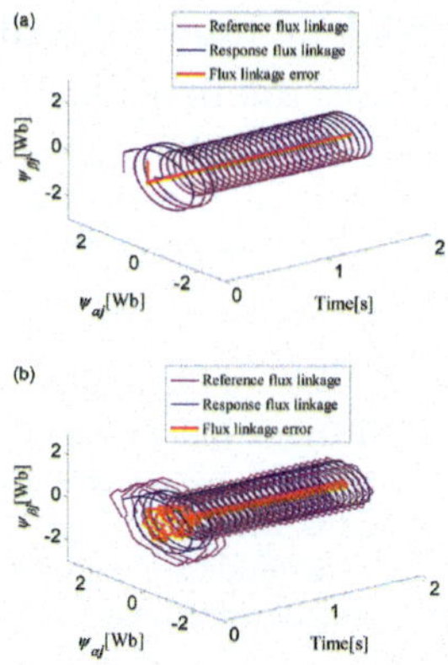

Fig. 10.16 Comparative simulation results of three-dimensional rotor flux trajectories under inductance parameter mismatch: (A) Proposed R-PTC, (B) Conventional PTC.

stator flux prediction and control in the conventional PTC algorithm, causing the α-β stator flux linkage of the modular multiphase PMSM to exhibit obvious pulsation. Conversely, Fig. 10.16A shows that the α-β stator flux linkage response value of the modular multiphase PMSM can accurately track the reference value when using the proposed R-PTC strategy. This is because the proposed R-PTC strategy effectively eliminates the influence of inductance parameter mismatch by adjusting proportional parameters.

10.6 Conclusion

In this chapter, a novel R-PTC strategy for modular multiphase PMSM is proposed for high-power traction applications. The proposed R-PTC method for the modular multiphase PMSM drive system has been proven to deliver excellent control performance, effectively managing both the electromagnetic torque and the stator flux. The stator flux and torque error-based cost function is replaced by the motor model-based voltage vector equation in the R-PTC strategy to avoid the adjustment of the weighting factor. Furthermore, the proposed R-PTC strategy utilizes a proportional controller to adjust the control voltage vector, thereby eliminating the influence of model parameter mismatch. Compared to the conventional PTC strategy, the proposed R-PTC strategy can significantly improve the predictive accuracy of the stator flux and torque and reduce the tracking errors under parameter mismatches. The simulation and experimental results obtained demonstrate the effectiveness of the proposed R-PTC strategy, which achieves the benefits of lower stator current THD, reduced torque ripple, and excellent dynamic performance under parameter mismatches.

References

[1] S. Morimoto, Y. Tong, et al., Loss minimization control of permanent magnet synchronous motor drives, IEEE Trans. Ind. Electron. 41 (05) (1994) 511–517.

[2] T. Jahns, G. Kliman, et al., Interior permanent-magnet synchronous motors for adjustable-speed drives, IEEE Trans. Ind. Appl. 22 (04) (1986) 738–747.

[3] Z. Chen, M. Tomita, et al., An extended electromotive force model for sensorless control of interior permanent-magnet synchronous motors, IEEE Trans. Ind. Electron. 50 (02) (2003) 288–295.

[4] S. Morimoto, Y. Takeda, et al., Current phase control methods for permanent magnet synchronous motors, IEEE Trans. Power Electron. 5 (02) (1990) 133–139.

[5] E. Levi, Multiphase electric machines for variable-speed applications, IEEE Trans. Ind. Electron. 55 (05) (2008) 1893–1909.

[6] Y. Hu, Z. Zhu, et al., Torque capability enhancement of dual three-phase PMSM drive with fifth and seventh current harmonics injection, IEEE Trans. Ind. Electron. 53 (05) (2017) 4526–4535.

[7] E. Levi, M. Jones, et al., A novel concept of a multiphase, multimotor vector controlled drive system supplied from a single voltage source inverter, IEEE Trans. Power Electron. 19 (02) (2004) 320–225.

[8] L. Shao, W. Hua, et al., Mathematical modeling of a 12-phase flux-switching permanent-magnet machine for wind power generation, IEEE Trans. Ind. Electron. 63 (01) (2016) 504–516.

[9] M. Onsal, Y. Demir, et al., A new nine-phase permanent magnet synchronous motor with consequent pole rotor for high-power traction applications, IEEE Trans. Magn. 53 (11) (2009) 1–6.

[10] H. Che, E. Levi, et al., Current control methods for an asymmetrical six-phase induction motor drive, IEEE Trans. Power Electron. 29 (01) (2014) 407–417.

[11] X. Wang, Z. Wang, et al., A hybrid direct torque control scheme for dual three-phase PMSM drives with improved operation performance, IEEE Trans. Power Electron. 34 (02) (2018) 1622–1634.

[12] J. Karttunen, S. Kallio, et al., Decoupled vector control scheme for dual three-phase permanent magnet synchronous machines, IEEE Trans. Ind. Electron. 61 (05) (2014) 2185–2196.

[13] Y. Demir, M. Aydin, A novel asymmetric and unconventional stator winding configuration and placement for a dual three-phase surface PM motor, IEEE Trans. Magn. 53 (11) (2017) 1–5.

[14] X. Han, D. Jiang, et al., Two-Segment Three-Phase PMSM Drive With Carrier Phase-Shift PWM for Torque Ripple and Vibration Reduction, IEEE Trans. Power Electron. 34 (01) (2019) 588–599.

[15] X. Han, D. Jiang, et al. Two-segment three-phase PMSM drive with carrier phase-shift PWM. IEEE APEC (2018) 848–854.

[16] S. Bhattacharya, D. Mascarella, et al., A dual three-level T-NPC inverter for high-power traction applications, IEEE J. Emerg. Sel. Top. Power Electron. 4 (02) (2016) 668–678.

[17] S. Bhattacharya, D. Mascarella, et al. Interleaved PWM control for neutral point balancing in dual 3-level traction drives. IEEE ECCE (2014) 1715–1721.

[18] S. Li, Z. Liu, Adaptive speed control for permanent magnet synchronous motor system with variations of load inertia, IEEE Trans. Ind. Electron. 56 (08) (2009) 3050–3059.
[19] C. Rojas, J. Rodriguez, et al., Predictive torque and flux control without weighting factors, IEEE Trans. Ind. Electron. 60 (02) (2013) 681–690.
[20] S. Davari, D. Khaburi, et al., An improved FCS–MPC algorithm for an induction motor with an imposed optimized weighting factor, IEEE Trans. Power Electron. 48 (03) (2012) 1540–1551.
[21] X. Zhang, B. Hou, et al., Double vectors model predictive torque control without weighting factor based on voltage tracking error, IEEE Trans. Power Electron. 33 (03) (2018) 2368–2380.
[22] J. Rodriguez, R.M. Kennel, et al., High-performance control strategies for electrical drives: an experimental assessment, IEEE Trans. Ind. Electron. 59 (02) (2012) 812–820.
[23] X. Zhang, Y. Zhang, et al., Model predictive current control for PMSM drives with parameter robustness improvement, IEEE Trans. Power Electron. 34 (02) (2018) 1645–1657.
[24] S. Huang, G. Wu, et al., Novel Predictive Stator Flux Control Techniques for PMSM drives, IEEE Trans. Power Electron. 34 (09) (2018) 8916–8929.
[25] C. Zhang, G. Wu, et al., Robust fault-tolerant predictive current control for permanent magnet synchronous motors considering demagnetization fault, IEEE Trans. Ind. Electron. 65 (07) (2017) 5324–5534.
[26] X. Zhang, B. Hou, et al., Deadbeat predictive current control of permanent-magnet synchronous motors with stator current and disturbance observer, IEEE Trans. Power Electron. 32 (05) (2017) 3818–3834.
[27] M. Yang, X. Lang, et al., A flux immunity robust predictive current control with incremental model and extended state observer for PMSM drive, IEEE Trans. Power Electron. 32 (12) (2017) 9267–9279.
[28] X. Zhang, Y. He, Direct Voltage-Selection Based Model Predictive Direct Speed Control for PMSM Drives without Weighting Factor, IEEE Trans. Power Electron. 34 (08) (2018) 7838–7851.
[29] S.A. Davari, D. Khaburi, et al., Using full order and reduced order observers for robust sensorless predictive torque control of induction motors, IEEE Trans. Power Electron. 27 (07) (2012) 3424–3433.
[30] F. Wang, S.A. Davari, et al., Finite control set model predictive torque control of induction machine with a robust adaptive observer, IEEE Trans. Ind. Electron. 64 (04) (2017) 2631–2641.
[31] J. Wang, F. Wang, et al., Design and implementation of disturbance compensation-based enhanced robust finite control set predictive torque control for induction motor systems, IEEE Trans. Ind. Inform. 13 (05) (2017) 2645–2656.
[32] M. Siami, D.A. Khaburi, et al., Torque ripple reduction of predictive torque control for PMSM drives with parameter mismatch, IEEE Trans. Power Electron. 32 (09) (2017) 7160–7168.
[33] J. Stumper, S. Kuehl, et al. Predictive torque control for AC drives: improvement of parametric robustness using two-degree-of-freedom control, IEEE Energy Convers. Congr. Expo. (2013) 1170–1175.
[34] E. Jung, H. Yoo, et al., A nine-phase permanent-magnet motor drive system for an ultrahigh-speed elevator, IEEE Trans. Ind. Appl. 48 (03) (2012) 987–995.

CHAPTER 11

Model-free predictive flux vector control for modular multiphase PMSM drives considering parameter mismatch

11.1 Introduction

Multiphase permanent magnet synchronous motors (PMSMs) have garnered wide acceptance as high-power railway traction and marine propulsion systems, owing to their high power density, high torque density, and better fault-tolerant capability [1–3]. Since fault-tolerance and reliability are major concerns in multiphase PMSMs, many new multiphase PMSM structures have been studied. To utilize well-established three-phase converter technologies, a novel multimodule three-phase PMSM (modular multiphase PMSM) with multiple three-phase winding sets is investigated in [4,5]. In this configuration, each three-phase winding set is driven by an independent three-phase VSI. The main advantage of the modular multiphase PMSM is that it can be extended in terms of fault tolerance to reconfigure the driver in case of an open-phase fault event [6]. However, the modular vector control scheme is the key to the industrial application of modular multiphase PMSMs. In [7,8], PI controllers were used to regulate the modular multiphase PMSM, which can achieve excellent dynamic and static control performance but is highly dependent on PI control gains.

Model predictive control (MPC) is a high-performance discrete closed-loop control algorithm that has been successfully applied to various motor drives [9]. Compared with the conventional PI controller, the MPC method has the advantage of accurately adjusting stator flux and torque, which makes it more suitable for modular multiphase PMSM. In [5], a novel model predictive torque and stator flux control method for modular multiphase PMSM was developed, which can achieve lower torque and stator flux ripple. In [10], a model predictive torque vector

Robust Model Predictive Control of Permanent Magnet Synchronous Motor Under Parameter Mismatch
https://doi.org/10.1016/B978-0-443-44209-4.00014-2

control scheme based on a multi-three-phase approach was proposed for multi-three-phase motor drives, where each three-segment stator winding set was independently controlled. However, all possible voltage vectors of the MPC method must depend on the prediction models of stator flux and torque. Therefore, when the parameters used in the prediction models do not match the actual modular multiphase PMSM parameters, the switching frequency of the MPC method becomes unstable, which directly affects the torque/flux control performance [11].

To improve the performance of a model predictive controller, the disturbance observer plays an important role in achieving strong robustness to parameter mismatch. In [12], an improved deadbeat predictive control method based on the extended disturbance observer was developed, which can reduce current tracking errors and fluctuations caused by parameter mismatch. In [13], an incremental prediction model with an inductance disturbance observer was introduced, which can improve the antiparameter disturbance capability of the model predictive controller. In [14], a robust model predictive current controller with constant switching frequency based on a Luenberger observer was presented, which can effectively minimize instantaneous current error in each sampling period. In [15–17], a robust predictive current control strategy based on an extended state observer was proposed by using the augmented system model, which can eliminate the impacts of parameter mismatches on the control performance of the model predictive controller. In [18], a robust cascaded deadbeat predictive controller with speed and disturbance observers was developed for the dual three-phase variable-flux PMSM drive system, which can improve the system's robustness under parameter variation and magnetization manipulation conditions. In [19,20], a composite robust current regulation method based on the adaptive disturbance observer was presented for the linear permanent-magnet synchronous motor, which can enhance the system's robustness against model parameter mismatch.

Recently, the model-free predictive control method based on observer compensation has been proposed to further enhance robustness against parameter mismatch. In [21–23], an improved model-free predictive controller based on the extended state observer was introduced, which eliminates the need for motor parameters and achieves better performance in terms of tracking error. In [24], a robust model-free predictive control method with parallel and cascade extended-state observers was presented, enhancing disturbance rejection attributes and eliminating the influence of parameter mismatch. In [25], a novel model-free predictive control strategy based on the hybrid cascade-parallel extended-state observer was proposed, addressing both parameter mismatch and noise amplification simultaneously. In [26], a finite control set model-free predictive controller based on the SMO was introduced, requiring only the input and output data of the system. In [27], a model-free predictive zero-sequence current control method was proposed for the dual inverter-fed open-end winding induction motor, improving the robustness of the entire control system and reducing the computational burden. Since the model-free predictive control method is implemented in the *d-q* axis current coordinate system [21–27], double current sampling and high sampling frequency are required in one control cycle, which increases the computational burden of the system. Furthermore, the vector control of a modular multiphase PMSM based on model-free control has not yet been reported. It is necessary to introduce the concept of model-free control into the vector control of modular multiphase PMSM to address the problem of model parameter dependence.

In this chapter, an improved model-free predictive flux vector control (MF-PFVC) mechanism, based on the *d-q* axis stator flux space, is studied to enhance the robustness of the modular multiphase PMSM system against model parameter mismatch. In addition, a multivariable unknown disturbance integral terminal SMO, based on the *d-q* axis stator flux, is designed to improve the stator flux tracking accuracy in the proposed MF-PFVC strategy. The performance of the proposed MF-PFVC strategy under parameter mismatch is significantly better than that of the conventional PTC method, especially in terms of stator flux tracking accuracy and torque/flux ripple. This chapter is organized as follows: The mathematical model and drive system of the modular multiphase PMSM are established in Section 11.2. In Section 11.3, the conventional PTC method for the modular multiphase PMSM is developed. The MF-PFVC method for the modular multiphase PMSM is proposed in Section 11.4. The effectiveness of the proposed MF-PFVC strategy is validated through simulations in Section 11.5. Finally, Section 11.6 concludes this chapter.

11.2 Mathematical model of modular multiphase PMSM

The structure diagram of the modular multiphase PMSM studied in this chapter is shown in Fig. 11.1. The modular multiphase PMSM is composed of three modular motor units with the same characteristics. Note that there are electrical, magnetic, and thermal isolation characteristics among the three modular motor units. The 12-slot windings of the modular

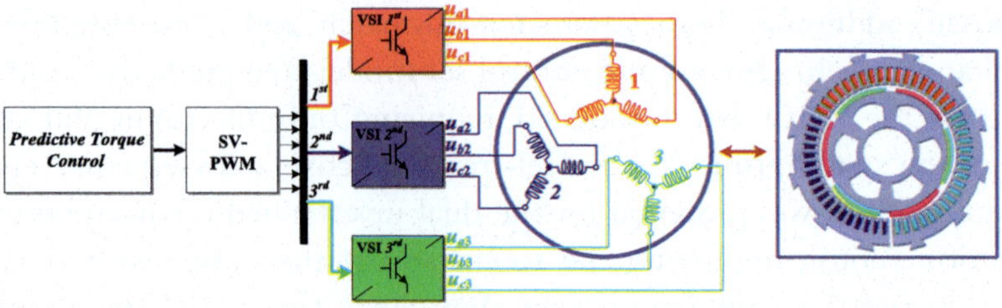

Fig. 11.1 Drive system structure diagram of a modular multiphase PMSM.

multiphase PMSM are divided into three parts. The stator winding of the modular multiphase PMSM consists of three modular motor windings, and each set of three-phase windings has a phase difference of 180 electrical angles, as shown in Fig. 11.1. Since the mutual inductance between windings in the motor units accounts for a small proportion of self-inductance, the d-q axis coordinate transformation matrix of the modular multiphase PMSM can be expressed as [2,8]

$$\mathbf{T}_{dq}^{abc} = \frac{2}{3}\begin{bmatrix} T_{11} & & \\ & T_{22} & \\ & & T_{33} \end{bmatrix} \tag{11.1}$$

With

$$\begin{cases} \mathbf{T}_{jj} = \dfrac{2}{3}\begin{bmatrix} \cos\theta & \cos(\theta - 120^{\circ}) & \cos(\theta + 120^{\circ}) \\ -\sin\theta & -\sin(\theta - 120^{\circ}) & -\sin(\theta + 120^{\circ}) \\ \dfrac{1}{2} & \dfrac{1}{2} & \dfrac{1}{2} \end{bmatrix} \\ j = 1,\quad 2,\quad 3 \end{cases} \tag{11.2}$$

where θ is the rotor position angle.

The voltage equations of the modular multiphase PMSM in a synchronous rotating frame are [5,8]

$$\begin{bmatrix} \mathbf{u}_{11} \\ \mathbf{u}_{22} \\ \mathbf{u}_{33} \end{bmatrix} = R_o \begin{bmatrix} \mathbf{i}_{11} \\ \mathbf{i}_{22} \\ \mathbf{i}_{33} \end{bmatrix} + \begin{bmatrix} \mathbf{L}_{11} & & \\ & \mathbf{L}_{22} & \\ & & \mathbf{L}_{33} \end{bmatrix} \left(\frac{d}{dt} \begin{bmatrix} \mathbf{i}_{11} \\ \mathbf{i}_{22} \\ \mathbf{i}_{33} \end{bmatrix} \right) + \omega_e \begin{bmatrix} \mathbf{G}_{11} & & \\ & \mathbf{G}_{22} & \\ & & \mathbf{G}_{33} \end{bmatrix} \begin{bmatrix} \mathbf{i}_{11} \\ \mathbf{i}_{22} \\ \mathbf{i}_{33} \end{bmatrix} + \omega_e \begin{bmatrix} \psi_f \\ \psi_f \\ \psi_f \end{bmatrix} \tag{11.3}$$

where j stands for any unit of the motor (j = 1, 2, 3); $\mathbf{u}_{jj}$ is the d- and q-axis voltages of each motor unit; $\mathbf{i}_{jj}$ is the d- and q-axis currents of each motor unit; $\mathbf{L}_{jj}$ is the stator inductance of each motor unit; $\mathbf{G}_{jj}$ is the cross coupling term of each motor unit; ψ_f is the flux linkage of permanent magnets; R is the stator resistance; and ω_e is the electrical rotor speed. The coefficient matrices of the state equations are

$$\mathbf{u}_{jj}=\begin{bmatrix} u_{dj} \\ u_{qj} \end{bmatrix}, \mathbf{i}_{jj}=\begin{bmatrix} i_{dj} \\ i_{qj} \end{bmatrix}, \mathbf{L}_{jj}=\begin{bmatrix} L_{do} & 0 \\ 0 & L_{qo} \end{bmatrix}, \mathbf{G}_{jj}=\begin{bmatrix} 0 & -L_{qo} \\ L_{do} & 0 \end{bmatrix}, \psi_f=\begin{bmatrix} 0 \\ \psi_{ro} \end{bmatrix}.$$

The electromagnetic torque of a modular multiphase PMSM can be expressed as:

$$T_e=\frac{3n_p}{2}\left[\psi_{ro}(i_{q1}+i_{q2}+i_{q3})+(L_{do}-L_{qo})(i_{d1}i_{q1}+i_{d2}i_{q2}+i_{d3}i_{q3})\right] \quad (11.4)$$

Where n_p is the number of pole pairs; J is the moment of inertia; and T_e and T_L are the electromagnetic torque and load torque of the modular multiphase PMSM, respectively.

Although the winding structure of modular multiphase PMSM is more complicated, the control method is similar to that of traditional three-phase PMSM. Through the characteristics analysis of modular multiphase PMSM, it can be observed that the VSI vector control of modular multiphase PMSM is similar to that of the three-phase PMSM [2,5,8]. Since the operating characteristics of each group of stator windings are very similar, the three-segment stator windings of modular multiphase PMSM can be excited by three VSIs in parallel, as shown in Fig. 11.1.

11.3 Conventional PTC method of modular multiphase PMSM

The purpose of the PTC method is to control the torque and stator flux of the modular multiphase PMSM. The modular PTC structural diagram of the modular multiphase PMSM drives is illustrated, as shown in Fig. 11.2. The control objectives of the PTC method are stator flux and torque. To minimize the absolute error between the predicted values and reference values, the cost function of predictive torque control can be defined as [2].

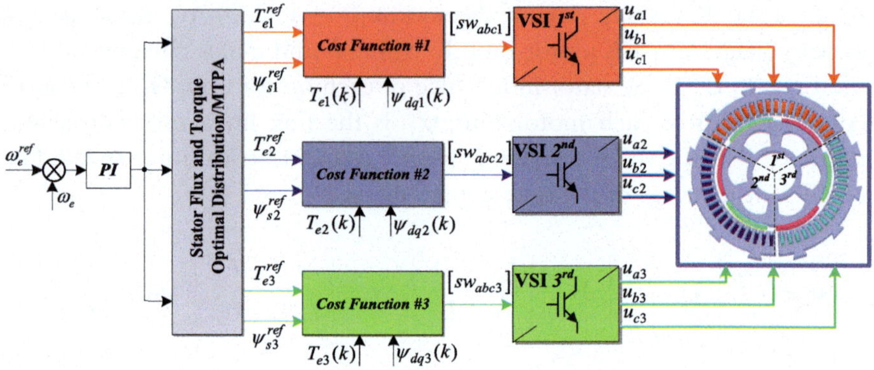

Fig. 11.2 The modular PTC structural diagram of the modular multiphase PMSM drives.

$$\zeta = \begin{bmatrix} g_{abc1} \\ g_{abc2} \\ g_{abc3} \end{bmatrix} = \begin{bmatrix} k_{\psi 1} & & \\ & k_{\psi 2} & \\ & & k_{\psi 3} \end{bmatrix} \begin{bmatrix} \left|\psi_{s1}^{\text{ref}} - \psi_{s1}^{p}(k+1)\right| \\ \left|\psi_{s2}^{\text{ref}} - \psi_{s2}^{p}(k+1)\right| \\ \left|\psi_{s3}^{\text{ref}} - \psi_{s3}^{p}(k+1)\right| \end{bmatrix} + \begin{bmatrix} \left|T_{e1}^{\text{ref}} - T_{e1}^{p}(k+1)\right| \\ \left|T_{e2}^{\text{ref}} - T_{e2}^{p}(k+1)\right| \\ \left|T_{e3}^{\text{ref}} - T_{e3}^{p}(k+1)\right| \end{bmatrix} \tag{11.5}$$

where $k_{\psi 1}$, $k_{\psi 2}$, and $k_{\psi 3}$ are the weighting factor; T_{ej}^{ref} and ψ_{sj}^{ref} are the torque reference value and flux-linkage reference value, respectively; $T_{ej}^{p}(k+1)$ and $\psi_{sj}^{p}(k+1)$ are the predicted torque value and predicted flux-linkage value, respectively.

If the PTC sampling period T_s is short enough, the torque and stator flux prediction equation of a surface-mounted modular multiphase PMSM ($L_{so} = L_{do} = L_{qo}$) can be expressed as

$$\begin{cases} T_{ej|sw}^{p}(k+1) = \dfrac{3}{2} n_p \psi_{ro} i_{qj|sw}^{p}(k+1) \\ \psi_{sj|sw}^{p}(k+1) = \psi_{sj}(k) - T_s R_o i_{sj}(k) + T_s u_{sw}(k) \\ i_{qj|sw}^{p}(k+1) = i_{qj}(k) + \dfrac{T_s}{L_{so}}\left[-L_{so}\omega_e(k) i_{dj}(k) - R_o i_{qj}(k) - \omega_e(k)\psi_{ro} + u_{qj|sw}(k)\right] \end{cases} \tag{11.6}$$

Eq. (11.6) can be used to predict the stator flux and torque for all seven voltage vectors of each VSI (sw = 1:7). From Eq. (11.6), it can be found that the prediction equations are absolutely dependent on the motor parameters (i.e., ψ_{ro}, R_o, and L_{so}), which means that the model parameter mismatch would lead to inaccurate prediction behavior and deteriorate the performance of the PTC algorithm.

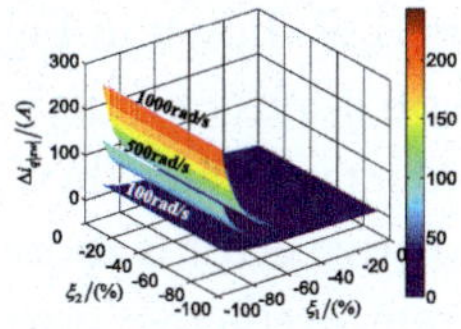

Fig. 11.3 The q-axis current predictive error under parameter mismatch.

To analyze parameter mismatch sensitivity of the conventional PTC method, introducing $L=L_{so}(1+\xi_1)$, $\psi_r=\psi_{ro}(1+\xi_2)$, $\xi_1=\Delta L_s / L_{so}$, and $\xi_2=\Delta\psi_r / \psi_{ro}$, where L_{so} and ψ_{ro} are actual values, and ΔL_s and $\Delta\psi_r$ are uncertain components of the stator inductance and flux linkage, respectively.

According to Eq. (11.6), the prediction equation can be obtained as

$$\breve{i}^{\,p}_{qj|sw}(k+1)=i_{qj}(k)+T_s\begin{bmatrix} -\omega_e(k)i_{dj}(k)-\dfrac{R_o}{L_{so}(1+\xi_1)}i_{qj}(k) \\ -\dfrac{\psi_{ro}(1+\xi_2)}{L_{so}(1+\xi_1)}\omega_e(k)+\dfrac{1}{L_{so}(1+\xi_1)}u_{qj|sw}(k) \end{bmatrix} \tag{11.7}$$

where $\breve{i}^{\,p}_{qj|sw}(k+1)$ is the prediction of q-axis current component under the parameter mismatch. According to Eq. (11.6) and Eq. (11.7), the current predictive error can be expressed as

$$\begin{aligned} \Delta i_{qj|sw} &= \breve{i}^{\,p}_{qj|sw}(k+1)-i^{p}_{qj|sw}(k+1) \\ &= T_s\left[\frac{R_o\xi_1}{L_{so}(1+\xi_1)}i_{qj}(k)-\frac{\psi_{ro}(\xi_2-\xi_1)}{L_{so}(1+\xi_1)}\omega_e(k)-\frac{\xi_1}{L_{so}(1+\xi_1)}u_{qj|sw}(k)\right] \end{aligned} \tag{11.8}$$

The q-axis voltage stationary equations can be expressed as follows when the modular multiphase PMSM reaches a stable state.

$$u_{qj|sw}(k)=R_o i_{qj}(k)+L_{so}\omega_e(k)i_{dj}(k)+\omega_e(k)\psi_{ro} \tag{11.9}$$

Under $i^{\mathrm{ref}}_{dj}(k)=0$, combining with Eqs. (11.8) and (11.9), the current predictive error can be simplified as

$$\Delta i_{qj|sw}=\breve{i}^{\,p}_{qj|sw}(k+1)-i^{p}_{qj|sw}(k+1)=-\frac{\psi_{ro}\xi_2\omega_e(k)}{L_{so}(1+\xi_1)}T_s \tag{11.10}$$

The main parameters of modular multiphase PMSM are given in Table II. According to Eq. (11.10), the q-axis current predictive error under parameter mismatch are plotted, as shown in Fig. 11.3. Fig. 11.3 analyzes the influence of different speeds on q-axis current predictive error under parameter mismatch. When the inductance and flux linkage parameter mismatch occur, the q-axis current predictive error is about 24A at 100 rad/s. At a speed of 400 rad/s, the q-axis current predictive error reaches 98A under inductance and flux linkage parameter mismatch ($\xi_1 = -90\%$, $\xi_2 = -90\%$). When the modular multiphase PMSM speed reaches 800 rad/s, the q-axis current predictive error is up to 195A under parameters mismatch ($\xi_1 = -90\%$, $\xi_2 = -90\%$). It can be inferred that the q-axis current predictive error increases with the increase of speed, which means that the high motor speed has a great impact on the switching state of the converter under parameter mismatch. If the parameter mismatch is not considered in the design of the predictive controller, the torque control performance of modular multiphase PMSM drive system will be deteriorated.

11.4 Novel MF-PFVC method of modular multiphase PMSM

Principle of MF-PFVC based on the ultra-local model

According to Eq. (11.3), the stator flux equations of the surface-mounted modular multiphase PMSM can be expressed as follows under parameter mismatch.

$$\underbrace{\frac{d}{dt}\begin{bmatrix}\psi_{11}\\ \psi_{22}\\ \psi_{33}\end{bmatrix}}_{\psi_\Lambda} = \underbrace{\begin{bmatrix}A_{11} & & \\ & A_{22} & \\ & & A_{33}\end{bmatrix}}_{E_\Lambda}\begin{bmatrix}\psi_{11}\\ \psi_{22}\\ \psi_{33}\end{bmatrix} + \underbrace{\begin{bmatrix}u_{11}\\ u_{22}\\ u_{33}\end{bmatrix}}_{U_\Lambda} - \underbrace{\begin{bmatrix}\delta_{11}\\ \delta_{22}\\ \delta_{33}\end{bmatrix}}_{\delta_\Lambda} \tag{11.11}$$

where $\boldsymbol{\psi}_{jj}$ is the d- and q-axis flux linkage; δ_{jj} is the d- and q-axis unknown disturbance caused by parameter mismatch; the matrixes of the stator flux equations are

$$\boldsymbol{\psi}_{jj} = \begin{bmatrix}\psi_{dj}\\ \psi_{qj}\end{bmatrix} = \begin{bmatrix}L_{so}i_{dj} + \psi_{ro}\\ L_{so}i_{qj}\end{bmatrix},\ \mathbf{A}_{jj} = \begin{bmatrix}-\dfrac{R_o}{L_{so}} & \omega_e\\ -\omega_e & -\dfrac{R_o}{L_{so}}\end{bmatrix},\ \boldsymbol{\delta}_{jj} = \begin{bmatrix}\delta_{dj}\\ \delta_{qj}\end{bmatrix}$$

With

$$\begin{cases} \delta_{dj} = -\dfrac{R_o}{L_{so}}\psi_{ro} - \dfrac{\xi_1 R_o}{(1+\xi_1)} i_{dj} + \dfrac{\xi_1}{(1+\xi_1)} u_{dj} \\ \delta_{qj} = \dfrac{\xi_1}{(1+\xi_1)}(u_{qj} - R_o i_{qj} - \psi_{ro}\omega_e - \psi_{ro}\xi_2\omega_e) + \psi_{ro}\xi_2\omega_e \end{cases} \tag{11.12}$$

According to [22] and [25], the changes in the input and output of highly nonlinear and time-varying systems can be approximated as finite-dimensional differential equations, specifically as:

$$M(t, y, \frac{dy}{dt}, \ldots, y^{(n)}, u, \frac{du}{dt}, \ldots, u^{(k)}) = 0 \tag{11.13}$$

where u and y are the system input and output, respectively; M is the derivable function of its arguments, and the superscripts (k) and (n) are system orders. From the implicit function theorem, we can get

$$y^{(n)} = \wp(t, y, \frac{dy}{dt}, \ldots, y^{(n)}, u, \frac{dy}{dt}, \ldots, u^{(k)}) \tag{11.14}$$

By setting $\wp = F + \alpha u$, the ultra-local model can be obtain as

$$y^{(n)} = \alpha u + F \tag{11.15}$$

where α is a nonphysical constant parameter; F contains the whole "system structure information", such as unknown disturbance caused by parameter mismatch.

According to Eqs. (11.11) and (11.15), the MF-PFVC, based on the ultra-local model of the modular multiphase PMSM, can be expressed as

$$\begin{cases} \dfrac{d}{dt}\psi_\Lambda = \alpha \mathbf{U}_\Lambda + \tilde{\mathbf{F}}_\Lambda \\ \psi_\Lambda = \mathbf{Q}_\Lambda \mathbf{x}_\Lambda + \mathbf{f}_\psi \end{cases} \tag{11.16}$$

With

$$\begin{cases} \alpha = \mathrm{diag}(1,1,1,1,1,1) \\ \tilde{\mathbf{F}}_\Lambda = \mathbf{E}_\Lambda \psi_\Lambda - \delta_\Lambda \end{cases} \tag{11.17}$$

where $\hat{\mathbf{F}}$ is the unknown disturbance caused by parameter mismatch, $\tilde{\mathbf{F}}_\Lambda = \begin{bmatrix} \tilde{\mathbf{F}}_{11} & \tilde{\mathbf{F}}_{22} & \tilde{\mathbf{F}}_{33} \end{bmatrix}^T$, $\tilde{\mathbf{F}}_{jj} = \begin{bmatrix} \tilde{F}_{dj} & \tilde{F}_{qj} \end{bmatrix}^T$, $\mathbf{Q}_\Lambda = L_{so} \cdot \alpha$; $\mathbf{f}_\psi = [\psi_{ro} \ \ 0 \ \ \psi_{ro} \ \ 0 \ \ \psi_{ro} \ \ 0]^T$, $\mathbf{x}_\Lambda = [\mathbf{i}_{11} \ \ \mathbf{i}_{22} \ \ \mathbf{i}_{33}]^T$.

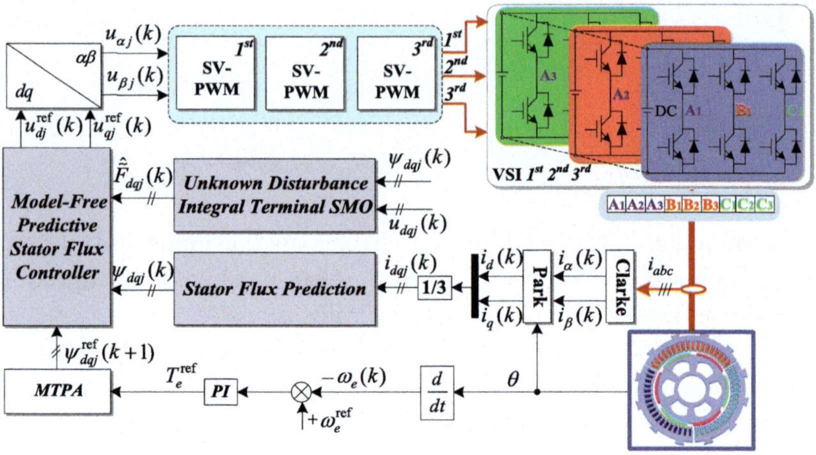

Fig. 11.4 Structural diagram of the modular multiphase PMSM drive system with the MF-PFVC strategy.

According to Eq. (11.16), the model-free predictive stator flux controller of the modular multiphase PMSM can be designed as follows when the parameter mismatches are considered

$$\mathbf{U}_{\Lambda}^{\text{ref}}(k) = \frac{1}{T_s}\boldsymbol{\psi}_{\Lambda}^{\text{ref}}(k+1) - \frac{1}{T_s}\widehat{\boldsymbol{\psi}}_{\Lambda}(k) - \hat{\hat{\mathbf{F}}}_{\Lambda}(k) \tag{11.18}$$

where $\mathbf{U}_{\Lambda}^{\text{ref}}(k)$ is the output voltage vector; $\boldsymbol{\psi}_{\Lambda}^{\text{ref}}(k+1)$ is the stator flux reference; $\hat{\boldsymbol{\psi}}_{\Lambda}(k)$ is the stator flux observation; and $\hat{\hat{\mathbf{F}}}_{\Lambda}(k)$ is unknown disturbance observation caused by parameter mismatch.

From Eq. (11.18), it can be observed that the novel model-free predictive stator flux controller is independent of motor parameters, which means that the MF-PFVC method is not affected by parameter mismatches. Fig. 11.4 shows the structural diagram of the modular multiphase PMSM drive system with the MF-PFVC strategy.

Design of multivariable unknown disturbance integral terminal SMO

In this section, the multivariable unknown disturbance integral terminal SMO is designed, which can accurately observe the unknown disturbance caused by parameter mismatch. Based on Eq. (11.11) and Eq. (11.16), the multivariable unknown disturbance integral terminal SMO is designed as follows:

$$\begin{cases} \dfrac{d}{dt}\widehat{\psi}_\Lambda = \mathbf{E}_\Lambda\widehat{\psi}_\Lambda + \mathbf{U}_\Lambda + \omega_e\mathbf{P}(\widehat{\psi}_\Lambda - \psi_\Lambda) - \mathbf{U}_\delta \\ \widehat{\psi}_\Lambda = \mathbf{Q}_\Lambda\hat{\mathbf{x}}_\Lambda + \mathbf{f}_\psi \end{cases} \tag{11.19}$$

where $\widehat{\psi}_\Lambda$ is the observed value of ψ_Λ; $\hat{\mathbf{x}}_\Lambda$ is the observed value of $\mathbf{x}_\Lambda$; $\mathbf{U}_\delta=[\mathbf{U}_{\delta 11}\quad \mathbf{U}_{\delta 22}\quad \mathbf{U}_{\delta 33}]^T$ is the sliding mode control function; and $\mathbf{U}_{\delta jj}=[\mathbf{U}_{\delta dj}\quad \mathbf{U}_{\delta qj}]^T$; $\mathbf{P}$ is the designed observer sliding gain.

The integral terminal sliding surface is designed as.

$$\mathbf{s}_\Lambda = \mathbf{e}_\Lambda + \rho\int_0^t \mathrm{sgn}(\mathbf{e}_\Lambda)d\tau \tag{11.20}$$

With

$$\mathbf{e}_\Lambda = \hat{\psi}_\Lambda - \psi_\Lambda = \mathbf{Q}_\Lambda(\hat{\mathbf{x}}_\Lambda - \mathbf{x}_\Lambda) = \begin{bmatrix} \mathbf{e}_{11} \\ \mathbf{e}_{22} \\ \mathbf{e}_{33} \end{bmatrix} = \mathbf{Q}_\Lambda \begin{bmatrix} \hat{\mathbf{i}}_{11} - \mathbf{i}_{11} \\ \hat{\mathbf{i}}_{22} - \mathbf{i}_{22} \\ \hat{\mathbf{i}}_{33} - \mathbf{i}_{33} \end{bmatrix} \tag{11.21}$$

where $\mathbf{e}_\Lambda = [\mathbf{e}_{11}\quad \mathbf{e}_{22}\quad \mathbf{e}_{33}]^T$ and $\mathbf{e}_{jj} = [e_{dj}\quad e_{qj}]^T$; $\mathbf{s}_\Lambda = [\mathbf{s}_{11}\quad \mathbf{s}_{22}\quad \mathbf{s}_{33}]^T$, and $\mathbf{s}_{jj} = [s_{dj}\quad s_{qj}]^T$; ρ is the designed parameters of unknown disturbance integral terminal SMO, and $\rho \succ 0$; sgn() is sign function, and $\mathrm{sgn}(\varepsilon)=\tanh(\varepsilon)=\dfrac{e^{2\varepsilon}-1}{e^{2\varepsilon}+1}$; ε is any function variable.

Taking the time derivative of Eq. (11.20) and combining it with Eqs. (11.16) and (11.19), one obtains:

$$\frac{d}{dt}\mathbf{s}_\Lambda = \mathbf{E}'_\Lambda\mathbf{e}_\Lambda + \omega_e(\mathbf{E}''_\Lambda + \mathbf{P})\mathbf{e}_\Lambda + \delta_\Lambda - \mathbf{U}_\delta + \rho\,\mathrm{sgn}(\mathbf{e}_\Lambda) \tag{11.22}$$

where $\mathbf{E}_\Lambda=\mathbf{E}'_\Lambda+\omega_e\mathbf{E}''_\Lambda$, $\mathbf{E}'_\Lambda=\begin{bmatrix} \mathbf{A}'_{11} & & \\ & \mathbf{A}'_{22} & \\ & & \mathbf{A}'_{33} \end{bmatrix}$, $\mathbf{A}'_{jj}=-\dfrac{R_o}{L_{so}}\begin{bmatrix} 1 & 0 \\ 0 & 1 \end{bmatrix}$,

$\mathbf{A}''_{jj}=\begin{bmatrix} 0 & 1 \\ -1 & 0 \end{bmatrix}$, $\mathbf{E}''_\Lambda=\begin{bmatrix} \mathbf{A}''_{11} & & \\ & \mathbf{A}''_{22} & \\ & & \mathbf{A}''_{33} \end{bmatrix}$.

To eliminate the influence of the speed, the gain matrixes of multivariable unknown disturbance integral terminal SMO is designed as $\mathbf{P}=-\mathbf{E}''_\Lambda$, where $\omega_e(\mathbf{E}''_\Lambda+\mathbf{P})\mathbf{e}_\Lambda=0$, the Eq. (11.22) is modified to

$$\frac{d}{dt}\mathbf{s}_{\Lambda} = \mathbf{E}'_{\Lambda}\mathbf{e}_{\Lambda} + \delta_{\Lambda} + \rho\,\mathrm{sgn}(\mathbf{e}_{\Lambda}) - \mathbf{U}_{\delta} \tag{11.23}$$

Combining Eqs. (11.20) and (11.23), the sliding mode control function is designed to guarantee the convergence of the multivariable unknown disturbance integral terminal SMO.

$$\mathbf{U}_{\delta} = \mathbf{E}'_{\Lambda}\mathbf{e}_{\Lambda} + \rho\,\mathrm{sgn}(\mathbf{e}_{\Lambda}) + \kappa_{\alpha}\mathbf{s}_{\Lambda} + \kappa_{\beta}\,\mathrm{sgn}(\mathbf{s}_{\Lambda}) \tag{11.24}$$

where $\kappa_{\alpha} = \kappa_{l}\mathrm{diag}(1,1,1,1,1,1)$ and $\kappa_{\beta} = \kappa_{\upsilon}\mathrm{diag}(1,1,1,1,1,1)$ are the designed observer sliding gain.

The following Lyapunov function is defined to prove the asymptotic stability of the multivariable unknown disturbance integral terminal SMO.

Combining with Eqs. (11.23) and (11.24), one yields

$$\begin{aligned}
\frac{d}{dt}\mathbf{V} &= \mathbf{s}_{\Lambda}^{T}(\mathbf{E}'_{\Lambda}\mathbf{e}_{\Lambda} + \delta_{\Lambda} + \rho\,\mathrm{sgn}(\mathbf{e}_{\Lambda}) - \mathbf{U}_{\delta}) \\
&= \mathbf{s}_{\Lambda}^{T}(\delta_{\Lambda} - \kappa_{\alpha}\mathbf{s}_{\Lambda} - \kappa_{\beta}\,\mathrm{sgn}(\mathbf{s}_{\Lambda})) \\
&= \mathbf{s}_{\Lambda}^{T}\delta_{\Lambda} - \kappa_{l}(s_{d1}^{2} + s_{q1}^{2} + s_{d2}^{2} + s_{q2}^{2} + s_{d3}^{2} + s_{q3}^{2}) - \mathbf{s}_{\Lambda}^{T}\kappa_{\beta}\,\mathrm{sgn}(\mathbf{s}_{\Lambda}) \\
&\le \|\mathbf{s}_{\Lambda}\|\,\|\delta_{\Lambda}\| - \kappa_{\upsilon}\|\mathbf{s}_{\Lambda}\| - \kappa_{l}(s_{d1}^{2} + s_{q1}^{2} + s_{d2}^{2} + s_{q2}^{2} + s_{d3}^{2} + s_{q3}^{2}) \\
&= \|\mathbf{s}_{\Lambda}\|(\|\delta_{\Lambda}\| - \kappa_{\upsilon}) - \kappa_{l}(s_{d1}^{2} + s_{q1}^{2} + s_{d2}^{2} + s_{q2}^{2} + s_{d3}^{2} + s_{q3}^{2})
\end{aligned} \tag{11.25}$$

In practical engineering, the unknown disturbance δ_{Λ} is bounded, that is $\delta_{\Lambda} \le N_{\delta}$, where N_{δ} is the boundary value. When κ_{l} and κ_{υ} are set to $\kappa_{l} \ge 0$ and $\kappa_{\upsilon} \ge N_{\delta} \ge \delta_{\Lambda}$, then $\frac{d}{dt}\mathbf{V} \le 0$. Therefore the stability and convergence of the proposed multivariable unknown disturbance integral terminal SMO is guaranteed.

If the sampling period is short enough, the discrete-time integral terminal SMO is given by Eq. (11.26) according to Eqs. (11.19) and (11.24).

$$\begin{cases}
\hat{\psi}_{dj}(k+1) = \left(1 - \dfrac{R_{o}}{L_{so}}T_{s}\right)\hat{\psi}_{dj}(k) + T_{s}u_{dj}(k) + \omega_{e}(k)\psi_{qj}(k)T_{s} - U_{\delta dj}(k)T_{s} \\
U_{\delta dj}(k) = -\dfrac{R_{o}}{L_{so}}e_{dj}(k) + \rho\,\mathrm{sgn}(e_{dj}(k)) + k_{l}s_{dj}(k) + k_{\upsilon}\,\mathrm{sgn}(s_{dj}(k)) \\
\hat{\psi}_{qj}(k+1) = \left(1 - \dfrac{R_{o}}{L_{so}}T_{s}\right)\hat{\psi}_{qj}(k) + T_{s}u_{qj}(k) - \omega_{e}(k)\psi_{dj}(k)T_{s} - U_{\delta qj}(k)T_{s} \\
U_{\delta qj}(k) = -\dfrac{R_{o}}{L_{so}}e_{qj}(k) + \rho\,\mathrm{sgn}(e_{qj}(k)) + k_{l}s_{qj}(k) + k_{\upsilon}\,\mathrm{sgn}(s_{qj}(k))
\end{cases} \tag{11.26}$$

According to Eqs. (11.22) and (11.17), when the multivariable unknown disturbance integral terminal SMO reaches the integral terminal sliding surface, the estimated value of the unknown disturbance is given as

$$\begin{cases} \hat{\hat{F}}_{dj}(k) = -\frac{R_o}{L_{so}}\psi_{dj}(k) + \omega_e\psi_{qj}(k) - \hat{\delta}_{dj}(k) \\ \hat{\delta}_{dj}(k) = U_{\delta dj}(k) \\ \hat{\hat{F}}_{qj}(k) = -\omega_e\psi_{dj}(k) - \frac{R_o}{L_{so}}\psi_{qj}(k) - \hat{\delta}_{qj}(k) \\ \hat{\delta}_{qj}(k) = U_{\delta qj}(k) \end{cases} \tag{11.27}$$

11.5 Simulations results

The modular multiphase PMSM drive system based on MF-PFVC strategy is established in MATLAB/Simulink.The main parameters of modular multiphase PMSM used in the MATLAB/Simulink and experiment are given in Table 11.1, and the sampling frequency is set as 10 kHz. The weight factor of the conventional PTC method is $k_{\psi 1}=k_{\psi 2}=k_{\psi 3}=4500$. In addition, the observer gains of the proposed multivariable unknown disturbance integral terminal SMO are set as $\lambda = 5$, $\eta = 15000$ and $\delta = 1000$, respectively. The speed reference is 100 rad/s, and the load torque reference is 2000 N.m.

Simulation results of modular multiphase PMSM under inductance parameter mismatch

The operating conditions of the modular multiphase PMSM in Case A are that the load torque is maintained at 2000 N·m, and the speed is maintained at 100 rad/s. At 0.8 s, there is a mismatch in the inductance parameter of the modular multiphase PMSM drive system, and the inductance value is stepped to 50% of its initial value, i.e., 1.446 mH. Comparative simulation

Table 11.1 Main parameters of the modular multiphase PMSM.

Parameter	Value	Parameter	Value
Inductance (L_o)	2.892/ mH	Stator phase resistance (R_o)	0.02/Ω
Flux linkage of PM (Ψ_{ro})	0.782/Wb	Rotational inertia (J)	80/kg m^2

results of the conventional PTC strategy and the proposed MF-PFVC strategy under inductance parameter mismatch are shown in Figs. 11.5–11.9.

Fig. 11.5 presents the comparative simulation results of the phase current and torque under inductance parameter mismatch. Simulation results of torque comparison illustrate that the conventional PTC method produces significant torque oscillation and stator current distortion under inductance parameter mismatch, as clearly shown in Fig. 11.5A. Since the stator flux response value of the proposed MF-PFVC strategy can strictly follow the reference value, and the inductance parameters are not required during control operation, the proposed PTSF method can significantly suppress torque oscillation and stator current distortion under inductance parameter mismatch, as shown in Fig. 11.5B. For the proposed MF-PFVC, note that there is significant overshoot in the torque waveform. The reason is that the unknown disturbance caused by inductance parameter mismatch is estimated by the multivariable unknown disturbance integral terminal SMO. Fig. 11.6 presents the comparative simulation results of

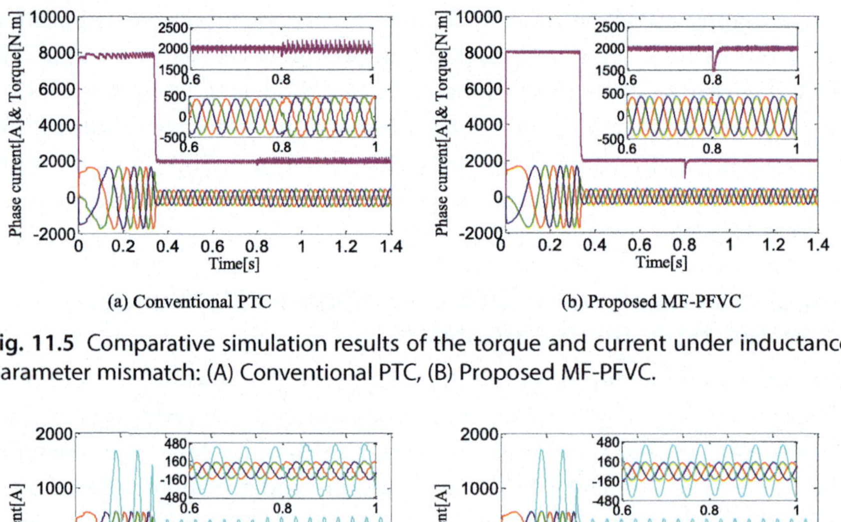

Fig. 11.5 Comparative simulation results of the torque and current under inductance parameter mismatch: (A) Conventional PTC, (B) Proposed MF-PFVC.

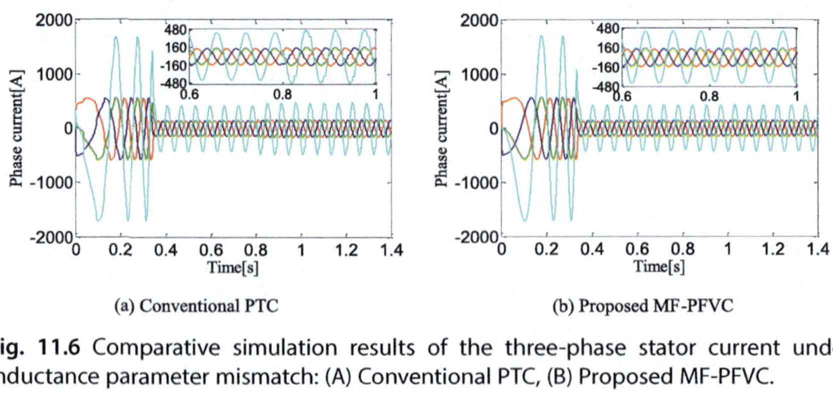

Fig. 11.6 Comparative simulation results of the three-phase stator current under inductance parameter mismatch: (A) Conventional PTC, (B) Proposed MF-PFVC.

the three-phase stator current under inductance parameter mismatch. As shown in Fig. 11.6, due to the modular multiphase PMSM consisting of three motor units with consistent characteristics, the three-phase stator current of the modular multiphase PMSM is three times that of the unit motor. Furthermore, compared with the conventional PTC method, the stator current distortion of the proposed MF-PFVC method is significantly reduced. This is because the proposed MF-PFVC method can effectively enhance robustness against inductance parameter mismatch by using the multivariable unknown disturbance integral terminal SMO.

Fig. 11.7 shows the comparative simulation results of the α-β stator flux linkage comparison between the proposed MF-PFVC and the conventional PTC under inductance parameter mismatch. From 0 to 0.8 s, it can be observed that the proposed MF-PFVC method exhibits the same excellent stator flux control performance as the conventional PTC when the modular multiphase PMSM operates normally. After 0.8 s, the α-β stator flux linkage response values of the modular multiphase PMSM show significant fluctuations when using the conventional PTC method under inductance parameter mismatch, as illustrated in Fig. 11.7A. Moreover, it can be seen that when the inductance parameters of the modular multiphase PMSM are mismatched, there is a noticeable oscillation in the α-β stator flux linkage wave form using the conventional PTC strategy, as clearly shown in Fig. 11.7A. The reason for this phenomenon is that the conventional PTC is highly sensitive to the inductance parameter of the modular multiphase PMSM. From Fig. 11.7B, it can be observed that the proposed MF-PFVC method significantly improves the α-β stator flux linkage control performance of the modular multiphase PMSM in the case

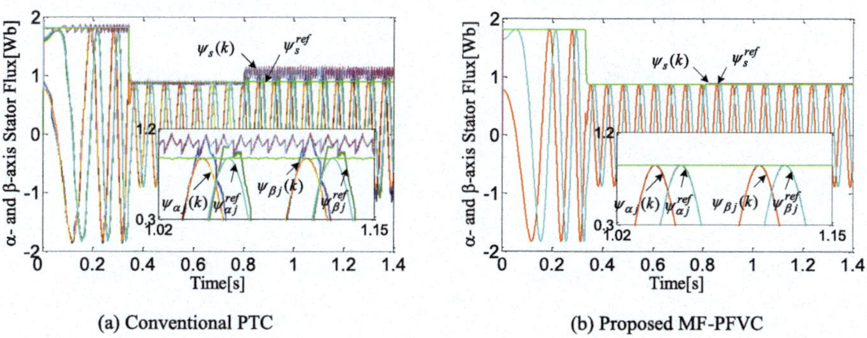

Fig. 11.7 Comparative simulation results of the α-β stator flux linkage under inductance parameter mismatch: (A) Conventional PTC, (B) Proposed MF-PFVC.

of inductance parameter mismatch. The α-β stator flux linkage oscillation caused by inductance parameter mismatch can be effectively suppressed by using the proposed MF-PFVC strategy. In addition, the α-β stator flux linkage response value of the modular multiphase PMSM can accurately track the reference value by adopting the proposed MF-PFVC strategy, as presented in Fig. 11.7B.

Fig. 11.8 shows the comparative simulation results of three-dimensional rotor flux trajectories between the proposed MF-PFVC strategy and the conventional PTC strategy under inductance parameter mismatch. From Fig. 11.8A, it can be observed that the three-dimensional rotor flux trajectories of the modular multiphase PMSM exhibit a large flux linkage fluctuation error when using the conventional PTC method under inductance parameter mismatch. However, the flux linkage fluctuation error of the three-dimensional rotor flux trajectories is drastically reduced because the proposed MF-PFVC method is immune to inductance parameter mismatches, as clearly shown in Fig. 11.8B. Fig. 11.9 depicts the observed simulation results of unknown disturbances using the multivariable unknown disturbance integral terminal SMO under inductance parameter mismatch. It can be seen that the multivariable unknown disturbance integral terminal SMO can simultaneously estimate the *d-q* stator flux linkage and unknown disturbances. Note that there is an overshoot when the inductance parameters are suddenly mismatched at 0.8 s; this occurs because the parameter mismatch impacts the multivariable unknown disturbance integral terminal SMO.

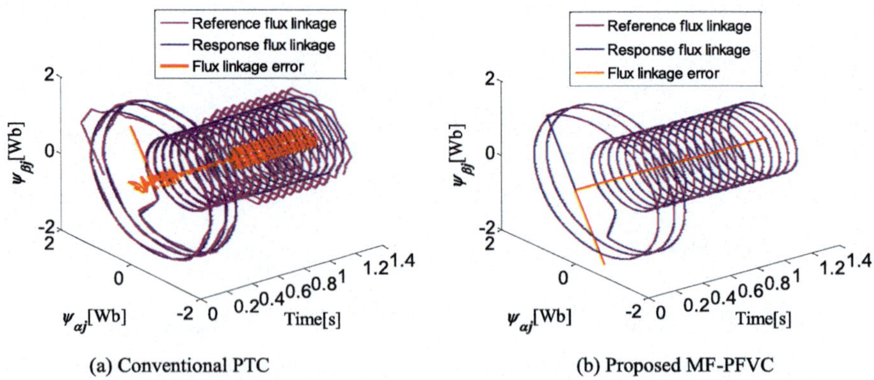

Fig. 11.8 Comparative simulation results of three-dimensional rotor flux trajectories under inductance parameter mismatch: (A) Conventional PTC, (B) Proposed MF-PFVC.

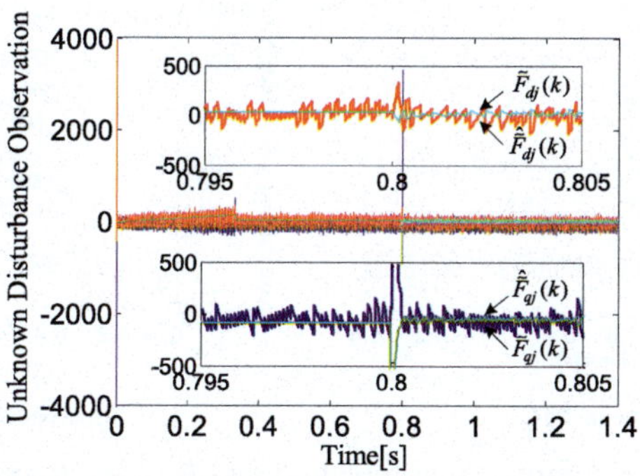

Fig. 11.9 Observation simulation results of an unknown disturbance under inductance parameter mismatch.

Simulation results of modular multiphase PMSM under flux linkage parameter mismatch

The operating conditions of the modular multiphase PMSM in Case B are that the load torque is maintained at 2000 N·m, and the speed is maintained at 100 rad/s. At 0.8 s, there is a mismatch in the flux linkage parameter of the modular multiphase PMSM drive system, and the flux linkage parameter value is stepped to 50% of its initial value, i.e., 0.391 Wb. Comparative simulation results of the conventional PTC strategy and the proposed MF-PFVC strategy under flux linkage parameter mismatch are shown in Figs. 11.10–11.14. Fig. 11.10 presents the comparative simulation results of the phase current and torque using the conventional PTC strategy and the proposed MF-PFVC strategy. The peak-to-peak torque ripple of the conventional PTC method is significantly increased under flux linkage parameter mismatch, while the peak-to-peak torque ripple of the modular multiphase PMSM is significantly reduced by using the proposed MF-PFVC method, as presented in Fig. 11.10. Fig. 11.10A demonstrates that the conventional PTC strategy provides a deteriorated torque wave form with large torque ripple in the presence of flux linkage parameter mismatch. The reason for the large torque ripple is that the conventional PTC strategy relies on accurate flux linkage parameters of the modular multiphase PMSM. In addition, it can be seen from Fig. 11.10A that the stator current suddenly increases to keep the torque constant under the flux linkage parameter mismatch. Compared with the conventional PTC strategy, the torque ripple of the modular multiphase PMSM is drastically decreased by using the proposed MF-PFVC strategy under flux linkage parameter mismatch, as depicted in Fig. 11.10B. The reason for the

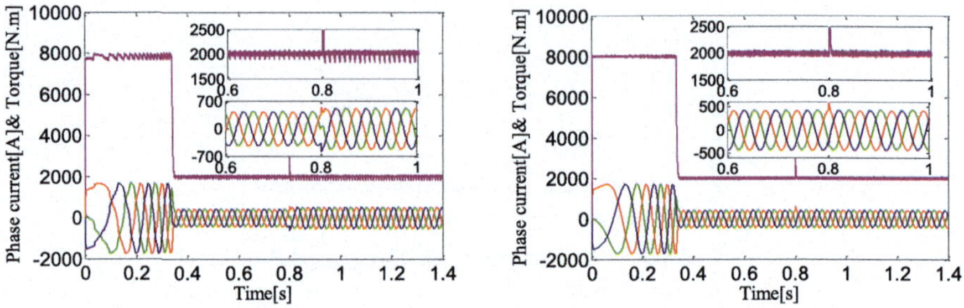

(a) Conventional PTC (b) Proposed MF-PFVC

Fig. 11.10 shows the comparative simulation results of the torque and current under flux linkage parameter mismatch: (A) Conventional PTC and (B) Proposed MF-PFVC.

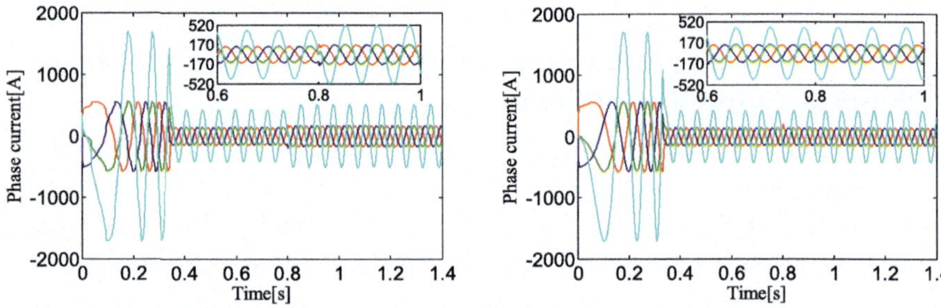

(a) Conventional PTC (b) Proposed MF-PFVC

Fig. 11.11 shows the comparative simulation results of the three-phase stator current under flux linkage parameter mismatch: (A) Conventional PTC and (B) Proposed MF-PFVC.

improvement in torque ripple is that the proposed MF-PFVC strategy is immune to the mismatch of the flux linkage parameter. It can also be noted in Fig. 11.10B that there is significant overshoot due to the operation of the multivariable unknown disturbance integral terminal SMO.

Fig. 11.11 presents the comparative simulation results of the three-phase stator current using the conventional PTC strategy and the proposed MF-PFVC strategy. In Fig. 11.11A, it can be seen that the three-phase stator current of the modular multiphase PMSM increases under the flux linkage parameter mismatch when using the conventional PTC strategy. The reason for this phenomenon is that the mismatch of the flux linkage parameter requires an increase in stator current to maintain the torque constant. Compared with the conventional PTC method, the proposed MF-PFVC method can significantly improve the stator current distortion and amplitude of the modular multiphase PMSM under the flux linkage

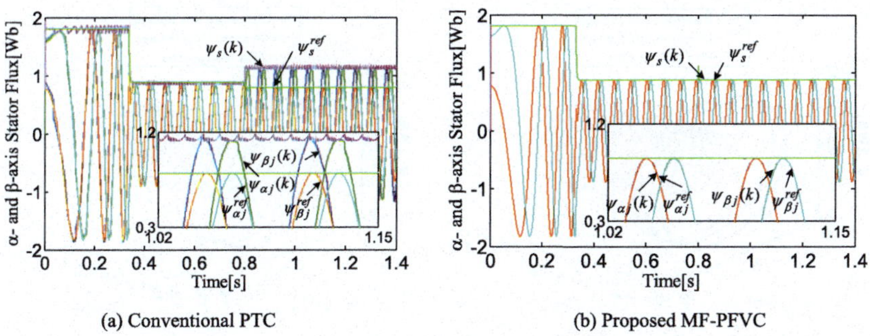

Fig. 11.12 Comparative simulation results of the α-β stator flux linkage under flux linkage parameter mismatch: (A) Conventional PTC, (B) Proposed MF-PFVC.

parameter mismatch, as clearly shown in Fig. 11.11B. This improvement is because the proposed MF-PFVC method can effectively enhance robustness against the flux linkage parameter mismatch by using the multivariable unknown disturbance integral terminal SMO.

Fig. 11.12 presents the comparative simulation results of α-β stator flux linkage comparison between the proposed MF-PFVC and conventional PTC under flux linkage parameter mismatch. From 0 to 0.8 s, it can be observed that both the proposed MF-PFVC method and the conventional PTC method achieve excellent α-β stator flux linkage control performance for modular multiphase PMSM in the absence of parameter mismatch. After 0.8 s, there is a significant error in the α-β stator flux linkage wave form under flux linkage parameter mismatch, and the response value of the α-β stator flux linkage fails to track its reference value, as shown in Fig. 11.12A. The reason for the significant stator flux linkage tracking error is that the conventional PTC strategy is highly affected by the flux linkage parameters of modular multiphase PMSM. Fig. 11.12B illustrates the simulation results that verify the performance of the proposed MF-PFVC strategy with a multivariable unknown disturbance integral terminal SMO under flux linkage parameter mismatch. The proposed MF-PFVC strategy significantly improves the stator flux control performance of modular multiphase PMSM, enabling the α-β stator flux linkage response value of modular multiphase PMSM to accurately track the reference value, as clearly shown in Fig. 11.12B. This improvement is attributed to the fact that the proposed MF-PFVC method does not rely on the inductance parameters of modular multiphase PMSM, allowing the response value of the stator flux linkage to accurately track its reference value.

Fig. 11.13 presents the comparative simulation results of three-dimensional rotor flux trajectories between the proposed MF-PFVC and the

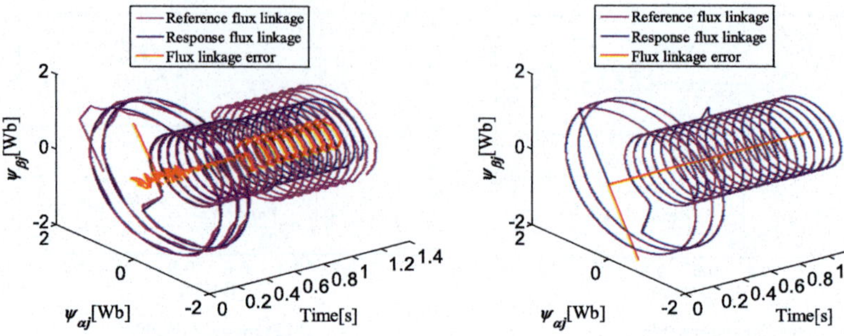

(a) Conventional PTC (b) Proposed MF-PFVC

Fig. 11.13 Comparative simulation results of three-dimensional rotor flux trajectories under flux linkage parameter mismatch: (A) Conventional PTC, (B) Proposed MF-PFVC.

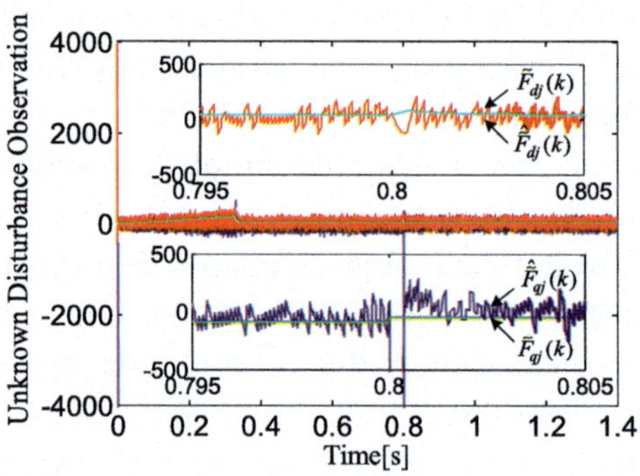

Fig. 11.14 Observation simulation results of an unknown disturbance under flux linkage parameter mismatch.

conventional PTC under flux linkage parameter mismatch. The three-dimensional rotor flux trajectories of the modular multiphase PMSM exhibit a large flux linkage static error under flux linkage parameter mismatch, as shown in Fig. 11.13A. The reason for the large flux linkage static error is that the flux linkage parameter mismatch deteriorates the control performance of the conventional PTC method. However, the flux linkage static error of the proposed MF-PFVC method can be effectively suppressed by using the multivariable unknown disturbance integral terminal SMO, as clearly illustrated in Fig. 11.13B. Fig. 11.14 depicts the observation simulation results of the unknown disturbance using the multivariable unknown disturbance integral terminal SMO under flux linkage parameter mismatch. The d-q stator flux linkage response value and the unknown disturbance predicted value of the modular multiphase PMSM can be accurately observed using the multivariable unknown disturbance integral terminal SMO, as clearly shown in Fig. 11.13A. Note that there is a sudden overshoot at 0.8 s, which is caused by flux linkage parameter mismatch.

11.6 Conclusion

This chapter proposes a novel model-free predictive flux vector control method (MF-PFVC) based on a multivariable unknown disturbance integral terminal SMO for a modular multiphase PMSM drive system that does not require motor parameters and demonstrates strong robustness. The *d-q* axis stator flux control is achieved by adopting model-free predictive control techniques, making it immune to any parameters of the modular multiphase PMSM. Furthermore, the proposed MF-PFVC strategy utilizes the multivariable unknown disturbance integral terminal SMO to obtain unknown disturbance information, thereby enhancing robustness against parameter mismatch. The proposed MF-PFVC strategy offers clear advantages in improving the tracking accuracy of *d-q* axis stator flux and reducing torque ripple compared to the conventional PTC strategy. Simulations and experimental results confirm that the proposed MF-PFVC method can significantly improve the operational behavior of the modular multiphase PMSM in the presence of significant parameter changes. In particular, the proposed MF-PFVC method achieves the benefits of lower stator flux tracking error, reduced torque ripple, and excellent dynamic performance under parameter mismatch conditions.

References

[1] E. Levi, Advances in converter control and innovative exploitation of additional degrees of freedom for multiphase machines, IEEE Trans. Ind. Electron. 63 (01) (2016) 433–448.

[2] J. Zhang, et al., Novel prescribed performance-tangent barrier Lyapunov function for neural adaptive control of the chaotic PMSM system by backstepping, Int J Electr Power Energy Syst (2020) 163–174. doi:10.1016/j.ijepes.2020.105991.

[3] S. Lin, W. Zhang, An adaptive sliding-mode observer with a tangent function-based PLL structure for position sensorless PMSM drives, Int J Electr Power Energy Syst (2017) 63–74. doi:10.1016/j.ijepes.2016.12.006.

[4] A. Abduallah, O. Dordevic, et al., Regenerative Test for Multiple Three-Phase Machines With Even Number of Neutral Points, IEEE Trans. Ind. Electron. 67 (03) (2020) 1684–1694.

[5] G. Wu, S. Huang, et al., Predictive Torque and Stator Flux Control for N*3-Phase PMSM Drives With Parameter Robustness Improvement, IEEE Trans. Power Electron. 36 (02) (2021) 1970–1983.

[6] F. Barrero, M.J. Duran, Recent Advances in the Design, Modeling, and Control of Multiphase Machines—Part I, IEEE Trans. Ind. Electron. 63 (01) (2016) 449–458.

[7] M. Onsal, Y. Demir, et al., A New Nine-Phase Permanent Magnet Synchronous Motor With Consequent Pole Rotor for High-Power Traction Applications, IEEE Trans. Magn. 53 (11) (2017) 1–6.

[8] X. Han, D. Jiang, et al., Two-Segment Three-Phase PMSM Drive With Carrier Phase-Shift PWM for Torque Ripple and Vibration Reduction, IEEE Trans. Power Electron. 34 (01) (2018) 588–599.

[9] Z. Zhou, C. Xia, et al., Torque Ripple Minimization of Predictive Torque Control for PMSM With Extended Control Set, IEEE Trans. Ind. Electron. 64 (09) (2017) 6930–6939.

[10] S. Rubino, R. Bojoi, et al., Model Predictive Direct Flux Vector Control of Multi-three-Phase Induction Motor Drives, IEEE Trans. Ind. Appl. 54 (05) (2018) 4394–4404.

[11] M. Siami, D.A. Khaburi, et al., Torque Ripple Reduction of Predictive Torque Control for PMSM Drives With Parameter Mismatch, IEEE Trans. Power Electron. 32 (09) (2017) 7160–7168.

[12] Y. Zhou, S. Zhang, et al., Current Prediction Error Based Parameter Identification Method for SPMSM With Deadbeat Predictive Current Control, IEEE Trans. Energy Convers. 36 (03) (2021) 1700–1710.

[13] X. Zhang, L. Zhang, et al., Model Predictive Current Control for PMSM Drives With Parameter Robustness Improvement, IEEE Trans. Power Electron. 34 (02) (2019) 1645–1657.

[14] C. Xia, M. Wang, et al., Robust Model Predictive Current Control of Three-Phase Voltage Source PWM Rectifier With Online Disturbance Observation, IEEE Trans. Ind. Inform. 8 (03) (2012) 459–471.

[15] X. Li, W. Tian, et al., A Generalized Observer-Based Robust Predictive Current Control Strategy for PMSM Drive System, IEEE Trans. Ind. Electron. 69 (02) (2022) 1322–1332.

[16] X. Yang, H. Hu, et al., A Quasi-Resonant Extended State Observer-Based Predictive Current Control Strategy for Three-Phase PWM Rectifier, IEEE Trans. Ind. Electron. 69 (12) (2022) 13910–13917.

[17] Z. Liu, Y. Wang, et al., An Approach to Suppress Low-Frequency Oscillation by Combining Extended State Observer With Model Predictive Control of EMUs Rectifier, IEEE Trans. Power Electron. 34 (10) (2019) 10282–10297.

[18] K. Yu, Z. Wang, et al., Robust Cascaded Deadbeat Predictive Control for Dual Three-Phase Variable-Flux PMSM Considering Intrinsic Delay in Speed Loop, IEEE Trans. Ind. Electron. 69 (12) (2022) 12107–12118.

[19] C. Bai, Z. Yin, et al., Multiple-Models Adaptive Disturbance Observer-Based Predictive Control for Linear Permanent-Magnet Synchronous Motor Vector Drive, IEEE Trans. Power Electron. 37 (08) (2022) 9596–9611.

[20] C. Bai, Z. Yin, et al., Robust Predictive Control for Linear Permanent Magnet Synchronous Motor Drives Based on an Augmented Internal Model Disturbance Observer, IEEE Trans. Ind. Electron. 69 (10) (2022) 9771–9782.

[21] Y. Zhang, J. Jin, et al., Model-Free Predictive Current Control of PMSM Drives Based on Extended State Observer Using Ultralocal Model, IEEE Trans. Ind. Electron. 68 (02) (2021) 993–1003.

[22] Y. Zhang, T. Jiang, et al., Model-Free Predictive Current Control of DFIG Based on an Extended State Observer Under Unbalanced and Distorted Grid, IEEE Trans. Power Electron. 35 (08) (2020) 8130–8139.

[23] X. Yuan, Y. Zuo, et al., Model-Free Predictive Current Control of SPMSM Drives Using Extended State Observer, IEEE Trans. Ind. Electron. 69 (07) (2022) 6540–6550.

[24] O. Babayomi, Z. Zhang, Model-Free Predictive Control of Power Converters with Multi-Frequency Extended State Observers, IEEE Trans. Ind. Electron. 70 (11) (2023) 11379–11389.

[25] O. Babayomi, Z. Zhang, et al., Model-Free Predictive Control of Power Converters with Cascade-Parallel Extended State Observers, IEEE Trans. Ind. Electron. 70 (10) (2023) 10215–10226.

[26] Z. Sun, Y. Deng, et al., Finite Control Set Model-Free Predictive Current Control of PMSM With Two Voltage Vectors Based on Ultralocal Model, IEEE Trans. Power Electron. 38 (01) (2023) 776–788.

[27] M.S. Mousavi, S.A. Davari, et al., Computationally Efficient Model-Free Predictive Control of Zero-Sequence Current in Dual Inverter Fed Induction Motor, IEEE J. Emerg. Sel. Top. Power Electron. 11 (02) (2023) 1332–1344.

CHAPTER 12

Predictive torque and stator flux control for modular multiphase PMSM drives with parameter robustness improvement

12.1 Introduction

Multiphase permanent magnet synchronous motors (multiphase PMSMs) have been widely adopted in many specialized applications, ranging from high-power traction vehicles to marine propulsion, due to their outstanding advantages, including better fault-tolerant capability and high power and torque densities [1–3]. To optimize the performance of multiphase PMSMs, a novel multi-three-phase PMSM (modular multiphase PMSM) with a special mechanical degree winding placement is investigated in [4–6]. The main benefits of this modular multiphase PMSM are its extremely low cogging and higher torque quality compared with conventional multiphase PMSMs. Moreover, the controller design of modular multiphase PMSMs is simpler and can be fed by several independent three-phase motor drives [4,5]. In [4–6], PI controllers were adopted to regulate the modular multiphase PMSM currents, which can reduce stator current harmonic distortion and achieve low torque ripple under steady-state conditions. However, the dynamic and static control performance of modular multiphase PMSMs is highly dependent on the PI controller gains.

Predictive torque control (PTC) offers the advantages of excellent dynamic and steady-state performance compared to the conventional PI controller, which makes it a powerful modular multiphase PMSM control strategy across the entire operating range [7–10]. In the PTC method, all possible voltage vectors of the converter are calculated using the predicted values of stator flux and torque. These predictions are determined by the mathematical model of the modular multiphase PMSM, which depends on the rotor flux linkage parameter and the inductance parameter. However, these parameters utilized in the PTC may not match the actual modular multiphase PMSM parameters due to parameter identification errors, or

Robust Model Predictive Control of Permanent Magnet Synchronous Motor Under Parameter Mismatch
https://doi.org/10.1016/B978-0-443-44209-4.00003-8

the parameters may change during motor operation [11]. Model parameter mismatches result in unstable switching frequency and inaccurate prediction behavior, which affect the performance of the PTC, i.e., torque/flux ripple increases [12].

Designing a disturbance observer is the main approach to improving the robustness of the model predictive controller under parameter mismatch. In [13,14], a robust cascaded nonlinear predictive controller based on a disturbance observer was proposed, which can achieve precise speed trajectory tracking and disturbance rejection in the presence of parameter mismatch and external perturbations. In [15,16], an improved deadbeat-based predictive current control method based on estimated disturbances was designed, which can achieve good dynamic performance and model parameter mismatch suppression. In [17], an improved predictive current control method with an incremental model was developed, which can operate without requiring knowledge of the rotor flux and improve robustness against inductance parameter mismatch by using an inductance disturbance observer. In [18], a novel linear quadratic model predictive control method based on a nonlinear disturbance observer was proposed, which can minimize torque ripple caused by parameter mismatch. In [19], a robust fault-tolerant predictive current control method based on a composite observer was proposed, which can eliminate the influence of permanent magnet demagnetization and parameter perturbation. In [13–19], the predictive control methods based on disturbance observer compensation rely on the accurate measurement of voltage and current.

Some modified predictive control methods without observer compensation have been presented to eliminate the influence of parameter mismatch. Instead of using disturbance observer-based voltage feedforward methods, a deadbeat predictive current control method with closed-form error compensation was proposed in [20], which can directly compensate for the current tracking errors caused by parameter mismatch. In [21], the robust model predictive control scheme with simplified repetitive control was presented, which can guarantee stability and maintain high-accuracy tracking under parameter mismatch and current distortion. In [22,23], a modified PTC algorithm with a proportional controller was proposed, which can improve prediction accuracy and achieve perfect prediction performance under inductance mismatch. In [24], a model-free predictive current control method based on a current difference detection technique was proposed, which is easy to implement due to its simplicity and independence from motor parameters. Although the model-free predictive control is quite independent of the motor model, it is strongly dependent on the accuracy of current references and current responses.

Since the parameter mismatch leads to inaccurate predictions of torque/flux and deteriorates the control performance, a novel parameter-immune predictive torque and stator flux control method is proposed in this chapter to guarantee the control performance of modular multiphase PMSM drives with parameter mismatch. The proposed PTSF method can enhance robustness against parameter mismatch by adopting the closed-loop control of *d-q* axis stator flux. Compared with the conventional PTC method, the proposed PTSF method can effectively eliminate the stator flux tracking errors and reduce the torque/flux ripple under parameter mismatch. This paper is organized as follows: the mathematical model of modular multiphase PMSM is established in Section 12.2. In Section 12.3, the robust predictive stator flux controller with improved parameter robustness is designed. The robust predictive torque controller with improved parameter robustness is proposed in Section 12.4. The effectiveness of the proposed PTSF algorithm is validated through simulations in Section 12.5. Section 12.6 concludes this chapter.

12.2 Mathematical model of modular multiphase PMSM

Taking a 6*3-phase PMSM as an example, the structure diagram of the modular multiphase PMSM studied in this chapter is shown in Fig. 12.1. The modular multiphase PMSM is composed of 6*3-phase PMSM units with the same characteristics. The stator winding of the modular multiphase PMSM consists of six segmented three-phase PMSM windings, and each set of three-phase windings has a phase difference of 180 electrical degrees. Note that there are electrical, magnetic, and thermal isolation characteristics among the three-phase PMSM unit windings.

The voltage equations of the modular multiphase PMSM in a synchronous rotating frame are as [4–6]

$$\begin{cases} \dfrac{di_{dj}}{dt} = -\dfrac{R}{L} i_{dj} + \omega_e i_{qj} + \dfrac{1}{L} u_{dj} \\ \dfrac{di_{qj}}{dt} = -\dfrac{R}{L} i_{qj} - \omega_e i_{dj} - \dfrac{\psi_r}{L} \omega_e + \dfrac{1}{L} u_{qj} \end{cases} \tag{12.1}$$

where j stands for any unit of the motor; u_{dj} and u_{qj} are the d- and q-axis voltages of the jth winding; i_{dj} and i_{qj} are the d- and q-axis currents of the

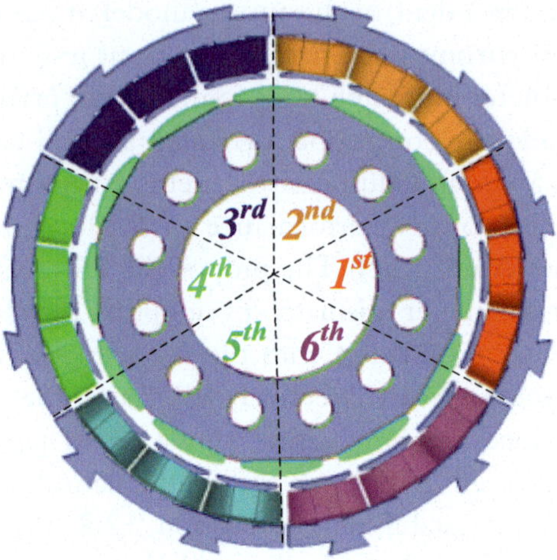

Fig. 12.1 The modular multiphase PMSM structure diagram.

jth winding; R and L are the stator resistance and stator inductance; ω_e is the electrical rotor speed; ψ_r is the flux linkage of permanent magnets.

The electromagnetic torque equation of a modular multiphase PMSM can be expressed as [4–6]

$$T_e = \frac{3n_p}{2}\sum_{j=1}^{N}(\psi_r i_{qj}) = \frac{3Nn_p}{2}\psi_r i_{qj} \tag{12.2}$$

The mechanical dynamic model of a modular multiphase PMSM can be described as

$$T_e - T_L = \frac{J}{n_p}\frac{d\omega_e}{dt} \tag{12.3}$$

where n_p is the number of pole pairs; J is the moment of inertia; and T_e and T_L are the electromagnetic torque and load torque of the N*3-phase PMSM, respectively.

12.3 Robust predictive stator flux control with parameter robustness improvement

According to the special winding structure and characteristics of the modular multiphase PMSM, the drive system structure is designed, as shown in Fig. 12.2. The modular multiphase PMSM is driven by 6*3-phase voltage source inverter structures (VSIs), and each set of windings is

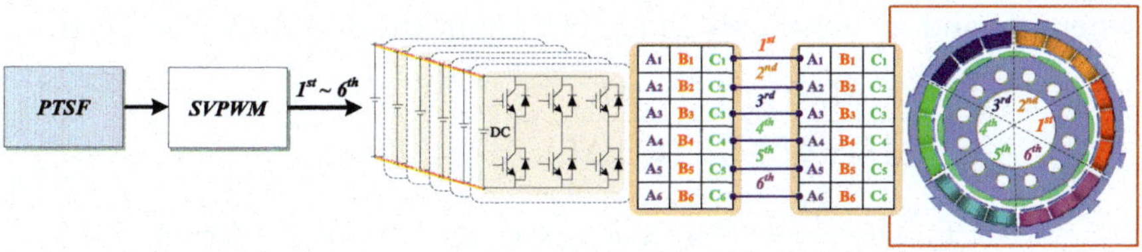

Fig. 12.2 Integral drive system diagram of a modular multiphase PMSM.

independently excited by a three-phase VSI structure. Through the characteristics analysis of the 6*3-phase PMSM, it can be observed that the vector control of the 6*3-phase PMSM is similar to that of the conventional three-phase PMSM [4–6]. In this chapter, a novel PTSF method is used to send out six groups of identical PWM driving signals, which can achieve the control of 6*3-phase VSI structures under parameter mismatches.

Design of predictive stator flux controller

This chapter mainly studies the novel robust predictive control under inductance and flux parameter mismatch. Introducing $L = L_o + \Delta L$ and $\psi_r = \psi_{ro} + \Delta\psi_r$, where L_o and ψ_{ro} are actual values, ΔL and $\Delta\psi_r$ are uncertain components of the corresponding parameter. According to Eq. (12.1), we can express the voltage equations of the modular multiphase PMSM as follows under inductance and flux parameters mismatch.

$$\begin{cases} \dfrac{di_{dj}}{dt} = -\dfrac{R_o}{L_o+\Delta L} i_{dj} + \omega_e i_{qj} + \dfrac{1}{L_o+\Delta L} u_{dj} \\ \dfrac{di_{qj}}{dt} = -\dfrac{R_o}{L_o+\Delta L} i_{qj} - \omega_e i_{dj} - \dfrac{\psi_{ro}+\Delta\psi_r}{L_o+\Delta L}\omega_e + \dfrac{1}{L_o+\Delta L} u_{qj} \end{cases} \tag{12.4}$$

According to Eq. (12.4), the stator flux equations of the modular multiphase PMSM can be obtained as follows under inductance and flux parameter mismatches.

$$\begin{cases} \dfrac{d\psi_{dj}}{dt} = u_{dj} - \dfrac{R_o}{L_o}\psi_{dj} + \omega_e \psi_{qj} + \dfrac{R_o}{L_o}\psi_{ro} + f_{dj} \\ \dfrac{d\psi_{qj}}{dt} = u_{qj} - \omega_e \psi_{dj} - \dfrac{R_o}{L_o}\psi_{qj} + f_{qj} \\ \psi_{dj} = L_o i_{dj} + \psi_{ro} \\ \psi_{qj} = L_o i_{qj} \end{cases} \tag{12.5}$$

where f_{dj} and f_{qj} are the d- and q-axis unknown disturbance, respectively.

$$\begin{cases} f_{dj} = \dfrac{\Delta L R_o}{(L_o + \Delta L)} i_{dj} - \dfrac{\Delta L}{(L_o + \Delta L)} u_{dj} \\ f_{qj} = \dfrac{\Delta L}{(L_o + \Delta L)} (-u_{qj} + R_o i_{qj} + \psi_{ro}\omega_e + \Delta\psi_r\omega_e) - \Delta\psi_r\omega_e \end{cases} \tag{12.6}$$

By discretizing the flux equation (Eq. 12.5), the discrete flux equations of the modular multiphase PMSM can be obtained as

$$\begin{cases} \boldsymbol{\psi}_j(k+1) = \mathbf{u}_j(k)T_s - \mathbf{A}(k)\boldsymbol{\psi}_j(k)T_s + \mathbf{B}T_s + \mathbf{f}(k)T_s \\ \boldsymbol{\psi}_j(k) = \mathbf{D}\mathbf{i}_j(k) + \mathbf{f}_\psi \end{cases} \tag{12.7}$$

where $\mathbf{u}_j(k) = \begin{bmatrix} u_{dj}(k) \\ u_{qj}(k) \end{bmatrix}$, $\psi_j(k) = \begin{bmatrix} \psi_{dj}(k) \\ \psi_{qj}(k) \end{bmatrix}$, $\mathbf{i}_j(k) = \begin{bmatrix} i_{dj}(k) \\ i_{qj}(k) \end{bmatrix}$,

$$\mathbf{A}(k) = \begin{bmatrix} \left(\dfrac{R_o}{L_o} - \dfrac{1}{T_s}\right) & -\omega_e(k) \\ \omega_e(k) & \left(\dfrac{R_o}{L_o} - \dfrac{1}{T_s}\right) \end{bmatrix}, \mathbf{B} = \begin{bmatrix} \dfrac{R_o}{L_o}\psi_{ro} \\ 0 \end{bmatrix}, \mathbf{f}_\psi = \begin{bmatrix} \psi_{ro} \\ 0 \end{bmatrix},$$

$$\mathbf{f}(k) = \begin{bmatrix} f_{dj}(k) \\ f_{qj}(k) \end{bmatrix}, \mathbf{D} = \begin{bmatrix} L_o & 0 \\ 0 & L_o \end{bmatrix}.$$

According to Eq. (12.7), the ideal output voltage vector of the predictive stator flux controller can be expressed as follows when inductance and flux parameter mismatches are considered.

$$\mathbf{u}_j^{ref}(k) = \frac{1}{T_s}\psi_j^{ref}(k+1) + \mathbf{A}(k)\psi_j(k) - \mathbf{B} - \mathbf{f}(k) \tag{12.8}$$

where $\mathbf{u}_j^{ref}(k)$ is the ideal output voltage vector of predictive stator flux controller and $\psi_j^{ref}(k+1)$ is the stator flux reference.

Parameter sensitivity analysis of predictive stator flux controller

The parameter sensitivity of the predictive stator flux controller is evaluated in this section. According to Eq. (12.7), the actual flux equations of the modular multiphase PMSM can be expressed as:

$$\psi_j(k+1) = \mathbf{u}_j(k)T_s - \mathbf{A}(k)\psi_j(k)T_s + \mathbf{B}T_s \tag{12.9}$$

The ideal output voltage vector of the predictive stator flux controller is applied to the actual flux equations of the modular multiphase PMSM under parameter mismatch. Therefore substituting Eq. (12.8) into Eq. (12.9) yields

$$\psi_j(k+1) = \psi_j^{\mathrm{ref}}(k+1) - \mathbf{f}(k)T_s \tag{12.10}$$

The stator flux predictive error is defined as.

$$\Delta\psi_j(k+1) = \psi_j(k+1) - \psi_j^{ref}(k+1) = -\mathbf{f}(k)T_s \tag{12.11}$$

with

$$\mathbf{f}(k)T_s = \begin{bmatrix} \Delta\psi_d \\ \Delta\psi_q \end{bmatrix} = \begin{bmatrix} f_{dj}(k)T_s \\ f_{qj}(k)T_s \end{bmatrix} \tag{12.12}$$

According to Eqs. (12.1) and (12.2), the voltage stationary equations can be expressed as follows when the modular multiphase PMSM reaches a stable state [19].

$$\begin{cases} u_{dj}(k) = R_o i_{dj}(k) - L_o\omega_e i_{qj}(k) \\ u_{qj}(k) = R_o i_{qj}(k) + L_o\omega_e i_{dj}(k) + \omega_e\psi_{ro} \\ T_e(k) = \dfrac{3Nn_p}{2}\psi_{ro} i_{qj}(k) \end{cases} \tag{12.13}$$

Under $i_{dj}^{ref}(k) = 0$, combining with Eqs. (12.6), (12.12), and (12.13), the *d*- and *q*-axis stator flux predictive errors can be simplified as

$$\begin{cases} \Delta\psi_d = -\dfrac{\chi_1 \cdot L_o}{(1+\chi_1)} \cdot \dfrac{2T_e(k)}{3Nn_p\psi_{ro}}\omega_e(k)T_s \\ \Delta\psi_q = \dfrac{\chi_2 \cdot \psi_{ro}}{(1+\chi_1)}\omega_e(k)T_s \\ \chi_1 = \Delta L / L_o \\ \chi_2 = \Delta\psi_r / \psi_{ro} \end{cases} \tag{12.14}$$

The modular multiphase PMSM parameters are shown in Table 12.1. The basic value of the flux linkage is 0.892 Wb. The simulation condition is: the load torque increases from 0 to 2000 Nm; the speed increases from 0 to 800 rad/s. According to Eq. (12.14), the *d*- and *q*-axis stator flux errors are plotted under parameters mismatch, as shown in Figs. 12.3 and 12.4. Fig. 12.3 shows the *d*- and *q*-axis stator flux predictive errors under inductance parameter mismatch. The *q*-axis stator flux predictive error is 0 Wb

Table 12.1 Main parameters of the modular multiphase PMSM.

Parameter	Value	Parameter	Value
Inductance (L_o)	3.572/mH	Stator phase resistance (R_o)	0.02/Ω
Flux linkage of PM (Ψ_{ro})	0.892/Wb	Rotational inertia (J)	100/kg.m2

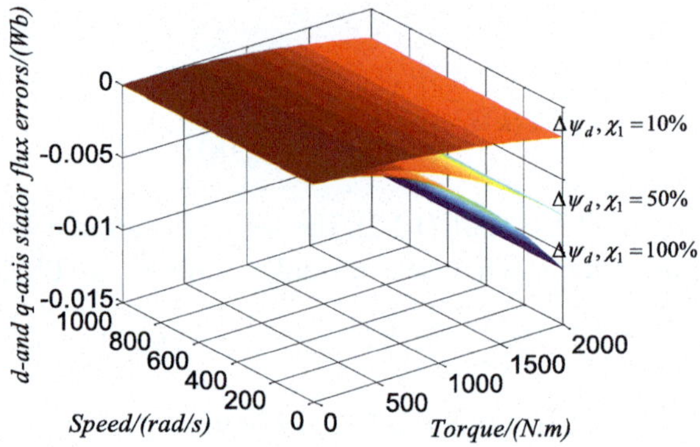

Fig. 12.3 The *d*- and *q*-axis stator flux predictive errors under inductance parameter mismatch.

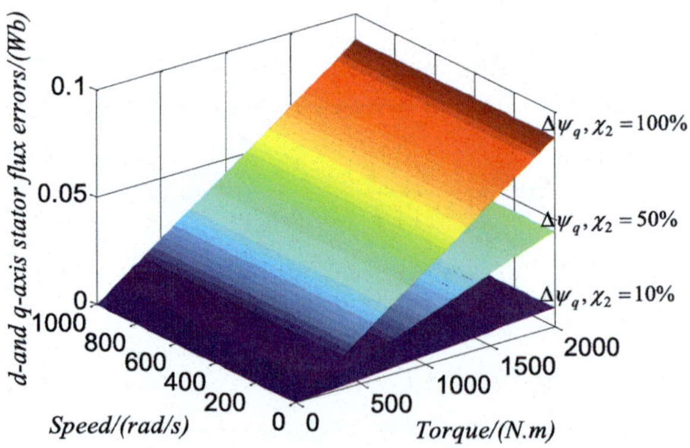

Fig. 12.4 The *d*- and *q*-axis stator flux predictive errors under flux linkage parameter mismatch.

under inductance parameter mismatch, which means that the inductance parameter mismatch does not affect the *q*-axis stator flux. At a speed of 800 rad/s, the *d*-axis stator flux predictive error is as low as $-8e^{-3}$Wb under inductance parameter mismatch ($\chi_1 = 100\%$). It can be inferred that the influence of the inductance parameter mismatch on the q-axis stator flux is very small. Fig. 12.4 shows the d- and q-axis stator flux predictive errors under flux parameter mismatch. When a flux parameter mismatch occurs, the d-axis stator flux predictive error is always 0 Wb, and the q-axis stator flux predictive error is 0.08 Wb. It can be inferred that flux parameter mismatch does not have considerable effect on d- and q-axis stator flux. Through the results described above, it is known that the predictive stator flux controller is immune to the modular multiphase PMSM parameters, and the parameters mismatch has little effect on the control performance of modular multiphase PMSM.

Predictive stator flux control with one-step delay compensation

Due to the control delay in the actual digital system, there is a one-beat delay between the actual output voltage vector and the ideal voltage vector of the predictive controller. To eliminate the influence of one beat delay on the predictive controller, it is necessary to calculate the output voltage vector one control cycle in advance. Fig. 12.5 demonstrates the one beat delay compensation in the predictive controller. The specific steps are as follows: First, at the kT_s moment, the stator flux value $\psi_j(k+1)$ of $(k+1)T_s$ moment should be predicted. Then, the predicted stator flux value $\psi_j(k+1)$ is used to calculate the ideal voltage vector $\mathbf{u}_j(k+1)$ at the kT_s moment.

At the kT_s moment, the stator flux value $\psi_j(k+1)$ of $(k+1)T_s$ moment can be predicted according to Eq. (12.9).

$$\psi_j(k+1) = \mathbf{u}_j(k)T_s - \mathbf{A}(k)\psi_i(k)T_s + \mathbf{B}T_s \tag{12.15}$$

According to Eq. (12.15), the voltage vector of $(k+1)T_s$ moment can be calculated as

$$\mathbf{u}_j(k+1) = \frac{1}{T_s}\psi_j(k+2) + \mathbf{A}(k+1)\psi_j(k+1) - \mathbf{B} \tag{12.16}$$

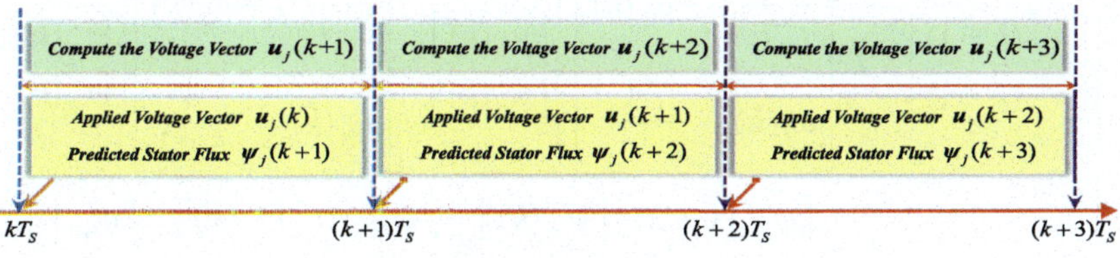

Fig. 12.5 Logic diagram of a beat delay compensation.

Note that the electrical time is far less than the mechanical time, the speed can be considered as unchanged in a control cycle (i.e., $\omega_e(k) \approx \omega_e(k+1)$), then $\mathbf{A}(k) \approx \mathbf{A}(k+1)$. At the kT_s moment, the output voltage vector of predictive stator flux controller can be obtained by substituting Eq. (12.15) into Eq. (12.16).

$$\mathbf{u}_j(k+1) = \frac{1}{T_s}\psi_j(k+2) + T_s\mathbf{A}(k)\left[\mathbf{u}_j(k) - \mathbf{A}(k)\psi_j(k) + \mathbf{B}\right] - \mathbf{B} \tag{12.17}$$

From Eq. (12.17), it is known that $\mathbf{u}_j(k)$, $\psi_j(k)$, $\mathbf{A}(k)$, and $\mathbf{B}$ are known quantities at the kT_s moment. Therefore the voltage vector of $(k+1)T_s$ moment can be calculated by equation Eq. (12.17).

There exists $\frac{R}{L} << \frac{1}{T_s}$, then $\mathbf{A}(k) \approx \begin{bmatrix} -\frac{1}{T_s} & -\omega_e(k) \\ \omega_e(k) & -\frac{1}{T_s} \end{bmatrix}$, one yields

$$\begin{bmatrix} u_{dj}(k+1) \\ u_{qj}(k+1) \end{bmatrix} = \begin{bmatrix} -1 & -\omega_e(k)T_s \\ \omega_e(k)T_s & -1 \end{bmatrix}$$
$$\left[\begin{bmatrix} u_{dj}(k) \\ u_{qj}(k) \end{bmatrix} - \frac{1}{T_s}\begin{bmatrix} -1 & -T_s\omega_e(k) \\ T_s\omega_e(k) & -1 \end{bmatrix}\begin{bmatrix} \psi_{dj}(k) \\ \psi_{qj}(k) \end{bmatrix} + \begin{bmatrix} \frac{R_o}{L_o}\psi_{ro} \\ 0 \end{bmatrix}\right]$$
$$+\frac{1}{T_s}\begin{bmatrix} \psi_{dj}(k+2) \\ \psi_{qj}(k+2) \end{bmatrix} - \begin{bmatrix} \frac{R_o}{L_o}\psi_{ro} \\ 0 \end{bmatrix} \tag{12.18}$$

In addition, the amplitude of the voltage source is much larger than the value of stator resistance voltage when the modular multiphase PMSM operates stably $\frac{R_o}{L_o}\psi_{ro} << u_{dj}(k)$. Therefore Eq. (12.18) is modified to

$$\begin{cases} \begin{bmatrix} u_{dj}^{ref}(k+1) \\ u_{qj}^{ref}(k+1) \end{bmatrix} = T_s\begin{bmatrix} -\frac{1}{T_s} & -\omega_e(k) \\ \omega_e(k) & -\frac{1}{T_s} \end{bmatrix}\left\{\begin{bmatrix} u_{dj}(k) \\ u_{qj}(k) \end{bmatrix} - \begin{bmatrix} -\frac{1}{T_s} & -\omega_e(k) \\ \omega_e(k) & -\frac{1}{T_s} \end{bmatrix}\begin{bmatrix} \psi_{dj}(k) \\ \psi_{qj}(k) \end{bmatrix}\right\} \\ \quad +\frac{1}{T_s}\begin{bmatrix} \psi_{dj}^{ref}(k+2) \\ \psi_{qj}^{ref}(k+2) \end{bmatrix}\begin{bmatrix} \psi_{dj}(k) \\ \psi_{qj}(k) \end{bmatrix} = \begin{bmatrix} L_o & 0 \\ 0 & L_o \end{bmatrix}\begin{bmatrix} i_{dj}(k) \\ i_{qj}(k) \end{bmatrix} + \begin{bmatrix} \psi_{ro} \\ 0 \end{bmatrix} \end{cases} \tag{12.19}$$

12.4 Robust PTC with parameter robustness improvement

Design of a robust predictive torque controller

According to Eq. (12.19), it is known that the key to realizing robust predictive stator flux control method is to obtain the d- and q-axis stator flux references value (i.e., $\psi_{dj}^{ref}(k+2), \psi_{qj}^{ref}(k+2)$). To calculate the accurate stator flux references value, the robust predictive torque controller is designed.

According to Eq. (12.2), the d- and q- axis stator flux linkage references can be calculated by using the Lagrange extrapolation under $i_{dj}^{ref}(k)=0$, expressed as follows

$$\begin{cases} \psi_{dj}^{ref}(k+2)=\psi_{dj}^{ref}(k+1)=\psi_{dj}^{ref}(k)=\psi_{ro}+L_o i_{dj}^{ref}(k)=\psi_{ro} \\ \psi_{qj}^{ref}(k+2)=3\psi_{qj}^{ref}(k)-3\psi_{qj}^{ref}(k-1)+\psi_{qj}^{ref}(k-2)=\dfrac{2L_o T_e^{ref}}{3Nn_p\psi_{ro}} \end{cases} \quad (12.20)$$

where T_e^{ref} is the torque reference.

According to Eq. (12.2), the electromagnetic torque equation can be rewritten as follows when parameter mismatches are considered.

$$\tilde{T}_e=\frac{3n_p}{2}\sum_{j=1}^{N}((\psi_{ro}+\Delta\psi_r)i_{qj})=T_e+\Delta T_e \quad (12.21)$$

where ΔT_e is the unknown torque disturbance caused by flux parameter mismatch, and $\Delta T_e=\dfrac{3Nn_p\Delta\psi_r}{2}i_{qj}$.

Substituting Eq. (12.21) into Eq. (12.3), we obtain

$$\frac{d\omega_e}{dt}=\frac{n_p}{J}(T_e+\Delta T_e-T_L) \quad (12.22)$$

According to Eq. (12.22), the speed equation in discrete-time form can be expressed as follows under parameter mismatch.

$$\omega_e(k+1)=\omega_e(k)+\frac{n_p}{J}T_d T_e(k)+\frac{n_p}{J}T_d\Delta T_e(k)-\frac{n_p}{J}T_d T_L \quad (12.23)$$

where T_d represents the speed sampling time, and $T_d=10T_s$.

The torque reference can be provided based on Eq. (12.23) as

$$T_e^{ref}=\frac{\omega_e^{ref}-\omega_e(k)-\frac{n_p}{J}T_d\Delta T_e(k)+\frac{n_p}{J}T_dT_L}{\frac{n_p}{J}T_d} \tag{12.24}$$

According to Eqs. (12.2) and (12.3), the actual speed equation can be expressed as

$$\omega_e(k+1)=\omega_e(k)+\frac{n_p}{J}T_dT_e(k)-\frac{n_p}{J}T_dT_L \tag{12.25}$$

Substituting Eq. (12.24) into Eq. (12.25), the speed predictive error can be obtained under steady-state operation.

$$\Delta\omega_e=\omega_e(k+1)-\omega_e^{ref}=-\frac{n_p}{J}T_d\Delta T_e(k)=-\frac{\chi_2 n_p}{J}T_dT_e(k) \tag{12.26}$$

The modular multiphase PMSM parameters are shown in Table 12.1. According to Eq. (12.26), the speed predictive error is plotted under flux linkage parameter mismatch, as shown in Fig. 12.6. From Fig. 12.6, it is evident that the influence of flux parameter mismatch on PTC is very small, with the maximum speed error being only 0.08 rad/s.

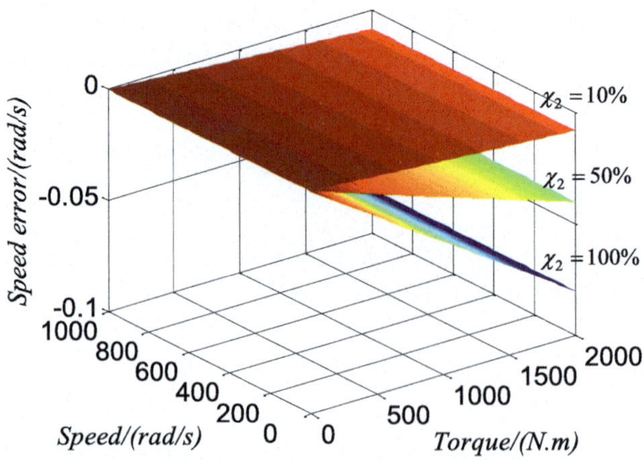

Fig. 12.6 Speed predictive error under flux parameter mismatch.

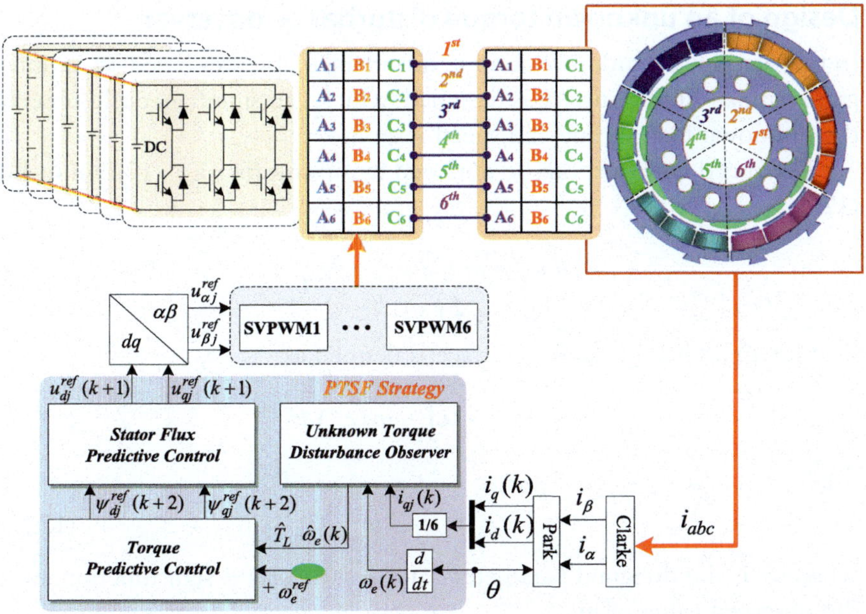

Fig. 12.7 Structural diagram of the modular multiphase PMSM drive system with the proposed PTSF method.

To enhance robustness against sudden changes in load torque, the speed feedback value is replaced by the observed speed value. Thus the robust predictive torque controller is designed as

$$T_e^{ref} = \frac{\omega_e^{ref} - \hat{\omega}_e(k) + \frac{n_p}{J} T_d \hat{T}_L}{\frac{n_p}{J} T_d} \tag{12.27}$$

where $\hat{T}_L$ is the observed values of T_L; $\hat{\omega}_e$ is the observed value of ω_e.

The structural diagram of the modular multiphase PMSM drive system with the proposed PTSF method is illustrated in Fig. 12.7. Through the above analysis, it is evident that the stator flux loop and torque loop of the modular multiphase PMSM control system are immune to motor parameters, enabling good control performance even under parameter mismatches.

Design of an unknown torque disturbance observer

In this section, the unknown torque disturbance observer based on an integral terminal sliding surface is designed, which can simultaneously observe the unknown load torque and speed response.

The speed equation can be obtained by substituting Eq. (12.2) into Eq. (12.3).

$$\frac{d\omega_e}{dt}=\frac{3Nn_p^2\psi_{ro}}{2J}i_{qj}-\frac{n_p}{J}T_L \tag{12.28}$$

The integral terminal sliding surface is defined as

$$s_\omega=e_\omega+\lambda\int_0^t \mathrm{sgn}(e_\omega)\,d\tau \tag{12.29}$$

with

$$e_\omega=\hat{\omega}_e-\omega_e \tag{12.30}$$

where λ is the designed parameters, and $\lambda>0$; *sgn*() is sign function; $\hat{\omega}_e$ is the observed value of ω_e.

To achieve chattering suppression in the sliding mode observer, a hyperbolic tangent sign function is given as

$$\mathrm{sgn}(v)=\tanh(v)=\frac{e^{2v}-1}{e^{2v}+1} \tag{12.31}$$

where v is any function variable.

According to Eq. (12.28), the sliding mode observer based on the integral terminal sliding surface is designed as

$$\frac{d\hat{\omega}_e}{dt}=\frac{3Nn_p^2\psi_{ro}}{2J}i_{qj}-U_\omega \tag{12.32}$$

where U_ω is the sliding mode control function.

Combining Eqs. (12.28) and (12.32), and taking the time derivative of Eq. (12.29), one yields

$$ds_\omega/dt=\frac{n_p}{J}T_L-U_\omega+\lambda\,\mathrm{sgn}(e_\omega) \tag{12.33}$$

According to Eq. (12.33), the sliding mode control function is designed as follows to guarantee the convergence of the sliding mode observer.

$$U_\omega=\lambda\,\mathrm{sgn}(e_\omega)+\eta s_\omega+\delta\,\mathrm{sgn}(s_\omega) \tag{12.34}$$

where η and δ are the designed observer sliding gain.

The following Lyapunov function is considered to prove the asymptotic stability of the proposed sliding-mode observer.

$$\frac{dV}{dt} = s_\omega \cdot \frac{ds_\omega}{dt} \tag{12.35}$$

Substituting Eqs. (12.33) and (12.34) into Eq. (12.35) yields

$$\begin{aligned} \frac{dV}{dt} &= s_\omega \cdot \frac{ds_\omega}{dt} = s_\omega\left(\frac{n_p}{J}T_L - U_\omega + \lambda\,\mathrm{sgn}(e_\omega)\right) \\ &= s_\omega\left(\frac{n_p}{J}T_L - \eta s_\omega - \delta\,\mathrm{sgn}(s_\omega)\right) \\ &= -\eta s_\omega^2 + \left(\frac{n_p}{J}s_\omega T_L - \delta \,\| s_\omega \|\right) \end{aligned} \tag{12.36}$$

The unknown load torque disturbance is bounded in engineer, that are $\|\hat{T}_L\| \leq N_\omega$, where N_ω is the boundary values. When η and δ are set to $\eta \geq 0$ and $\delta \geq \frac{n_p}{J}N_\omega$, then $\frac{dV}{dt} \leq 0$. Therefore the proposed sliding-mode observer can reach the integral terminal sliding surface and maintain stable, that is, $e_\omega = \frac{de_\omega}{dt} = 0$, $s_\omega = \frac{ds_\omega}{dt} = 0$. According to Eq. (12.33), the observed value of unknown disturbance can be obtained as

$$\hat{T}_L = \frac{J}{n_p}U_\omega(k) \tag{12.37}$$

If the sampling period T_s is short enough, the discrete forms of Eq. (12.32) can be deduced as according to Eq. (12.34).

$$\begin{aligned} \hat{\omega}_e(k+1) = \hat{\omega}_e(k) + \frac{3Nn_p^2\psi_{ro}}{2J}T_s i_{qj}(k) - \lambda T_s\,\mathrm{sgn}(e_\omega(k)) \\ -\eta T_s s_\omega(k) - \delta T_s\,\mathrm{sgn}(s_\omega(k)) \end{aligned} \tag{12.38}$$

12.5 Simulation results

The modular multiphase PMSM drive system based on PTSF method is established in MATLAB/Simulink. The main parameters of modular multiphase PMSM used in the MATLAB/Simulink are given in Table 12.1, and the sampling frequency is set as 10 vkHz. In addition, to guarantee the stability and convergence of the unknown torque disturbance observer, the observer gains are set as $\lambda = 5$, $\eta = 1000$, and $\delta = 200$.

Simulation for modular multiphase PMSM under inductance parameter mismatch

The operating conditions of the modular multiphase PMSM in Case A are that the load torque is maintained at 2000 N·m and the speed is maintained at 150 rad/s. At 1 s, there is a mismatch in the inductance parameter of the modular multiphase PMSM drive system, and the inductance value is stepped to 50% of its initial value, i.e., 1.786 mH. Comparative simulation results of the conventional PTC strategy and the proposed PTSF strategy under inductance parameter mismatch are shown in Figs. 12.8–12.12. Fig. 12.8 presents the comparative simulation results of the speed response under inductance parameter mismatch. From 0 to 1 s, it can be seen from Fig. 12.8 that the response speed value of the conventional PTC strategy is as good as that of the proposed PTSF strategy, and the speed static errors of the two methods are about 0.2 rad/s. After 1 s, the conventional PTC strategy exhibits the same superior speed control performance as the proposed PTSF strategy under inductance parameter mismatch, as shown clearly in Fig. 12.8B. In summary, it can be seen from Fig. 12.8 that the inductance parameter mismatch has little impact on the speed control performance of both the conventional PTC strategy and the proposed PTSF strategy.

Fig. 12.9 presents the comparative simulation results of the phase current and torque using the conventional PTC strategy and the proposed PTSF strategy under inductance parameter mismatch. From Fig. 12.9A, it can be observed that the conventional PTC strategy causes torque oscillation and stator current distortion in modular multiphase PMSM under

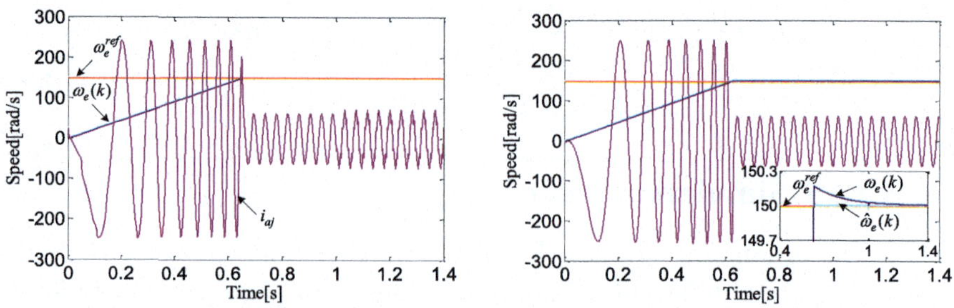

(a) Conventional PTC strategy (b) Proposed PTSF strategy

Fig. 12.8 Simulation results of the speed response under inductance parameter mismatch: (A) Conventional PTC strategy, (B) Proposed PTSF strategy.

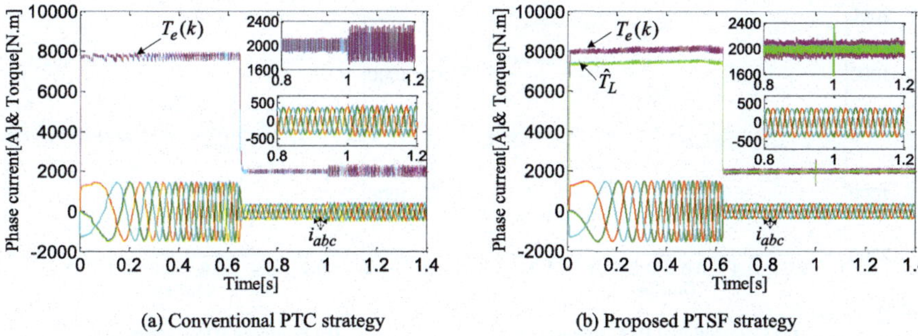

Fig. 12.9 Comparative simulation results of the torque and current under inductance parameter mismatch. (A) Conventional PTC strategy. (B) Proposed PTSF strategy.

inductance parameter mismatch. Since the stator flux response value of the proposed PTSF strategy can strictly follow the reference value and the inductance parameters are not required during control operation, the proposed PTSF strategy can significantly suppress torque oscillation and stator current distortion in modular multiphase PMSM under inductance parameter mismatch, as shown in Fig. 12.9B. From 0 to 0.4 s, it can be seen from Fig. 12.9B that there is a large estimation error between the observed value of load torque and the actual value of electromagnetic torque. The reason for the large estimation error is that the unknown torque disturbance observer does not reach the convergence state during the motor start-up process. After 0.4 s, it can be observed that the unknown torque disturbance observer based on the integral terminal sliding surface can accurately estimate the load torque, regardless of whether there is inductance parameter mismatch or the system is in normal condition. For the proposed PTSF strategy, note that there is a significant overshoot in the torque waveform of the modular multiphase PMSM. The reason for this phenomenon is that the unknown disturbance caused by inductance parameter mismatch is estimated by the unknown torque disturbance observer.

Fig. 12.10 presents the comparative simulation results of the three-phase stator current using the conventional PTC strategy and the proposed PTSF strategy under inductance parameter mismatch. From Fig. 12.10, it can be observed that the stator current fundamental value of the modular multiphase PMSM is approximately 312.2A, while that of the motor unit is about 52.04A, which is exactly six times. As shown in Fig. 12.10, due to the modular multiphase PMSM consisting of three motor units with consistent characteristics, the three-phase stator current of the modular multiphase PMSM is six times that of the unit motor. The reason for this

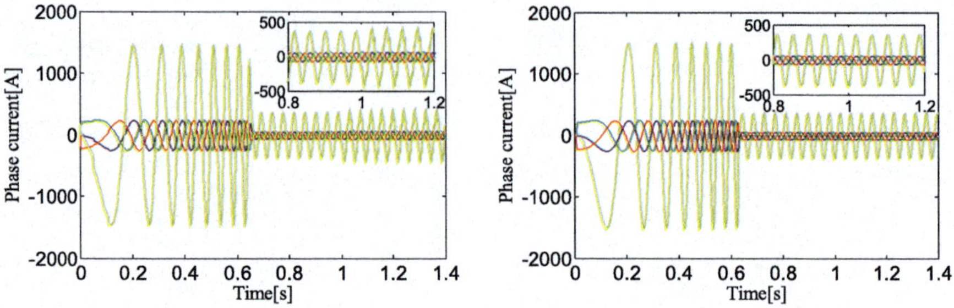

(a) Conventional PTC strategy (b) Proposed PTSF strategy

Fig. 12.10 Comparative simulation results of three-phase stator current under inductance parameter mismatch. (A) Conventional PTC strategy. (B) Proposed PTSF strategy.

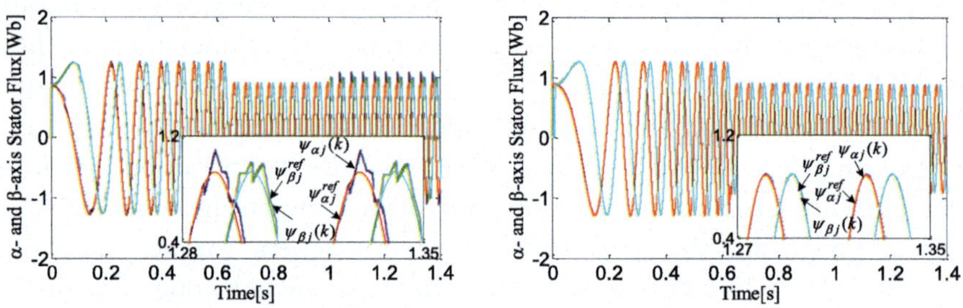

(a) Conventional PTC strategy (b) Proposed PTSF strategy

Fig. 12.11 Comparative simulation results of the α-β stator flux linkage under inductance parameter mismatch. (A) Conventional PTC strategy. (B) Proposed PTSF strategy.

mathematical relationship is that the modular multiphase PMSM adheres to the motor superposition characteristics. From Fig. 12.10A, it can be observed that the inductance parameter mismatch can cause significant distortion of the stator current in the modular multiphase PMSM when using the conventional PTC strategy. However, compared with the conventional PTC strategy, the torque ripple and stator current distortion of the modular multiphase PMSM are drastically reduced by adopting the proposed PTSF strategy in the case of parameter mismatch, as shown in Fig. 12.10B.

Fig. 12.11 presents the comparative simulation results of the α-β stator flux linkage using the conventional PTC strategy and the proposed PTSF strategy under inductance parameter mismatch. From 0 to 1 s, the conventional PTC exhibits the same superior stator flux control performance as the proposed PTSF when parameter mismatch does not occur. Comparative experimental results illustrate that the conventional

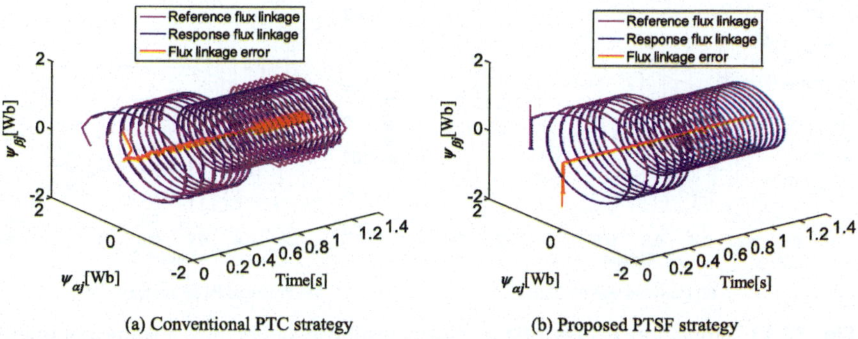

Fig. 12.12 Comparative simulation results of three-dimensional rotor flux trajectories under inductance parameter mismatch. (A) Conventional PTC strategy. (B) Proposed PTSF strategy.

PTC strategy produces an obvious static tracking error in the α-β stator flux linkage under inductance parameter mismatch, as clearly shown in Fig. 12.11A. The stator flux linkage response value of the proposed PTSF strategy strictly follows the reference value, and the flux linkage parameters are not needed in the control operation process. Thus the proposed PTSF strategy can significantly suppress the influence of flux linkage parameter mismatch, as shown in Fig. 12.11B.

Fig. 12.12 presents the comparative simulation results of the three-dimensional rotor flux trajectories using the conventional PTC strategy and the proposed PTSF strategy under inductance parameter mismatch. The conventional PTC strategy exhibits a large flux linkage fluctuation error between the α-β stator flux linkage response value and the reference value under inductance parameter mismatch. In contrast, the proposed PTSF strategy drastically reduces the flux linkage fluctuation error of the modular multiphase PMSM because it is immune to inductance parameter variations. In summary, the conventional PTC strategy is highly sensitive to the inductance parameter, which leads to flux linkage fluctuations, stator current distortion, and torque ripple under inductance parameter mismatch. However, these defects can be completely avoided because the proposed PTSF strategy is immune to the inductance parameter.

Simulation for modular multiphase PMSM under flux-linkage parameter mismatch

The operating conditions of the modular multiphase PMSM in Case B are that the load torque is maintained at 2000 N·m and the speed is maintained at 150 rad/s. At 1 s, there is a mismatch in the flux linkage parameter of the modular multiphase PMSM drive system, and the flux linkage value steps

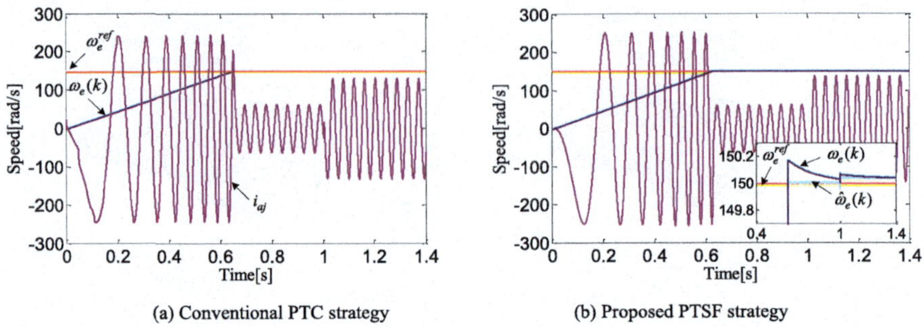

Fig. 12.13 Simulation results of the speed response under flux linkage parameter mismatch: (A) Conventional PTC strategy, (B) Proposed PTSF strategy.

to 50% of its initial value, i.e., 0.446 Wb. Comparative simulation results of the conventional PTC strategy and the proposed PTSF strategy under flux linkage parameter mismatch are shown in Figs. 12.13–12.17. Fig. 12.13 presents the comparative simulation results of the speed response using the conventional PTC strategy and the proposed PTSF strategy under flux linkage parameter mismatch. The response speed value of the conventional PTC is as good as that of the proposed PTSF under flux linkage parameter mismatch, as shown clearly in Fig. 12.13. From Fig. 12.13B, it can be observed that the unknown torque disturbance observer can accurately estimate the speed at both low and high speeds. However, the speed feedback value exhibits an overshoot under the sudden change of load torque, but the overshoot value is very small, as shown clearly in Fig. 12.13B. The simulation results demonstrate that the predictive control of the speed loop does not lead to a degradation of the overall performance under flux linkage parameter mismatch.

Fig. 12.14 presents the comparative simulation results of the phase current and torque using the conventional PTC strategy and the proposed PTSF strategy under flux linkage parameter mismatch. From 0 to 0.4 s, there is a large estimation error because the unknown torque disturbance observer does not reach the convergence state during the motor start-up process. Note that there is a large overshoot when the flux linkage parameters are suddenly mismatched at 1 s; the reason is that the unknown torque disturbance observer is affected by the parameter mismatch. After 1 s, it can be observed that the unknown torque disturbance observer can accurately estimate the load torque under flux linkage parameter mismatch. Fig. 12.14B demonstrates that the proposed PTSF strategy provides a satisfactory torque waveform with small torque ripples in the presence of flux linkage

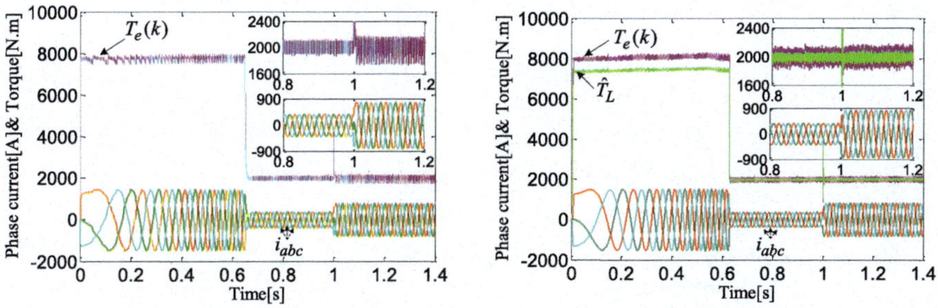

(a) Conventional PTC strategy (b) Proposed PTSF strategy

Fig. 12.14 Comparative simulation results of the torque and phase current under flux linkage parameter mismatch. (A) Conventional PTC strategy. (B) Proposed PTSF strategy.

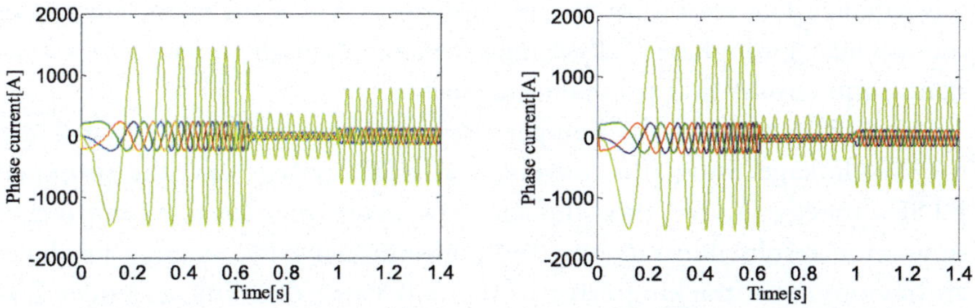

(a) Conventional PTC strategy (b) Proposed PTSF strategy

Fig. 12.15 Comparative simulation results of three-phase stator current under flux linkage parameter mismatch.

parameter mismatch. Compared with the proposed PTSF method, the conventional PTC method exhibits significant torque ripple under flux linkage parameter mismatch, as shown in Fig. 12.14A. The reason for the torque ripple is that the predictions of the conventional PTC method depend on the precise mathematical model of the modular multiphase PMSM.

Fig. 12.15 presents the comparative simulation results of the three-phase stator current using the conventional PTC strategy and the proposed PTSF strategy under flux linkage parameter mismatch. From Fig. 12.15, it can be observed that the stator current waveform of the proposed PTSF strategy is significantly superior to that of the conventional PTC strategy under flux linkage parameter mismatch. The reason for the large stator flux static error and torque ripple is that the conventional PTC strategy must strictly depend on the precise flux linkage parameter. In addition, it can be observed from Fig. 12.15 that the stator current of the modular multiphase PMSM will suddenly increase when the flux linkage parameters are

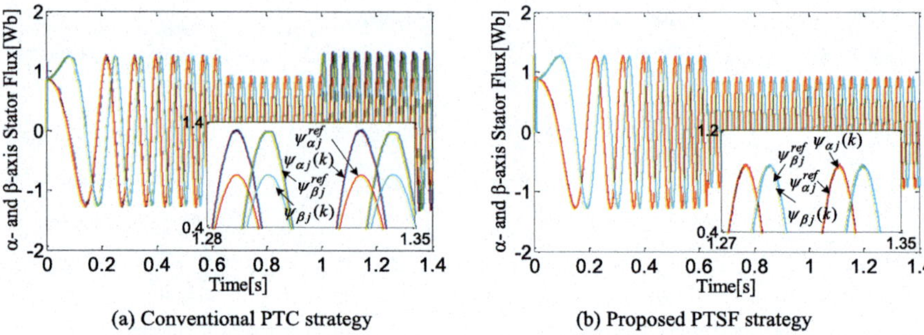

Fig. 12.16 Comparative simulation results of the α-β stator flux linkage under flux linkage parameter mismatch: (A) Conventional PTC strategy, (B) Proposed PTSF strategy.

mismatched. The reason for this phenomenon is that the stator current of the modular multiphase PMSM must be large enough to maintain a constant torque under flux parameter mismatch.

Fig. 12.16 presents the comparative simulation results of the α-β stator flux linkage using the conventional PTC strategy and the proposed PTSF strategy under flux linkage parameter mismatch. Comparative simulation results illustrate that the conventional PTC strategy produces an obvious static tracking error in the α-β stator flux linkage under flux linkage parameter mismatch, as clearly shown in Fig. 12.16A. The reason for this phenomenon is that the conventional PTC strategy strictly relies on the mathematical model and flux linkage parameter of the modular multiphase PMSM. Compared with the conventional PTC strategy, it can be seen from Fig. 12.16B that the stator flux linkage response value of the modular multiphase PMSM can strictly follow the reference value by adopting the proposed PTSF strategy under flux linkage parameter mismatch. The reason is that the flux linkage parameters are not needed in the control operation process, and the proposed PTSF method can significantly suppress the influence of flux linkage parameter mismatch, as shown in Fig. 12.16B.

Fig. 12.17 presents the comparative simulation results of the three-dimensional rotor flux trajectories using the conventional PTC strategy and the proposed PTSF strategy under flux linkage parameter mismatch. From Fig. 12.17A, it can be seen that there is a significant α-β stator flux linkage tracking error in the modular multiphase PMSM between the stator flux linkage response value and the reference value under flux linkage parameter mismatch. The reason for this phenomenon is that the flux linkage

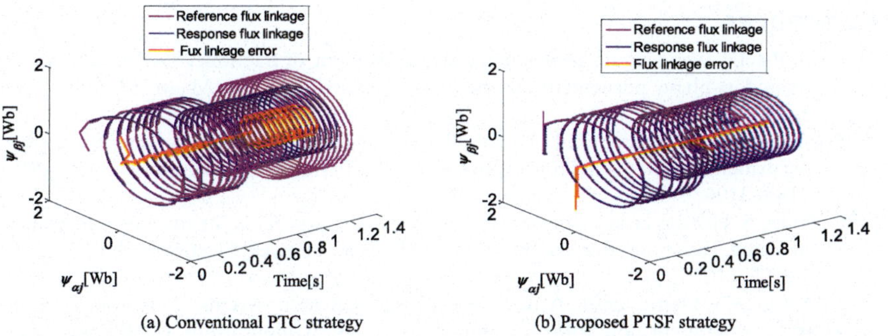

Fig. 12.17 Comparative simulation results of three-dimensional rotor flux trajectories under flux linkage parameter mismatch. (A) Conventional PTC strategy. (B) Proposed PTSF strategy.

parameter mismatch deteriorates the control performance of the conventional PTC strategy at 1 s. However, compared with the conventional PTC strategy, the α-β stator flux linkage tracking error in the modular multiphase PMSM can be effectively suppressed by using the proposed PTSF strategy, as shown clearly in Fig. 12.17B. The reason is that the proposed PTSF strategy can eliminate the impact of flux linkage parameter mismatch on the control performance of the modular multiphase PMSM.

12.6 Conclusion

In this chapter, a novel robust PTSF control method for a modular multiphase PMSM drive system, called PTSF, has been proposed, which can effectively enhance robustness against parameter mismatch. The robust predictive stator flux controller is formulated in the dq stator flux space, which is independent of the motor parameters and compensates for the one-step delay. Furthermore, the robust predictive torque controller is designed based on an unknown torque disturbance observer, which enhances robustness against parameter mismatch and load torque disturbance. Compared with the conventional PTC strategy, the proposed PTSF method effectively enhances robustness against parameter mismatch by adopting the closed-loop control of the *d-q* axis stator flux. Comparative simulation results demonstrate that the proposed PTSF strategy outperforms the conventional PTC strategy in yielding high stator flux precision tracking and satisfactory parameter immunity under parameter mismatch while achieving smaller stator current THD and lower torque/flux ripple.

References

[1] V.I. Patel, J. Wang, et al., Demagnetization assessment of fractional-slot and distributed wound 6-phase permanent magnet machines, IEEE Trans. Magn. 51 (06) (2015) 1–11.

[2] J. Karttunen, S. Kallio, et al., Decoupled vector control scheme for dual three-phase permanent magnet synchronous machines, IEEE Trans. Ind. Electron. 61 (05) (2014) 2185–2196.

[3] Y. Hu, S. Huang, et al., Control of dual three-phase permanent magnet synchronous machine based on five-leg inverter, IEEE Trans. Power Electron. 34 (11) (2019) 11071–11079.

[4] M. Onsal, Y. Demir, et al., A new nine-phase permanent magnet synchronous motor with consequent pole rotor for high-power traction applications, IEEE Trans. Magn. 53 (11) (2017) 1–6.

[5] X. Han, D. Jiang, et al., Two-segment three-phase PMSM drive with carrier phase-shift PWM, in: IEEE APEC, 2018, pp. 848–854.

[6] X. Han, D. Jiang, et al., Two-Segment Three-Phase PMSM Drive With Carrier Phase-Shift PWM for Torque Ripple and Vibration Reduction, IEEE Trans. Power Electron. 34 (01) (2018) 588–599.

[7] W. Xie, X. Wang, et al., Finite-control-set model predictive torque control with a deadbeat solution for PMSM drives, IEEE Trans. Ind. Electron. 62 (09) (2015) 5402–5410.

[8] Z. Xiang, X. Zhu, et al., Multi-level design optimization and operation of a brushless double mechanical ports flux-switching permanent magnet motor, IEEE Trans. Ind. Electron. 63 (10) (2016) 6042–6054.

[9] C. Shan, L. Wang, et al., A cascade MPC control structure for a PMSM with speed ripple minimization, IEEE Trans. Ind. Electron. 60 (08) (2013) 2978–2987.

[10] C. Garcia, J. Rodriguez, et al., Abu-Rub, H. Full predictive cascaded speed and current control of an induction machine, IEEE Trans. Energy Convers. 31 (03) (2016) 1059–1067.

[11] X. Zhang, Z. Liang, et al., Model predictive current control for PMSM drives with parameter robustness improvement, IEEE Trans. Power Electron. 34 (02) (2018) 1645–1657.

[12] L. Tong, X. Zhou, et al., An SRF-PLL-based sensorless vector control using the predictive deadbeat algorithm for the direct-driven permanent magnet synchronous generator, IEEE Trans. Power Electron. 29 (06) (2014) 2837–2849.

[13] R. Errouissi, M. Ouhrouche, et al., Robust cascaded nonlinear predictive control of a permanent magnet synchronous motor with antiwindup compensator, IEEE Trans. Ind. Electron. 59 (08) (2011) 3078–3088.

[14] C. Garcia, J. Rodriguez, et al., Full predictive cascaded speed and current control of an induction machine, IEEE Trans. Energy Convers. 31 (03) (2016) 1059–1067.

[15] L. He, F. Wang, et al., Zynq implemented lunenberger disturbance observer based predictive control scheme for PMSM drives, IEEE Trans. Power Electron. 35 (02) (2020) 1770–1778.

[16] X. Zhang, B. Hou, et al., Deadbeat predictive current control of permanent-magnet synchronous motors with stator current and disturbance observer, IEEE Trans. Power Electron. 32 (05) (2016) 3818–3834.

[17] M. Yang, X. Lang, et al., Flux immunity robust predictive current control with incremental model and extended state observer for PMSM drive, IEEE Trans. Power Electron. 32 (12) (2017) 9267–9279.

[18] T. Shi, Y. Yan, et al., Linear quadratic regulator control for PMSM drive systems using nonlinear disturbance observer, IEEE Trans. Power Electron. 35 (05) (2019) 5093–5101.

[19] C. Zhang, G. Wu, et al., Robust fault-tolerant predictive current control for permanent magnet synchronous motors considering demagnetization fault, IEEE Trans. Ind. Electron. 65 (07) (2017) 5324–5334.
[20] C. Xu, Z. Han, et al., Deadbeat predictive current control for permanent magnet synchronous machines with closed form error compensation, IEEE Trans. Power Electron. 35 (05) (2020) 5018–5030.
[21] Y. Liu, S. Cheng, et al., Robust model predictive control with simplified repetitive control for electrical machine drives, IEEE Trans. Power Electron. 34 (05) (2019) 4524–4535.
[22] J.F. Stumper, S. Kuehl, et al., Predictive torque control for AC drives: improvement of parametric robustness using two degree of freedom control, in: IEEE Energy Convers, Congress and Exposition, 2013, pp. 15–19.
[23] M. Siami, D.A. Khaburi, et al., Torque ripple reduction of predictive torque control for PMSM drives with parameter mismatch, IEEE Trans. Power Electron. 32 (09) (2016) 7160–7168.
[24] C.K. Lin, T.H. Liu, et al., Model-free predictive current control for interior permanent-magnet synchronous motor drives based on current difference detection technique, IEEE Trans. Ind. Electron. 61 (02) (2013) 667–681.

CHAPTER 13

Robust dual-loop cascade predictive tolerant control for modular multiphase PMSM with multiparameter mismatch fault diagnosis

13.1 Introduction

Although three-phase permanent magnet synchronous motors (PMSM) are commonly used [1,2], multiphase PMSM are more suitable for special applications that require high power and redundancy, such as ship propulsion, vehicle traction, and wind power generation [3]. Based on their stator winding structure, multiphase PMSM can be categorized into conventional and modular multiphase structures [4]. The modular multiwinding structure has several advantages, such as convenient controller design, simple topology, and enhanced fault-tolerant control performance; thus it can better meet the operation quality requirements of large torque, wide speed range, and high error tolerance compared with conventional multiphase structures [5,6].

In the modular multiwinding stator structure configuration, N-segment multimodule three-phase PMSM (modular multiphase PMSM) are ideal for the current energy production and transportation industry [7,8]. This drive topology with multiple three-phase units can fully integrate three-phase technology and expand the fault tolerance of the module. The modular multiphase PMSM are primarily studied for their motor body design and topology optimization. Their controller uses the traditional proportional–integral (PI) control [9–11]. The modular multiphase PMSM drive systems [9–11] were adjusted using a PI controller to reduce the current harmonics and torque ripple. However, their response quality was considerably impacted by the controller gain.

The key indexes for judging the modular multiphase PMSM performance are its response and tracking accuracy of speed and current.

Robust Model Predictive Control of Permanent Magnet Synchronous Motor Under Parameter Mismatch
https://doi.org/10.1016/B978-0-443-44209-4.00010-5

Compared with PI control, predictive control has the advantages of a simple principle and easy debugging; thus it is an efficient control scheme for modular multiphase PMSM [12–15]. However, predictive control relies on system parameters [16], and when the parameters employed by the predictive controller deviate from those of the motor, i.e., a mismatch parameter fault occurs, its predictive behavior will be inaccurate. This behavior will affect its speed and current response accuracy, as well as cause system divergence [17,18].

During a parameter mismatch fault, system robustness can be improved by designing a disturbance observer for feedback compensation. Yu et al. [19] proposed a cascade predictive control algorithm, which was integrated with a composite disturbance observer. Li et al. [20] presented an enhanced predictive control strategy that incorporates a generalized observer designed to tackle challenges including model discrepancies, digital delays, and load disturbances. Wang et al. [21] proposed a strategy that integrates observer and predictive models. Wu et al. [22] developed a method that integrated an observer designed to suppress tracking errors and fluctuations in stator flux and torque resulting from complex parameter mismatches. In [19–22], predictive control methods combined with disturbance observers eliminated control errors; however, the dynamic response was not ideal due to parameter mismatch fault.

The robustness of predictive control can be enhanced via parameter online fault diagnosis. Yao et al. [23] presented a novel method for identifying resistance and inductance parameters that relies on current injection. Zhou et al. [24] proposed a novel predictive control strategy that employs Kalman filter identification. Zhang et al. [25] proposed a predictive control scheme that utilizes a parameter extraction algorithm. Wang et al. [26] designed a predictive controller whose parameters were updated in real-time using online-identified parameters, effectively eliminating control offset caused by parameter mismatch faults. In [23–26], the proposed predictive control methods based on parameter identification effectively suppressed the impact of parameter mismatch faults; however, they had high requirements for parameter identification performance and did not consider the impact of unknown load disturbances.

This chapter is organized as follows. The mathematical model and drive topology of the modular multiphase PMSM are established in Section 13.2. Building on this foundation, the robust dual-loop cascade predictive tolerant controller is proposed in Section 13.3. Section 13.4 presents the modular multiparameter fault diagnosis system, including parameter identification model construction, ANN structure design, and an improved weight learning algorithm. Section 13.5 validates the effectiveness of the proposed CTC-MFD strategy. The conclusion is presented in Section 13.6.

13.2 Mathematical model and driving topology of modular multiphase PMSM

Fig. 13.1 illustrates the stator winding configuration and drive topology of the modular multiphase PMSM. In the stator, the three-phase winding sets A_j–B_j–C_j and A_n–B_n–C_n are in phase with A_1–B_1–C_1, and all three winding sets are distributed along the stator circumference at 120 mechanical degrees. Meanwhile, to enhance system fault tolerance, a drive topology consisting of N parallel inverters with independent DC power sources has been designed.

For motors with n three-phase winding units, the d-q axis voltage transformation of any winding unit j ($j = 1, 2, \ldots, n$) can be expressed as

$$\begin{bmatrix} u_{dj} \\ u_{qj} \\ u_0 \end{bmatrix} = \frac{2}{3} \begin{bmatrix} \cos\theta & \cos\left(\theta - \frac{2\pi}{3}\right) & \cos\left(\theta + \frac{2\pi}{3}\right) \\ \sin\theta & \sin\left(\theta - \frac{2\pi}{3}\right) & \sin\left(\theta + \frac{2\pi}{3}\right) \\ \frac{1}{2} & \frac{1}{2} & \frac{1}{2} \end{bmatrix} \begin{bmatrix} u_{aj} \\ u_{bj} \\ u_{cj} \end{bmatrix} \tag{13.1}$$

where u_{dj} and u_{qj} denote the d- and q-axis voltages of unit j, respectively; u_{aj}, u_{bj}, and u_{cj} denote the a-axis, b-axis, and c-axis voltages of unit j, respectively; and θ is the rotor angle.

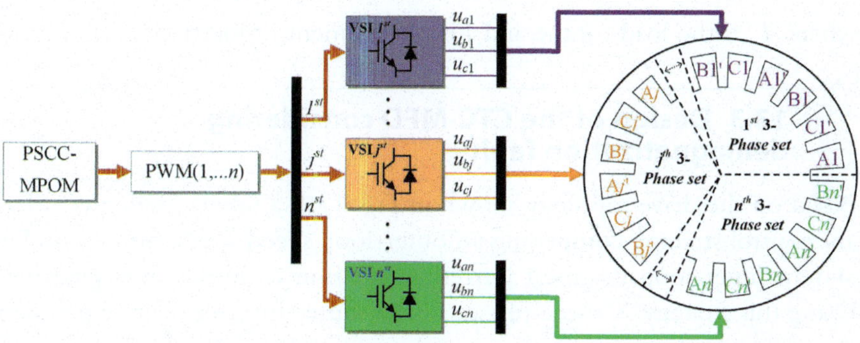

Fig. 13.1 The stator winding configuration and drive topology schematic of a modular multiphase PMSM.

In the d-q coordinate system, the mathematical model of a modular multiphase PMSM is equivalent to the superposition of a conventional three-phase PMSM. Therefore the state equation of any unit j of a surface-mounted modular multiphase PMSM can be expressed as

$$\frac{dx_j}{dt} = A_j x_j + B_j u_j + D_j \tag{13.2}$$

with $x_j = \begin{bmatrix} i_{dj} \\ i_{qj} \end{bmatrix}$, $u_j = \begin{bmatrix} u_{dj} \\ u_{qj} \end{bmatrix}$, $A_j = \begin{bmatrix} -\frac{R_j}{L_{sj}} & \omega_e \\ -\omega_e & \frac{-R_j}{L_{sj}} \end{bmatrix}$, $B_j = \begin{bmatrix} \frac{1}{L_{sj}} & 0 \\ 0 & \frac{1}{L_{sj}} \end{bmatrix}$,

$D_j = \begin{bmatrix} 0 \\ \frac{-\psi_r \omega_e}{L_{sj}} \end{bmatrix}$, where i_{dj} and i_{qj} are the d- and q-axis currents of unit j, respectively; R_j and L_{sj} are the resistance and inductance of unit j, respectively; ω_e is the electrical angular velocity; ψ_r is the permanent magnet flux linkage.

The total output torque of the modular multiphase PMSM can be obtained from the sum of the torque generated by each unit.

$$T_e = \frac{3p}{2}\sum_{j=1}^{n}\left(\psi_r i_{qj}\right) = \frac{3p}{2}\psi_r\left(i_{q1} + \cdots i_{qj} + \cdots i_{qn}\right) \tag{13.3}$$

where p is the number of pole pairs.

The mechanical motion equation for the modular multiphase PMSM is

$$T_e - T_L = \frac{J}{p}\frac{d\omega_e}{dt} \tag{13.4}$$

where T_L is the load torque and J is the moment of inertia.

13.3 Design of the CTC-MFD considering demagnetization fault

Based on the drive topology shown in Fig. 13.1, a robust predictive tolerant control strategy comprising an outer-loop speed-prediction controller and an inner-loop deadbeat current-prediction controller was proposed. Using this strategy, N three-phase voltage source inverters can be precisely controlled during parameter mismatch faults and high-power impact loads by sending N sets of PWM drive signals with model parameter correction.

Design of the deadbeat current predictive controller

The general solution of the state Eq. (13.2) for the modular multiphase PMSM unit j can be expressed as

$$x_j(t) = e^{A_j(t-t_o)}x_j(t_o) + \int_{t_o}^{t} e^{A_j(t-\tau)}\left(B_j u_j(\tau) + D_j\right)d\tau \tag{13.5}$$

Let $t_0 = kT_s$ and $t = (k + 1)T_s$, where T_s is the control period. When T_s is extremely small, the following expression can be derived as

$$e^{-\frac{R_j T_s}{L_{sj}}} \approx 1 - \frac{R_j T_s}{L_{sj}}, \cos(\omega_e T_s) \approx 1, \sin(\omega_e T_s) \approx \omega T_s \tag{13.6}$$

From Eqs. (13.2) and (13.6), the discrete state equation of the three-phase unit j can be obtained as

$$u_j(k) = F_j^{-1}\left[i_j(k+1) - E_j(k)i_j(k) - P_j(k)\right] \tag{13.7}$$

where $i_j(k) = \begin{bmatrix} i_{dj}(k) \\ i_{qj}(k) \end{bmatrix}$, $u_j(k) = \begin{bmatrix} u_{dj}(k) \\ u_{qj}(k) \end{bmatrix}$, $P_j(k) = \begin{bmatrix} 0 \\ -\frac{\psi_r}{L_{sj}} T_s \omega_e(k) \end{bmatrix}$,

$$E_j(k) = \begin{bmatrix} 1 - \frac{R_j}{L_{sj}} T_s & T_s \omega_e(k) \\ -T_s \omega_e(k) & 1 - \frac{R_j}{L_{sj}} T_s \end{bmatrix}, \quad F_j = \begin{bmatrix} \frac{T_s}{L_{sj}} & 0 \\ 0 & \frac{T_s}{L_{sj}} \end{bmatrix}.$$

An irreversible one-sample delay exists in the output voltage of the controller in a practical digital control system; this results in the voltage vector calculated at kT_s not being applied until $(k + 1)T_s$. To suppress the effect of the one-beat delay on the system response, the sampled current $i_j(k + 1)$ at kT_s was obtained using a current predictor. Then, $i_j(k + 1)$ was used to calculate the voltage vector $u_j(k + 1)$ for the next control cycle.

As the mechanical time constant of the motor was considerably larger than its electrical time constant, the speed was approximately constant within a control cycle. Therefore $E_j(k) \approx E_j(k + 1)$ and $P_j(k) \approx P_j(k + 1)$. At $(k + 1)T_s$, the voltage vector output by the deadbeat current controller of the three-phase unit j was obtained from Eq. (13.7).

$$u_j^{ref}(k+1) = F_j^{-1}\left[i^{ref}(k+2) - E_j(k)i_j(k+1) - P_j(k)\right] \tag{13.8}$$

where $u_j^{ref}(k+1) = \begin{bmatrix} u_{dj}^{ref}(k+1) \\ u_{qj}^{ref}(k+1) \end{bmatrix}$ and $i^{ref}(k+2) = \begin{bmatrix} i_d^{ref}(k+2) \\ i_q^{ref}(k+2) \end{bmatrix}$ are the output voltage of the deadbeat current controller and the current vector given by the speed controller, respectively. $i_j(k+1) = [i_{dj}(k+1)\ i_{qj}(k+1)]^T$ was the predicted current of the next moment.

According to Eq. (13.8), the output voltage of a deadbeat current controller depends on the acquisition of the sampled current $i_j(k+1)$ and the given current $i^{ref}(k+2)$ of the speed controller. Herein, the current predictor was used to predict the sampled current value $i_j(k+1)$ as

$$\begin{cases} i_{dj}(k+1) = \left(1 - \dfrac{R_j}{L_{sj}} T_s\right) i_{dj}(k) + T_s \omega_e(k) i_{qj}(k) + \dfrac{T_s}{L_{sj}} u_{dj}(k) \\ i_{qj}(k+1) = \left(1 - \dfrac{R_j}{L_{sj}} T_s\right) i_{qj}(k) - T_s \omega_e(k) i_{dj}(k) - \dfrac{\psi_r}{L_{sj}} T_s \omega_e(k) + \dfrac{T_s}{L_{sj}} u_{qj}(k) \end{cases} \tag{13.9}$$

Design of speed prediction controller

Using the speed prediction control, the response speed can accurately track the command speed and resist the high-power impact load. Due to the limitations of the rotational inertia and maximum torque driving ability of the system, T_s of the current loop was less than the sampling time T_{ds} of the speed loop. Although the traditional first-order Euler discretization method employs a simple algorithm, it may experience oscillations when achieving high-bandwidth predictive speed control and exhibit poor resistance to high-power impact loads. Therefore, to avoid increasing the computational complexity considerably, the second-order Euler discretization method was used to design the speed prediction controller.

By employing the control strategy under $i_d = 0$, the motion equation of the modular multiphase PMSM was obtained from Eqs. (13.3) and (13.4) as

$$\frac{d\omega_e}{dt} = \frac{p}{J}\left(\frac{3p}{2}\sum_{j=1}^{n}(\psi_r i_{qj}) - T_L\right) \tag{13.10}$$

As $dT_L/dt = 0$, taking the derivative of Eq. (13.10) yields

$$\frac{d^2\omega_e}{dt^2}=\frac{3p^2}{2J}\frac{d\sum_{j=1}^{n}(\psi_r i_{qj})}{dt} \tag{13.11}$$

By the second-order expansion of speed, we can obtain

$$\omega_e(k+1)=\omega_e(k)+\frac{d\omega_e}{dt}\cdot T_{ds}+\frac{d^2\omega_e}{dt^2}\cdot\frac{T_{ds}^2}{2} \tag{13.12}$$

By substituting Eqs. (13.10) and (13.11) into Eq. (13.12), the q-axis current command value, output by the speed prediction controller, can be obtained as

$$\sum_{j=1}^{n}(\psi_r i_{qj}(k+1))=\frac{\omega_e(k+1)-\omega_e(k)+\frac{p}{J}T_{ds}\left(T_L-\frac{57p}{40}\sum_{n=1}^{n}(\psi_r i_{qj}(k))\right)}{\frac{3p^2}{40J}T_{ds}} \tag{13.13}$$

As each independent stator winding of the modular multiphase PMSM is the same and the three-phase unit is controlled according to the power-sharing principle, $\sum_{j=1}^{n}(\psi_r i_{qj}(k))=n\psi_r i_{qj}(k)$. Let $i_{qj}(k)=i_q^{ref}(k)$, $\omega_e(k)=\omega_e^{ref}$, then according to Eq. (13.13), we can get

$$i_q^{ref}(k)-\frac{\omega_e^{ref}-\omega_e(k-1)+\frac{pT_L}{J}T_{ds}-\frac{57np^2}{40J}T_{ds}\psi_r i_q(k-1)}{\frac{3np^2}{40J}\psi_r T_{ds}} \tag{13.14}$$

According to Lagrange's theorem, the command values of d- and q-axis currents at $(k+2)T_s$ are

$$\begin{cases} i_d^{ref}(k+2)=i_d^{ref}(k+1)=i_d^{ref}(k)=0 \\ i_q^{ref}(k+2)=3i_q^{ref}(k)-3i_q^{ref}(k-1)+i_q^{ref}(k-2) \end{cases} \tag{13.15}$$

By inputting the d-q axis current commands $i_d^{ref}(k+2)$ and $i_q^{ref}(k+2)$, as well as the vector $i_j(k+1)$ output by the current predictor, into the deadbeat current controller, the voltage vector required to accurately follow the command for the modular multiphase PMSM current is calculated.

Robust double-loop predictive tolerant control

According to Eqs. (13.8) and (13.14), the current loop and speed loop predictive controllers comprise the resistance, inductance, flux linkage, and load torque (R_j, L_{sj}, ψ_r, and T_L). The control performance of these controllers depends on the actual values of motor parameters and the load torque. Therefore a robust predictive control strategy with a modular multiparameter online monitoring function was designed, wherein the fixed parameters in the controller were updated using online monitoring parameters. Fig. 13.4 shows the robust predictive control system for modular multiphase PMSM. To obtain robust dual-loop control, the online monitoring of the electromagnetic parameters of three-phase winding units and unknown load torque disturbances must be performed.

13.4 Modular multiparameter fault diagnosis system

The modular parameter online fault diagnosis system mainly involves three processes: the construction of a parameter identification model, the design of a neural network structure, and the improvement of neural network learning algorithms.

Construction of a full-rank identification model

The discrete model of the three-phase winding unit j was developed using Eq. (13.7) as follows:

$$\begin{cases} i_{dj}(k+1) = i_{dj}(k) + \dfrac{T_s}{L_{sj}}\left[u_{sdj}(k) - R_j i_{dj}(k) + D_{dj}(k)\right] \\ i_{qj}(k+1) = i_{qj}(k) + \dfrac{T_s}{L_{sj}}\left[u_{sqj}(k) - R_j i_{qj}(k) + D_{qj}(k)\right] \end{cases} \tag{13.16}$$

where $D_{dj}(k) = L_{sj}\omega_e(k)i_{qj}(k)$ and $D_{qj}(k) = -L_{sj}\omega_e(k)i_{dj}(k)$-$\omega_e(k)\psi_r$ are the components of the d-q axis back electromotive force at kT_s and $u_{sdj}(k)$ and $u_{sqj}(k)$ are the d-q axis reconstruction voltages of three-phase winding unit j. If $U_{ANj} = (2S_{aj}\ S_{bj}\ S_{cj})U_{dcj}/3$, $U_{BNj} = (2S_{bj}\ S_{aj}\ S_{cj})U_{dcj}/3$, and $U_{CNj} = (2S_{cj}\ S_{aj}\ S_{bj})$ $U_{dcj}/3$ are the AC-side phase voltages of the three-phase winding unit j and S_{aj}, S_{bj}, and S_{cj} are the inverter switch states, then $u_{sdj}(k)$ and $u_{sqj}(k)$ can be specifically represented as

$$\begin{cases} u_{sdj}(k) = \dfrac{2}{3}\left(U_{ANj}\cos(\theta) + U_{BNj}\cos\left(\theta - \dfrac{2\pi}{3}\right) + U_{CNj}\cos\left(\theta + \dfrac{2\pi}{3}\right)\right) \\ u_{sqj}(k) = \dfrac{2}{3}\left(-U_{ANj}\sin(\theta) - U_{BNj}\sin\left(\theta - \dfrac{2\pi}{3}\right) - U_{CNj}\sin\left(\theta + \dfrac{2\pi}{3}\right)\right) \end{cases} \tag{13.17}$$

The system identification model, Eq. (13.16), is a second-order under-rank system with three unknown motor parameters (R_j, L_{sj}, and $\psi_p r$). To avoid poor convergence resulting from an under-ranked system, the system identification model was extended by adding the current increment state equation. According to Eq. (13.16), the equations at kT_s and $(k\text{-}1)T_s$ are

$$\begin{cases} \dfrac{\Delta i_{dj}(k)}{T_s} = \dfrac{u_{sdj}(k-1) - R_j i_{dj}(k-1) + D_{dj}(k-1)}{L_{sj}} \\ \dfrac{\Delta i_{qj}(k)}{T_s} = \dfrac{u_{sqj}(k-1) - R_j i_{qj}(k-1) + D_{qj}(k-1)}{L_{sj}} \end{cases} \tag{13.18}$$

$$\begin{cases} \dfrac{\Delta i_{dj}(k-1)}{T_s} = \dfrac{u_{sdj}(k-2) - R_j i_{dj}(k-2) + D_{dj}(k-2)}{L_{sj}} \\ \dfrac{\Delta i_{qj}(k-1)}{T_s} = \dfrac{u_{sqj}(k-2) - R_j i_{qj}(k-2) + D_{qj}(k-2)}{L_{sj}} \end{cases} \tag{13.19}$$

where $\Delta i_{dj}(k) = i_{dj}(k) - i_{dj}(k-1)$ and $\Delta i_{qj}(k) = i_{qj}(k) - i_{qj}(k-1)$ are the d-axis and q-axis current increments at kT_s, respectively; $\Delta i_{dj}(k-1) = i_{dj}(k-1) - i_{dj}(k-2)$ and $\Delta i_{qj}(k-1) = i_{qj}(k-1) - i_{qj}(k-2)$ are the *d*- and *q*-axis current increments at $(k-1)T_s$, respectively.

As the speed can be considered constant during adjacent control cycles, subtracting Eq. (13.19) from Eq. (13.18) yields the current increment state equation as

$$\begin{cases} \dfrac{\Delta i_{dj}(k) - \Delta i_{dj}(k-1)}{T_s} = \dfrac{\Delta u_{sdj}(k) - R_j \Delta i_{dj}(k-1) + L_{sj}\omega_e(k)\Delta i_{qj}(k-1)}{L_{sj}} \\ \dfrac{\Delta i_{qj}(k) - \Delta i_{qj}(k-1)}{T_s} = \dfrac{\Delta u_{sqj}(k) - R_j \Delta i_{qj}(k-2) - L_{sj}\omega_e(k)\Delta i_{dj}(k-1)}{L_{sj}} \end{cases} \tag{13.20}$$

where $\Delta u_{sdj}(k) = u_{sdj}(k-1) - u_{sdj}(k-2)$ and $\Delta u_{sqj}(k) = u_{sqj}(k-1) - u_{sqj}(k-2)$ denote the d- and q-axis voltage increments at kT_s, respectively. Based on Eq. (13.20), the current increment state equation has a rank of 2 and does not include the flux linkage parameter. Thus it can be used to establish an adjustable model for identifying the stator resistance and inductance parameters.

When using the control strategy with $i_d = 0$, it can further be obtained by Eq. (13.20).

$$\begin{cases} L_{sj} = \dfrac{\Delta u_{sdj}(k)}{-\omega_e(k)\Delta i_{qj}(k-1)} \\ R_{sj} = \dfrac{\Delta u_{sdj}(k)\left(\Delta i_{qj}(k) - \Delta i_{qj}(k-1)\right)}{T_s\omega_e(k)\Delta i_{qj}(k-1)\Delta i_{qj}(k-2)} + \dfrac{\Delta u_{sqj}(k)}{\Delta i_{qj}(k-2)} \end{cases} \tag{13.21}$$

After obtaining the inductance and resistance parameters of the three-phase winding unit, an identification model for ψ_r can be established based on Eq. (13.16) as

$$\psi_r = \frac{L_{sj}\left(i_{qj}(k+1) - i_{qj}(k)\right) - T_s\left(u_{sqj}(k) - R_j i_{qj}(k)\right)}{-T_s\omega_e(k)} \tag{13.22}$$

By combining Eqs. (13.3) and (13.4), we obtain

$$\frac{d\omega_e}{dt} = \frac{3p^2}{2J}\sum_{j=1}^{n}(\psi_r i_{qj}) - \frac{p}{J}T_L \tag{13.23}$$

Using Eq. (13.23), the identification model for load torque can be presented as

$$T_L = \frac{3np^2T_s\psi_r i_{qj}(k) - 2J(\omega_e(k) - \omega_e(k-1))}{2pT_s} \tag{13.24}$$

By combining Eqs. (13.21), (13.22), and (13.23), a full-rank identification model of $R_{sj}, L_{sj}, \psi_r, T_L$ can be obtained.

Design of the neural network structure

The ANN parameter identification method considered the actual motor and the identification model as the reference and adjustable model, respectively. Meanwhile, the parameters to be identified are designed as weights for the ANN structure. Fig. 13.2 shows the ANN structure obtained for the three-phase winding unit j.

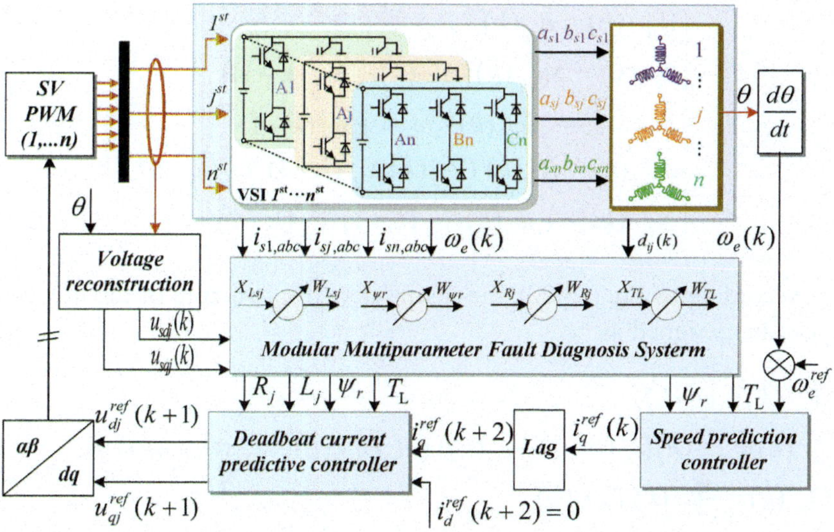

Fig. 13.2 One-beat delay compensation logic diagram based on predicted current.

The ANN adjustable model is expressed as

$$O(Wij, Xij) = \sum_{i=0}^{n} W_{ij} X_{ij} \tag{13.25}$$

where W_{ij}, X_{ij}, and O_{ij} are the neural network weights, input, and output activation functions, respectively, of the three-phase winding unit *j*. If $d_{ij}(k)$ represents the target output of the actual reference model, the ANN parameter identification structure can be constructed using the full-rank identification model.

Specifically, the neural network structure of identifiers L_{sj} and R_{sj} can be calculated using Eq. (13.21).

$$\begin{cases} W_{L_{sj}}(k) = L_{sj}(k) \\ d_{L_{sj}}(k) = \Delta u_{sdj}(k) \\ O_{L_{sj}}(k) = -L_{sj}(k)\,\omega_e(k)\,\Delta i_{qj}(k-1) \\ X_{L_{sj}}(k) = -\omega_e(k)\,\Delta i_{qj}(k-1) \end{cases} \tag{13.26}$$

$$\begin{cases} W_{R_{sj}}(k) = R_{sj}(k) \\ d_{R_{sj}}(k) = \Delta u_{sdj}(k)\left(\Delta i_{qj}(k) - \Delta i_{qj}(k-1)\right) \\ \qquad + \Delta u_{sqj}(k) T_s \omega_e(k) \Delta i_{qj}(k-1) \\ O_{R_{sj}}(k) = R_{sj}(k) T_s \omega_e(k) \Delta i_{qj}(k-1) \Delta i_{qj}(k-2) \\ X_{R_{sj}}(k) = T_s \omega_e(k) \Delta i_{qj}(k-1) \Delta i_{qj}(k-2) \end{cases} \tag{13.27}$$

According to Eq. (13.22), the neural network structure of the identifier ψ_r can be designed as

$$\begin{cases} W_{\psi_r}(k) = \psi_r(k) \\ d_{\psi_r}(k) = L_{sj}(k)\left(i_{qj}(k+1) - i_{qj}(k)\right) - T_s\left(u_{sqj}(k) - R_j(k) i_{qj}(k)\right) \\ O_{\psi_r}(k) = -\psi_r(k) T_s \omega_e(k) \\ X_{\psi_r}(k) = -T_s \omega_e(k) \end{cases} \tag{13.28}$$

According to Eq. (13.24), the neural network structure of the identifier T_L can be calculated as

$$\begin{cases} W_{T_L}(k) = T_L(k) \\ d_{T_L}(k) = 3np^2 i_{qj}(k)\left(L_{sj} \Delta i_{qj}(k+1) - T_s\left(u_{sqj}(k) - R_j \times i_{qj}(k)\right)\right) \\ \qquad + 2J\omega_e(k)(\omega_e(k) - \omega_e(k-1)) \\ O_{T_L}(k) = -2T_L(k) p T_s \omega_e(k) \end{cases} \tag{13.29}$$

Improvement of the weight-learning algorithm

The conventional ANN weight-learning algorithm, based on the least mean square (LMS), can be expressed as

$$\begin{cases} W_{ij}(k+1) = W_{ij}(k) + 2\eta_{ij} X_{ij}(k) \varepsilon_{ij}(k) \\ \varepsilon_{ij}(k) = d_{ij}(k) - O_{ij}(k) \end{cases} \tag{13.30}$$

where $\varepsilon_{ij}(k)$ and η_{ij} are the output error and convergence step size of the three-phase winding unit j, respectively. To ensure the convergence of the ANN identification system, η_{ij} must meet the following requirements [27].

$$0 < 2\eta_{ij} \left| X_{ij}(k) \right|^2 < 1 \tag{13.31}$$

The inherent feature of conventional LMS algorithms is that their step size must be carefully tuned. To improve their ability to track time-varying system parameters, momentum terms can be added to the weight adjustment algorithm [27,28]; however, this approach encounters overshoot and oscillation issues. The LMS algorithm can be improved by changing the step size according to specific features during its adaptation [29,30]. However, most variable step size methods are highly complex and introduce more parameters that require adjustment; this is not suitable for modular multi-three-phase PMSM systems, which require a large number of identification parameters.

For the modular multiparameter online monitoring of the modular multiphase PMSM, the conventional ANN weight-learning algorithm was modified. The improved two-stage LMS algorithm (TS-LMS) is

$$\begin{cases} W_{ij}(k+1) = W_{ij}(k) W_{ij}(k) + 2\eta_{ij}(k) X_{ij}(k) \varepsilon_{ij}(k) \\ \eta_{ij}(k) = \begin{cases} \eta_{ijss} & if \left| A_{new} - A_{old} \right| < \lambda_{ij} \\ \eta_{ijcs} & otherwise \end{cases} \end{cases} \tag{13.32}$$

where η_{ijcs} is the step size of the convergence stage, η_{ijss} is the step size of the steady-state stage, A_{new} is the latest sum of m weights, A_{old} is the value of the last sum of m weights, and λ_{ij} is a normal number with a very small value.

In the improved weight learning algorithm, the learning of neural network weights was divided into convergence and steady-state stages based on the characteristics of motor system parameter identification. The convergence stage uses a larger step size η_{ijcs}, whereas the steady-state stage uses a smaller step size η_{ijss}. As $|A_{new} - A_{old}|$ decreases with the gradual convergence of the network, the convergence stage of the identification result at the current time can be determined by setting the value of λ_{ij}. Note that the improved weight learning algorithm only introduced a new parameter λ_{ij}, contrary to the conventional LMS algorithm; this parameter was easily selected by varying the weight values in the steady-state stage. η_{ijcs} and η_{ijss} are multiples; if η_{ijcs} satisfies Eq. (13.31), the convergence of the improved weight learning algorithm can be guaranteed.

13.5 Simulation results

Taking the modular multiphase PMSM as an example ($N = 3$), the modular multiphase PMSM driving system based on the CTC-MFD strategy was constructed in Matlab. Then, some experimental results were compared with those of the conventional SCPC strategy. Table 13.1 lists the main motor parameters. The DC-side bus voltage, speed, sampling frequency, and initial load torque are 1500 V, 100 rad/s, 5 kHz, and 1600 N·m, respectively. In addition, the step sizes of the proposed TS-LMS algorithm are set as shown in Table 13.2.

Simulation results of modular multiphase PMSM under flux linkage mismatch fault

Fig. 13.3 presents the comparison results of multiparameter online monitoring based on the ANN under the mismatch fault of flux linkage. The modular multiparameter monitoring system begins at 0.25 s and simulates a sudden change in the flux linkage parameters from 0.892 to 0.446 Wb at 0.6 s. Fig. 13.3A shows that the mismatch fault of flux linkage causes

Table 13.1 Main parameters of PMSM.

Parameter	Value	Parameter	Value
Inductance (L_{sj})	3.572/mH	Stator phase resistance (R_j)	0.02/Ω
Flux linkage of PM (Ψ_r)	0.892/Wb	Rotational inertia (J)	100/kg m^2

Table 13.2 Control parameters of PMSM.

Parameter	Value	Parameter	Value
η_{Lsjss}	3×10^{-11}	η_{Rjss}	1×10^{-7}
$\eta_{\psi rss}$	3×10^{-7}	η_{TLss}	2×10^{-2}
η_{Lsjcs}	3×10^{-12}	η_{Rjcs}	4×10^{-3}
$\eta_{\psi rcs}$	1×10^{-7}	η_{TLcs}	2×10^{-4}
λ_{Lsj}	1.3×10^{-4}	λ_{Rj}	2×10^{-3}
$\lambda_{\psi r}$	1.3×10^{-2}	λ_{TL}	20

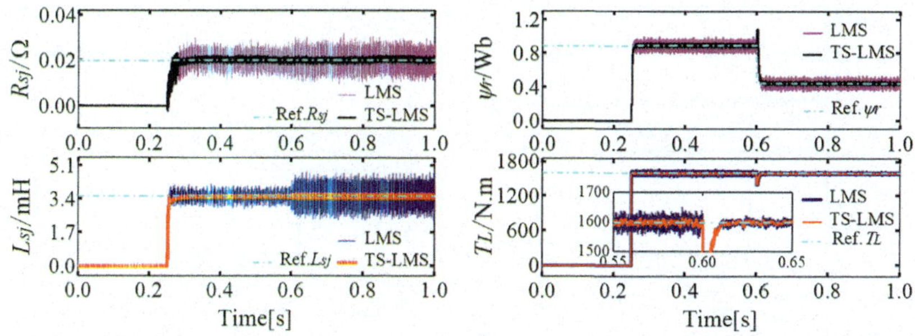

Fig. 13.3 Multiparameter online monitoring based on the ANN under the mismatch fault of flux linkage: (A) Resistance and inductance and (B) Flux linkage and torque.

oscillations in resistance and inductance under the LMS algorithm as high as ±0.015 Ω and ±1.16 mH, respectively. The oscillations in resistance and inductance under the proposed TS-LMS algorithm are only ±0.002 Ω and ±0.12 mH, respectively. Fig. 13.3B shows that the parameter monitoring system can quickly and accurately identify flux linkage and the unknown load disturbance. Compared with the LMS algorithm, the proposed TS-LMS algorithm has lower oscillation in steady-state identification results under similar convergence speeds.

Fig. 13.4 presents the comparison results between conventional SCPC and proposed CTC-MFD strategies under the mismatch fault of flux linkage. Fig. 13.4A shows that when a flux linkage mismatch fault occurs at 0.6 s, the deviation between the response and reference current of the q-axis under the conventional SCPC method is about 6A; however, the effect of the flux linkage mismatch fault on the d-axis response current is not obvious. Fig. 13.4B shows that using the CTC-MFD strategy, the response deviations in the q-axis current and speed are slightly reduced because the load torque and flux linkage in the CTC-MFD controller are corrected online.

Simulation results of modular multiphase PMSM under inductance mismatch fault

Fig. 13.5 presents the comparison results of multiparameter online monitoring based on the ANN under the mismatch fault of inductance. The modular multiparameter monitoring system starts at 0.25 s and simulates a sudden change in the inductance parameters from 3.572 mH to 1.786 mH at 0.6 s. Fig. 13.5A shows that the parameter monitoring system can quickly

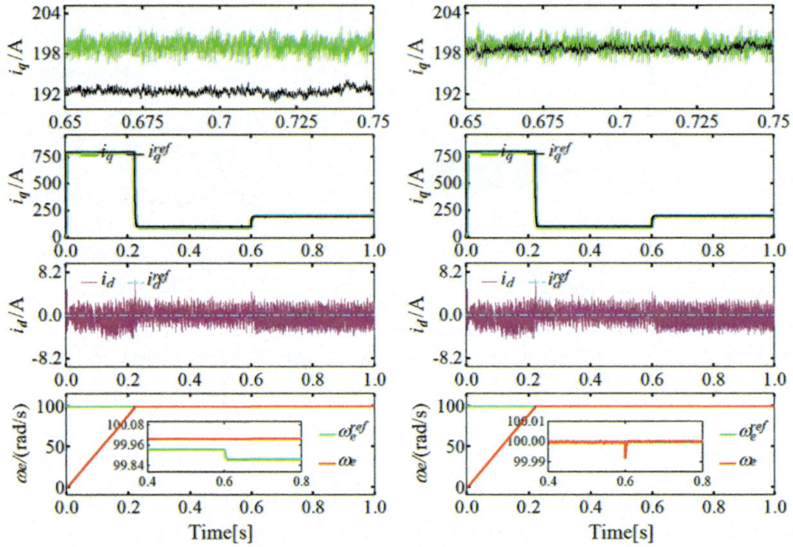

Fig. 13.4 Simulation results under the mismatch fault of flux linkage: (A) Conventional SCPC strategy and (B) Proposed CTC-MFD strategy.

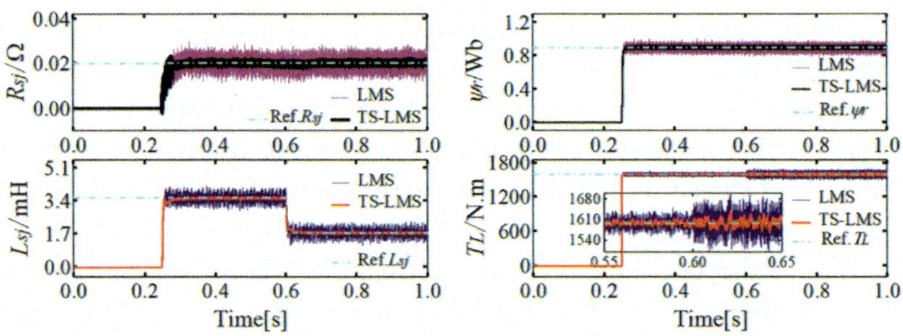

Fig. 13.5 Multiparameter online monitoring based on the ANN under the mismatch fault of inductance: (A) Resistance and inductance and (B) Flux linkage and torque.

and accurately identify the changes in inductance parameters. Compared with the LMS algorithm, TS-LMS reduces the oscillation of resistance and inductance by ±0.005 Ω and ±0.4 mH, respectively. Fig. 13.5B shows that the oscillation of load torque determined using the LMS algorithm increases from ±20 to ±80 N·m under the mismatch fault of inductance, and the oscillation in the flux linkage identification results remains relatively large at around ±0.13 Wb. However, the oscillations in the identification results of flux linkage and load torque are very small.

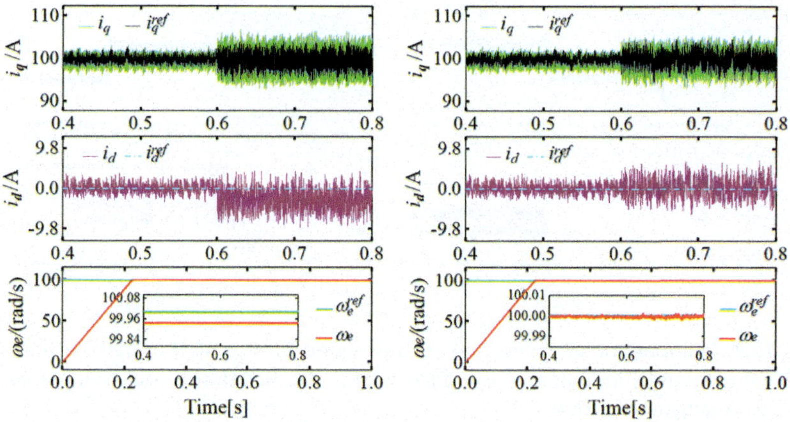

Fig. 13.6 Simulation results under the mismatch fault of inductance: (A) Conventional SCPC strategy and (B) Proposed CTC-MFD strategy.

Fig. 13.6 presents the comparison results between conventional SCPC and proposed CTC-MFD strategies under the mismatch fault of inductance. As shown in Fig. 13.6A, under the conventional SCPC strategy, the inductance mismatch primarily affects the *d*-axis response current, resulting in a deviation of up to 6 A between the response current and the reference current. Due to the lack of accurate load torque feedback values, the static speed error under the conventional SCPC strategy reaches 0.09 rad/s. As shown in Fig. 13.6B, under the proposed CTC-MFD strategy, the inductance and load torque parameters in the controller can be corrected online, thereby significantly improving the *d*-axis current deviation and static speed error.

Modular multiphase PMSM simulation under resistance, inductance, and flux linkage mismatch faults

Fig. 13.7 presents the comparison results of multiparameter online monitoring based on the ANN under mismatch faults of resistance, inductance, and flux linkage. The modular multiparameter monitoring system starts at 0.25 s and simulates sudden changes in resistance from 0.02 Ω to 0.04 Ω, inductance from 3.572 mH to 1.786 mH, and flux linkage from 0.892 Wb to 0.446 Wb at 0.6 s, respectively. Fig. 13.7 demonstrates that the parameter monitoring system can rapidly and accurately identify variations in resistance, inductance, and flux linkage parameters. Fig. 13.7A indicates that, compared to the LMS algorithm, the TS-LMS algorithm reduces oscillations in resistance and inductance by ±0.003 Ω and ±0.2 mH, respectively. Fig. 13.7B shows that under simultaneous multiparameter mismatch

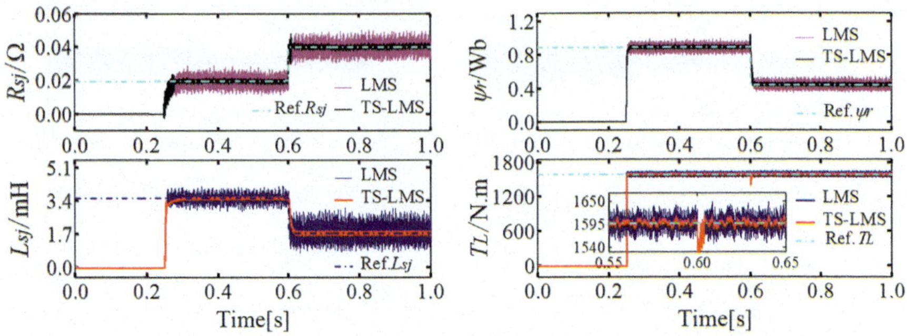

Fig. 13.7 Multiparameter online monitoring based on the ANN under the mismatch fault of resistance, inductance, and flux linkage: (A) Resistance and inductance and (B) Flux linkage and torque.

faults, the identification results of flux linkage and load torque using the LMS algorithm still maintain oscillations of approximately ±0.15 Wb and ±40 N · m, respectively, whereas the oscillations in both flux linkage and load torque identification results are significantly smaller with the TS-LMS algorithm. Therefore the results demonstrate that the modular multiparameter monitoring system can quickly and accurately track changes in each parameter at 0.6 s.

Fig. 13.8 presents the comparison results between conventional SCPC and the proposed CTC-MFD strategies under the mismatch fault of resistance, inductance, and flux linkage. Fig. 13.8A shows that the *d-q* axis response currents and speed deviate from the reference values under the conventional SCPC strategy by 5.4 A and 8 A, and 0.14 rad/s, respectively. The reason is that the controller uses fixed parameter values, which cannot be updated in real-time with changes in operating conditions. However, as shown in Fig. 13.8B, when the proposed CTC-MFD strategy is applied, the fault diagnosis system monitors parameter changes in real-time and updates the fixed parameter values in the controller. This effectively avoids parameter mismatch issues in the control system, reducing the deviations in current and speed static errors to negligible levels.

Modular multiphase PMSM simulation under different impact loads

During the modular multiphase PMSM simulation, the efficacy of the CTC-MFD strategy was verified against the high-power impact load. To this end, impact load disturbances of 4000 and 8000 N·m were added at 0.6 s, and the load was unloaded at 0.8 s.

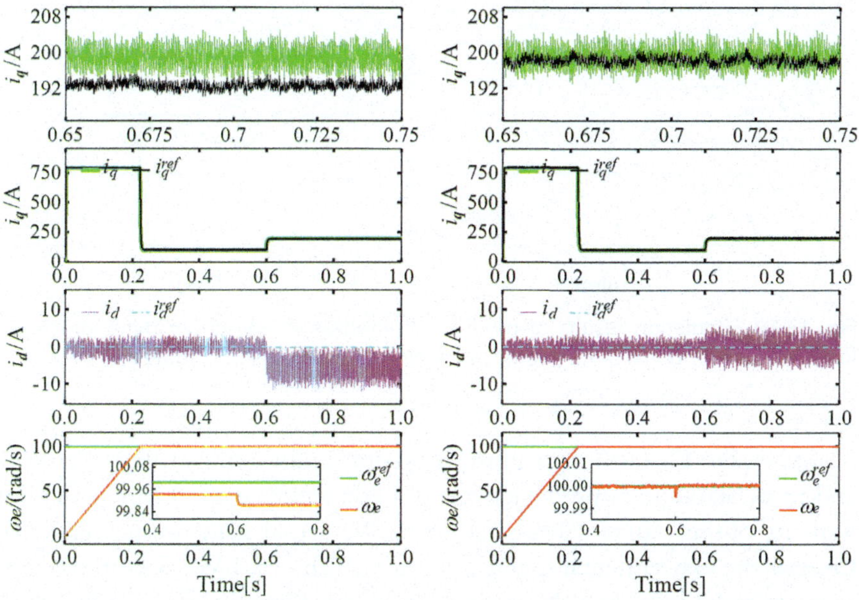

Fig. 13.8 Simulation results under the mismatch fault of resistance, inductance, and flux linkage: (A) Conventional SCPC strategy and (B) Proposed CTC-MFD strategy.

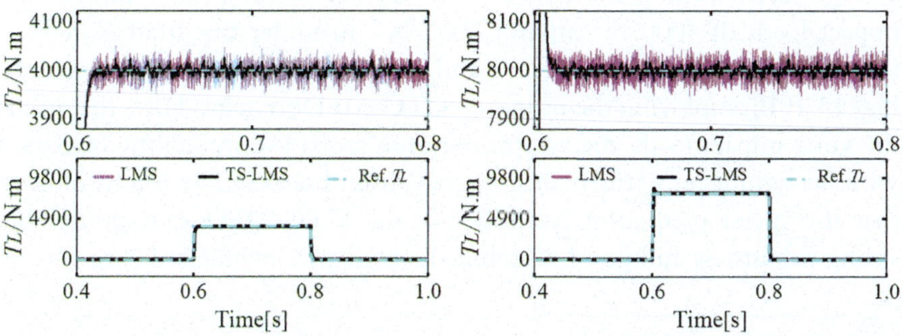

Fig. 13.9 Online monitoring of impact load based on the ANN: (A) Impact load is 4000 N·m and (B) Impact load is 8000 N·m.

Fig. 13.9 presents the comparison results of the online monitoring of load torque under different impact loads. As shown in Fig. 13.9A, both the TS-LMS and LMS algorithms can effectively track the dynamic changes of load torque, with the TS-LMS algorithm demonstrating smoother tracking trajectories and a faster recovery to steady state during sudden load variations. Fig. 13.9B further reveals that, when subjected to an 8000 N · m impact load, the tracking results of both algorithms exhibit slight overshoot.

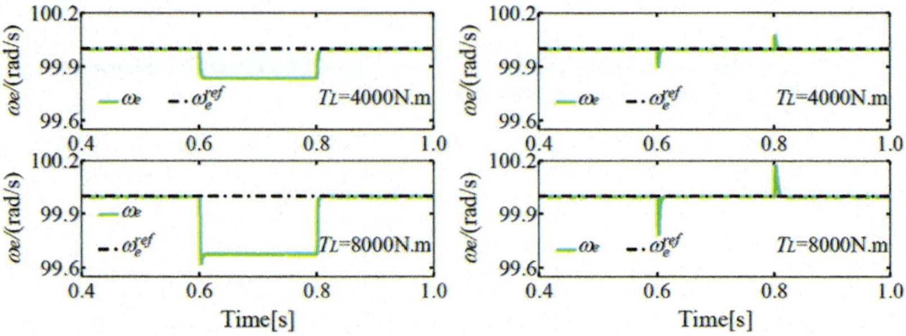

Fig. 13.10 Simulation results of modular multiphase PMSM under different impact loads: (A) Conventional SCPC strategy and (B) Proposed CTC-MFD strategy.

However, the TS-LMS algorithm maintains significantly smaller overshoot and more stable response characteristics compared to the LMS algorithm, with its output curve staying closer to the actual load torque reference. Overall, the experimental results indicate that the TS-LMS algorithm offers superior stability and tracking accuracy under dynamic load conditions.

Fig. 13.10 presents the comparison results of conventional SCPC and proposed CTC-MFD strategies under different impact loads. As shown in Fig. 13.10A, using the conventional SCPC strategy, the application of impact loads of 4000 N · m and 8000 N · m under this strategy reduces the response speed by 0.21 rad/s and 0.37 rad/s, respectively. In contrast, Fig. 13.10B, employing the proposed CTC-MFD strategy, shows that under the same impact loads, the system response speed is only slightly increased, with no significant performance degradation observed. The results indicate that the speed prediction controller of the CTC-MFD strategy exhibits strong robustness and good resistance to high-power impact loads.

13.6 Conclusion

In this chapter, a novel cascade predictive speed and current-tolerant control scheme was proposed for modular multiphase PMSM. This scheme considerably improved the robustness of modular multiphase PMSM against multiparameter faults. An improved two-stage LMS algorithm was designed to train the ANN, which accurately monitored the unknown load torque disturbances and electromagnetic parameters of the modular multiphase PMSM. The parameters monitored online were used to correct the fixed parameters in the speed and current prediction controller in real time, thereby avoiding the fault effects of unknown load disturbances and parameter mismatches. Compared with the conventional SCPC strategy, the proposed CTC-MFD strategy considerably optimized the speed and current prediction accuracies and reduced the tracking errors. It also efficiently overcame the mismatch fault effects. These findings indicated the feasibility of the proposed CTC-MFD strategy, which exhibited excellent dynamic and static performance under parameter mismatch faults and high-power impact loads.

References

[1] M.S. Khajueezadeh, S. FeizHoseini, et al., Analysis of Torsional Vibrations on the Resolver Under Eccentricity in PMSM Drive System, IEEE Sens. J. 22 (22) (2022) 21592–21599.

[2] M.S. KhajueeZadeh, M. Emadaleslami, et al., Comprehensive Investigation of the Resolver's Eccentricity Effect on the Field-Oriented Control of PMSM, IEEE Sens. J. 23 (17) (2023) 19145–19152.

[3] S.V. Nair, H. P, et al., Six-step operation of a symmetric dual three-phase PMSM with minimal circulating currents for extended speed range in electric vehicles, IEEE Trans. Ind. Electron. 69 (08) (2022) 7651–7662.

[4] A. Tessarolo, F. Luise, Design for improved fault tolerance in large synchronous machines, in: 2015 IEEE Workshop on Electrical Machines Design, Control and Diagnosis, 2015, pp. 45–52.

[5] S. Rubino, I.R. Bojoi, et al., Modular stator flux and torque control of multi-three-phase induction motor drives, IEEE Trans. Ind. Appl. 56 (06) (2020) 6507–6525.

[6] T. Zhao, S. Wu, et al., Multiphase PMSM With Asymmetric Windings for More Electric Aircraft, IEEE Trans. Transp. Electrific. 6 (04) (2020) 1592–1602.

[7] A. Salem, M. Narimani, A review on multiphase drives for automotive traction applications, IEEE Trans. Transp. Electrific. 5 (04) (2019) 1329–1348.

[8] S. Rubino, O. Dordevic, et al., A novel matrix transformation for decoupled control of modular multiphase PMSM drives, IEEE Trans. Power Electron. 36 (07) (2021) 8088–8101.

[9] M. Onsal, Y. Demir, et al., A new nine-phase permanent magnet synchronous motor with consequent pole rotor for high-power traction applications, IEEE Trans. Magn. 53 (11) (2017) 1–6.

[10] D. Jiang, K. Liu, et al., Four-module three-phase PMSM drive for suppressing vibration and common-mode current, IEEE Trans. Ind. Appl. 57 (05) (2021) 4874–4883.

[11] X. Han, D. Jiang, et al., Two-segment three-phase PMSM drive with carrier phase-shift pwm for torque ripple and vibration reduction, IEEE Trans. Power Electron. 34 (01) (2018) 588–599.

[12] M. Kashif, B. Singh, Reduced-sensor-based multistage model reference adaptive control of pv-fed PMSM drive for water pump, IEEE Trans. Ind. Electrific. 70 (04) (2023) 3782–3792.

[13] R. Vadivel, Y.H. Joo, Reliable fuzzy H∞ control for permanent magnet synchronous motor against stochastic actuator faults, IEEE Transactions on Systems, Man, and Cybernetics: Systems 51 (04) (2021) 2232–2245.

[14] M. Saeed, W. Song, et al., Low-complexity deadbeat model predictive current control for open-winding PMSM drive with zero-sequence current suppression, IEEE Trans. Transp. Electrific. 7 (04) (2021) 2671–2682.

[15] G. Wu, S. Huang, et al., Robust predictive torque control of N*3-phase PMSM for high-power traction application, IEEE Trans. Power Electron. 35 (10) (2020) 10799–10809.

[16] Y. Xu, S. Li, et al., Integral sliding mode control based deadbeat predictive current control for PMSM drives with disturbance rejection, IEEE Trans. Power Electron. 37 (03) (2022) 2845–2856.

[17] J. He, R. Tang, et al., Robust predictive current control of permanent magnet synchronous motor using voltage coefficient matrix update, Int. J. Electr. Power Energy Syst. 159 (2024) 109999. doi:10.1016/j.ijepes.2024.109999.

[18] C. Zhang, et al., Robust Fault-Tolerant Predictive Current Control for Permanent Magnet Synchronous Motors Considering Demagnetization Fault, IEEE Trans. Ind. Electrific. 65 (07) (2018) 5324–5334.

[19] K. Yu, Z. Wang, et al., Robust Cascaded Deadbeat Predictive Control for Dual Three-Phase Variable-Flux PMSM Considering Intrinsic Delay in Speed Loop, IEEE Trans. Ind. Electrific. 69 (12) (2022) 12107–12118.

[20] X. Li, W. Tian, et al., A generalized observer-based robust predictive current control strategy for PMSM drive system, IEEE Trans. Ind. Electrific. 69 (02) (2022) 1322–1332.

[21] F. Wang, D. Ke, et al., Enhanced predictive model based deadbeat control for PMSM drives using exponential extended state observer, IEEE Trans. Ind. Electrific. 69 (03) (2022) 2357–2369.

[22] G. Wu, et al., Predictive torque and stator flux control for N*3-phase PMSM drives with parameter robustness improvement, IEEE Trans. Power Electron. 36 (02) (2021) 1970–1983.

[23] Y. Yao, Y. Huang, et al., An improved deadbeat predictive current control with online parameter identification for surface-mounted PMSM, IEEE Trans. Ind. Electrific. 67 (12) (2020) 10145–10155.

[24] Y. Zhou, S. Zhang, et al., Current prediction error based parameter identification method for SPMSM with deadbeat predictive current control, IEEE Trans. Energy Convers. 36 (03) (2021) 1700–1710.

[25] X. Zhang, Z. Zhao, et al., Robust model predictive current control based on inductance and flux linkage extraction algorithm, IEEE Trans. Veh. Technol. 69 (12) (2020) 14893–14902.

[26] Z. Wang, J. Chai, et al., A novel online parameter identification algorithm designed for deadbeat current control of the permanent-magnet synchronous motor, IEEE Trans. Ind. Appl. 58 (02) (2022) 2029–2041.

[27] S. Roy, J.J. Shynk, Analysis of the momentum LMS algorithm, IEEE Transactions on Acoustics, Speech, and Signal Processing 38 (12) (1990) 2088–2098.

[28] B. Sun, Z. Teng, et al., A novel LMS-based SANC for conveyor belt-type checkweigher, IEEE Trans. Instrum. Meas. 70 (2021) 1–10.

[29] S.S. Kumar, N. Mukundan, et al., Modified LMS control for a grid interactive PV–fuel cell–electrolyzer hybrid system with power dispatch to the grid, IEEE Trans. Ind. Appl. 58 (06) (2022) 7907–7918.

[30] A. Kumar, N. Patel, et al., L2 norm enabled adaptive LMS control for grid-connected photovoltaic converters, IEEE Trans. Ind. Appl. 58 (04) (2022) 5328–5339.

CHAPTER 14

Predictive power control of modular multiphase PM energy storage motor for urban rail transit

14.1 Introduction

Urban public transport is the lifeblood of urban economic life and an important infrastructure related to the national economy and social stability. As an efficient, safe, and environmentally friendly new means of transportation, urban rail transit has gradually become the mainstream of urban public transport development [1,2]. The research on the regenerative energy storage technology of train braking is not only of great significance to the safe and stable operation of urban rail transit but also of great value in energy conservation and sustainable development. Flywheel energy storage technology has the advantages of high power density, fast response speed, and green environmental protection [3]. It has unique advantages in stabilizing the traction grid voltage and improving the regenerative energy recovery of train braking. It is especially suitable for urban rail transit braking energy storage power generation.

The permanent magnet (PM) motor has become the most commonly used and ideal type of energy storage motor in flywheel energy storage systems due to its high power density, high operating efficiency, and good dynamic and static performance [4,5]. To overcome the capacity limitation of a single power electronic converter, the flywheel energy storage system typically employs multiple sets of multiphase PM motors as the main generators. Multiphase PM motors have two significant advantages: First, the amplitude of electromagnetic torque ripple decreases as the motor phase number increases [6,7]. Second, the short-circuit and open-circuit faults of the multiphase PM motor drive system can be collectively referred to as motor phase failures, which can be addressed through fault isolation [8,9]. The multiphase PM motor can operate smoothly under phase failure conditions and exhibits strong fault tolerance and reliability [10]. Therefore, in

Robust Model Predictive Control of Permanent Magnet Synchronous Motor Under Parameter Mismatch
https://doi.org/10.1016/B978-0-443-44209-4.00013-0

the field of urban rail transit, which demands higher and stricter reliability requirements, the novel modular multiphase PM energy storage motor with strong fault tolerance and high reliability offers more advantages than the conventional three-phase PM motor.

A great deal of the literature has reported and described different multiphase PM motor topologies, modeling and design methods, and control technologies. In [11], a novel multiphase PM motor design with a dual three-phase asymmetric stator winding was proposed, which can achieve higher torque density levels and excellent torque quality. In [12], a novel six-phase PM motor with a new asymmetric stator AC winding scheme and placement was proposed, which can achieve better motor performance than conventional dual three-phase PM motors. To improve the fault tolerance of motor drive systems, a dual three-phase 12-slot 10-pole PM motor was designed in [13], which can avoid excessive torque ripple and unbalanced radial forces under faulty operating conditions. In [14,15], a novel modular fault-tolerant multiphase PM motor was developed, which can solve the problem of full-load testing for a single AC PM electric machine. In [16], a new nine-phase permanent magnet synchronous motor with a consequent pole rotor was proposed for high-power traction applications, which can take advantage of both simpler and cheaper construction in addition to wider flux-weakening capability. In [17], a 20-pole–24-slot surface PM synchronous motor with a consequent pole rotor was proposed for in-wheel direct drive, which can avoid all of the problematic features of direct-drive in-wheel motors. In [18], a nine-phase PM synchronous motor drive system based on multiple three-phase voltage source inverters (VSIs) was presented for an ultrahigh-speed elevator, which can ensure the fault-tolerance capability of the ultrahigh-speed elevator system. However, the design of multiphase PM motor topologies was only considered in [11–18].

The effective control strategy can greatly improve the performance of the modular multiphase PM energy storage motor drive system. During train braking and starting, the flywheel energy storage system uses the modular multiphase PM energy storage motor as a generator and motor, respectively. The controller of the modular multiphase PM energy storage motor can be divided into two parts: the energy storage controller and the energy release controller. The modular multiphase PM energy storage motor control system employs a three-loop cascade control structure. In the field of urban rail transit applications, a fast electromagnetic torque response is required to ensure the high dynamic performance of the entire flywheel energy storage system. Compared to conventional PI control, the predictive control method offers better dynamic and steady-state performance, enabling the modular multiphase PM energy storage motor to achieve excellent control performance across the entire operating range [19–21]. In [22], a robust fault-tolerant predictive current control method was proposed for PM synchronous motors, which improves the robustness of the current control loops and maintains a quick dynamic response. In [23], a predictive stator flux control algorithm was introduced for PM synchronous motor drives, achieving low torque ripple, low stator current harmonics, and excellent steady-state performance. In [24], a robust nonlinear predictive current control algorithm was developed for PM synchronous motor drives, optimizing the current control loop performance under model parameter perturbation. In [25], a cascaded robust fault-tolerant predictive control strategy based on an integral terminal sliding mode observer was presented for PM synchronous motor drives, delivering high-performance speed and current loops. In [22–25], the predictive control methods were designed for three-phase PM synchronous motors, achieving outstanding control performance. However, there are essential differences between modular multiphase PM energy storage motors and conventional three-phase PM synchronous motors. Therefore designing a predictive controller for modular multiphase PM energy storage motors remains a significant challenge.

In this chapter, a predictive power control (PPC) strategy for the modular multiphase PM energy storage motor is proposed, which can achieve lower torque ripple and zero-error tracking of stator flux. The predictive control algorithm is applied to the modular multiphase PM energy storage motor for the first time. First, the output characteristics of the modular multiphase PM energy storage motor are analyzed in detail, and the mathematical model is presented. Second, the operation principle and topological structure of the modular multiphase PM energy storage motor are developed. Thirdly, a novel predictive direct power control algorithm is proposed for the modular multiphase PM energy storage motor, which can obtain excellent power control performance. This chapter is organized as follows: the novel modular multiphase PM energy storage motor is designed in Section 14.2, the predictive direct power control of the novel modular multiphase PM energy storage motor is proposed in Section 14.3, the simulation results are given in Section 14.4, and Section 14.5 concludes this chapter.

14.2 The design of a novel modular multiphase PM energy storage motor

Mathematical model of modular multiphase PM energy storage motor

To overcome the capacity limitations of a single power electronic converter, the modular multiphase PM energy storage motor is designed in this chapter. The modular multiphase PM energy storage motor consists of N three-phase PMSM units with repetitive characteristics. Taking the modular 6*3-phase PM energy storage motor as an example, the finite element simulation model of the motor is established using finite element software, and the output characteristics of the motor are obtained through finite element calculations. The structural diagram of the modular 6*3-phase PM energy storage motor is shown in Fig. 14.1.

According to the characteristics analysis of the modular multiphase PM energy storage motor, the d- and q-axis voltage equations of the modular multiphase PM energy storage motor can be obtained as [10,16]

$$\begin{cases} \dfrac{di_{dj}}{dt} = -\dfrac{R_o}{L_o} i_{dj} + \omega_e i_{qj} + \dfrac{1}{L_o} u_{dj} \\ \dfrac{di_{qj}}{dt} = -\dfrac{R_o}{L_o} i_{qj} - \omega_e i_{dj} - \dfrac{\psi_{ro}}{L_o} \omega_e + \dfrac{1}{L_o} u_{qj} \end{cases} \tag{14.1}$$

where j stands for any unit of the motor, i_{dj} and i_{qj} are the d- and q-axis currents, respectively; R_o and L_o are the stator resistance and stator inductance, respectively; ω_e is the electrical rotor speed, ψ_{ro} is the flux linkage of permanent magnets, u_{dj} and u_{qj} represent the d- and q-axis voltages of the jth winding, respectively.

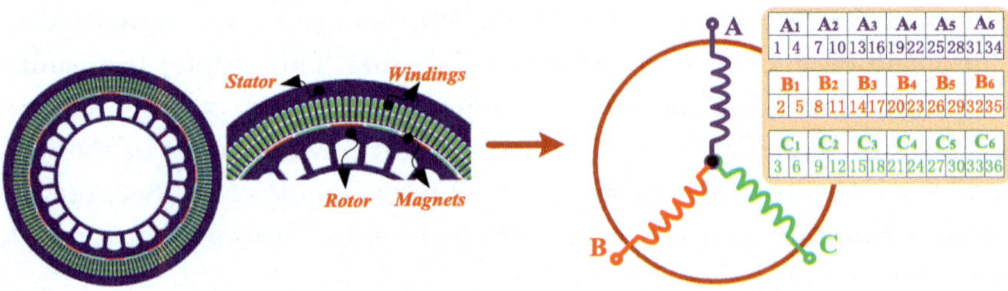

Fig. 14.1 The structural diagram of the modular 3*3-phase PM energy storage motor.

The modular multiphase PM energy storage motor is composed of three-phase SPMSM units with identical characteristics. Therefore the stator resistance and stator inductance of each three-phase SPMSM unit are equal. The electromagnetic torque produced by each three-phase SPMSM unit is equal. The output electromagnetic torque of the modular multiphase PM energy storage motor is the sum of the electromagnetic torque produced by each three-phase SPMSM unit. The electromagnetic torque produced by the modular multiphase PM energy storage motor can be expressed as [16]

$$T_e = \frac{3n_p}{2}\sum_{j=1}^{N}(\psi_{ro} i_{qj}) \tag{14.2}$$

The mechanical dynamic model of the modular multiphase PM energy storage motor can be described as:

$$T_e - T_L = \frac{J}{p}\frac{d\omega_e}{dt} \tag{14.3}$$

where n_p is the number of pole pairs, J is the moment of inertia; T_e and T_L are the electromagnetic torque and load torque of the modular multiphase PM energy storage motor, respectively.

Topological structure of modular multiphase PM energy storage motor drive system

With the increasing power load of high-speed railway electric multiple units (EMUs), the total capacity of the converter is also increasing. The capacity of a single power electronics converter module cannot meet the demands of the high-speed railway EMU system. Thus a multi-module parallel power electronics converter energy storage system based on a modular multiphase PM energy storage motor is proposed in this chapter. Due to the limitations of flywheel safety and the space constraints of high-speed railway EMUs, the application of flywheel energy storage technology in rail transit is restricted to ground-based systems. The topological structure of the modular multiphase PM energy storage motor system for ground rail transit is illustrated in Fig. 14.2. The topology consists of a back-to-back converter, and the flywheel is connected to the converter through the modular multiphase PM energy storage motor to facilitate energy transfer.

The modular multiphase PM energy storage motor system can be divided into charging and discharging working states, as shown in Fig. 14.3.

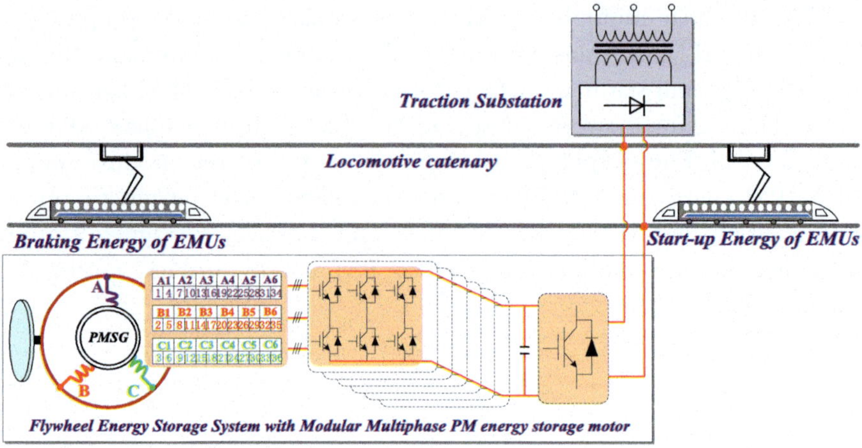

Fig. 14.2 Topological structure of a modular multiphase PM energy storage motor system for ground rail transit.

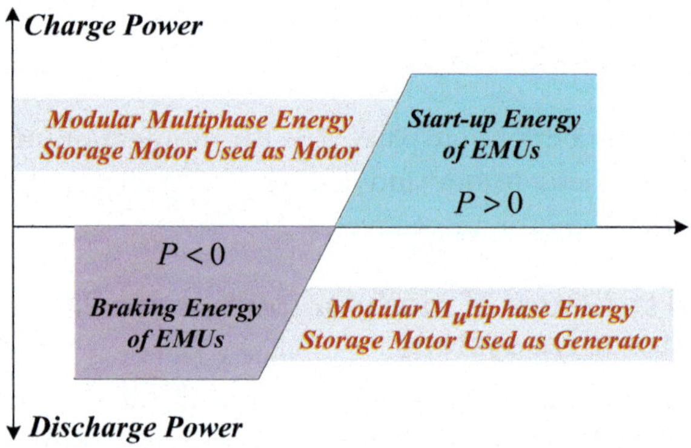

Fig. 14.3 Energy recovery process of a modular multiphase PM energy storage system.

When the high-speed railway EMUs accelerates and leaving the station, the flywheel energy storage system is charged state. The modular multiphase PM energy storage motor is used as motor (i.e., power of modular multiphase PM energy storage motor$P > 0$). When the high-speed railway EMUs decelerates and brakes into the station, the flywheel energy storage system is discharged state. The modular multiphase PM energy storage motor is used as generator (i.e., power of modular multiphase PM energy storage motor$P < 0$). Through the above analysis, the charging and discharging control of modular multiphase PM energy storage motor system is essentially the power control of modular multiphase PM energy storage motor.

14.3 PPC of modular multiphase PM energy storage motor

Drive system structure of a modular multiphase PM energy storage motor

Taking the modular 6⋆3-phase PM energy storage motor as an example, the modular 6⋆3-phase PM energy storage motor is driven by six drive units to obtain the required power. The 18 stator winding wirings of the modular 6⋆3-phase PM energy storage motor are connected with six VSIs in parallel, as shown in Fig. 14.4. It can be seen from Fig. 14.4 that each group of stator windings is excited independently with three-phase symmetrical voltage generated by a VSI. The volt-second characteristics of any two VSIs are the same in any subcycle. The mathematical model and characteristics of each motor unit are identical to those of a regular PMSM, and all the characteristics of each motor unit are repetitive. The control of a VSI can be analogous to that of a conventional inverter. In this chapter, a novel predictive direct power control method is proposed for the modular multiphase PM energy storage motor.

PPC of modular multiphase PM energy storage motor

The PPC method is a discrete control algorithm based on the precise mathematical model of a modular multiphase PM energy storage motor. It is necessary to discretize the mathematical model of the modular multiphase PM energy storage motor and then design the predictive power controller.

The general solution of the voltage equations of a modular multiphase PM energy storage motor can be expressed as

$$\mathbf{x}_j(t) = e^{\mathbf{A}(t-t_o)}\mathbf{x}_j(t_o) + \int_{t_o}^{t} e^{\mathbf{A}(t-\tau)}(\mathbf{B}\mathbf{u}_j(\tau) + \mathbf{D})d\tau \tag{14.4}$$

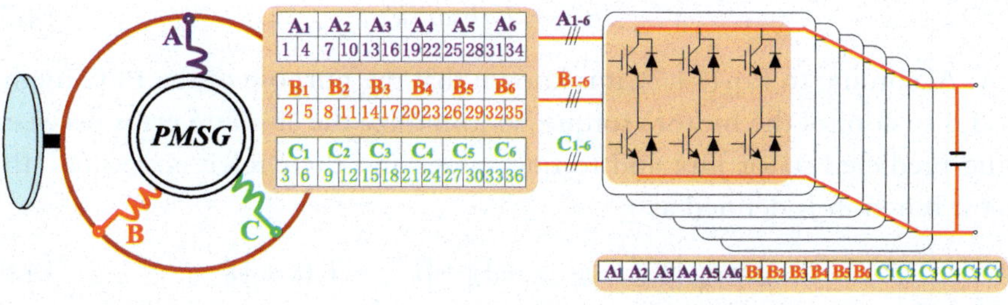

Fig. 14.4 Drive system structure of modular 6*3-phase PM energy storage motor drive.

where $x_j(t)=\begin{bmatrix} i_{dj} \\ i_{qj} \end{bmatrix}$, $u_j(t)=\begin{bmatrix} u_{dj} \\ u_{qj} \end{bmatrix}$, $A(t)=\begin{bmatrix} -\frac{R_o}{L_o} & \omega_e \\ -\omega_e & -\frac{R_o}{L_o} \end{bmatrix}$, $B=\begin{bmatrix} \frac{1}{L_o} & 0 \\ 0 & \frac{1}{L_o} \end{bmatrix}$,

$D=\begin{bmatrix} 0 \\ -\frac{\omega_e}{L_o}\psi_{ro} \end{bmatrix}$.

Let $t_o = kT_s$, $t_o = (k+1)T_s$, and T_s is the sampling period. When the sampling period is short enough, we can get

$$\begin{cases} e^{-\frac{R_o T_s}{L_o}} \approx 1-\frac{R_o T_s}{L_o} \\ e^{-\frac{R_o T_s}{L_o}} \approx 1-\frac{R_o T_s}{L_o} \\ \cos(\omega_e T_s) \approx 1 \\ \sin(\omega_e T_s) \approx \omega_e T_s \end{cases} \tag{14.5}$$

Thus, according to Eq. (14.1), the discrete state equation of the modular multiphase PM energy storage motor can be obtained as

$$\begin{cases} u_{dj}(k)=\frac{L_o}{T_s}i_{dj}(k+1)+\left(R_o-\frac{L_o}{T_s}\right)i_{dj}(k)-L_o\omega_e(k)i_{qj}(k) \\ u_{qj}(k)=\frac{L_o}{T_s}i_{qj}(k+1)+\left(R_o-\frac{L_o}{T_s}\right)i_{qj}(k)+L_o\omega_e(k)i_{dj}(k)+\psi_{ro}\omega_e(k) \end{cases} \tag{14.6}$$

The power discrete expression of a modular multiphase PM energy storage motor can be expressed as

$$P(k)=T_e(k)\omega_e(k) \tag{14.7}$$

According to Eq. (14.7), it is known that the purpose of the PPC method is to control the motor's torque. To minimize the absolute error between the predicted stator flux and torque values and their reference values, the cost function is defined as

$$g=k_\psi\left|\psi_s^{ref}-\psi_s(k+1)\right|+\left|T_e^{ref}-T_e(k+1)\right| \tag{14.8}$$

where k_ψ is a weighting factor; T_e^{ref} and ψ_s^{ref} are the torque reference value and flux linkage reference value, respectively; $T_e(k+1)$ and $\psi_s(k+1)$ are the predicted torque value and predicted flux linkage value, respectively.

If the PPC sampling period T_s is short enough, the discrete form of modular multiphase PM energy storage motor can be modeled by the first-order Taylor expansion. The torque and stator flux prediction equation of modular multiphase PM energy storage motor can be expressed as

$$\begin{cases} \psi_s(k+1) = \sqrt{\psi_{dj}^2(k+1) + \psi_{qj}^2(k+1)} \\ T_e(k+1) = \dfrac{3n_p}{2} \displaystyle\sum_{j=1}^{N} \left[\psi_{ro} i_{qj}(k+1)\right] \end{cases} \tag{14.9}$$

where $\psi_{dj}(k+1)$ and $\psi_{qj}(k+1)$ are the d- and q-axis stator flux components, respectively; $i_{qj}(k+1)$ is the q-axis current response.

According to Eqs. (14.2) and (14.6), the torque prediction equation of the modular multiphase PM energy storage motor can be expressed as:

$$T_e(k+1) = \frac{3Nn_p}{2} \left[\psi_{ro} i_{qj}(k+1)\right] \tag{14.10}$$

where

$$i_{qj}(k+1) = \frac{1}{L_o} T_s u_{qj}(k) - \left(\frac{R_o}{L_o} T_s - 1\right) i_{qj}(k) - T_s \omega_e(k) i_{dj}(k) - \frac{1}{L_o} T_s \psi_{ro} \omega_e(k) \tag{14.11}$$

According to Eq. (14.6), the stator flux prediction equation of the modular multiphase PM energy storage motor can be expressed as

$$\begin{cases} \psi_{dj}(k+1) = T_s u_{dj}(k) + \left(1 - \dfrac{R_o}{L_o} T_s\right) \psi_{dj}(k) + T_s \omega_e(k) \psi_{qj}(k) + \dfrac{R_o}{L_o} \psi_{ro} T_s \\ \psi_{qj}(k+1) = T_s u_{qj}(k) + \left(1 - \dfrac{R_o}{L_o} T_s\right) \psi_{qj}(k) - T_s \omega_e(k) \psi_{dj}(k) \end{cases} \tag{14.12}$$

where

$$\begin{cases} \psi_{dj}(k) = L_o i_{dj}(k) + \psi_{ro} \\ \psi_{qj}(k) = L_o i_{qj}(k) \end{cases} \tag{14.13}$$

According to Eq. (14.7), the torque reference value can be obtained as

$$T_e^{ref} = \frac{P^{ref}}{\omega_e(k)} \tag{14.14}$$

Using the maximum torque per ampere (MTPA) control strategy, the relationship between the stator flux reference and the torque reference can be expressed as

$$\psi_s^{ref} = \sqrt{\psi_d^2 + \psi_q^2} = \sqrt{(\psi_{ro})^2 + \left(L_o \cdot \frac{T_e^{ref}}{\frac{3Nn_p}{2}\psi_{ro}}\right)^2} \tag{14.15}$$

The structural diagram of the modular multiphase PM energy storage motor system with PPC strategy is shown in Fig. 14.5. According to Fig. 14.5, the operating principle of the modular multiphase PM energy storage motor drive system is as follows: According to Eqs. (14.10) and (14.12), the predicted values of stator flux and torque at the $(k+1)T_s$ moment can be calculated. Then, the switching signal of the VSI can be obtained by substituting the predicted values and the response values into the cost function (Eq. 14.8). Because six VSIs are connected with six sets

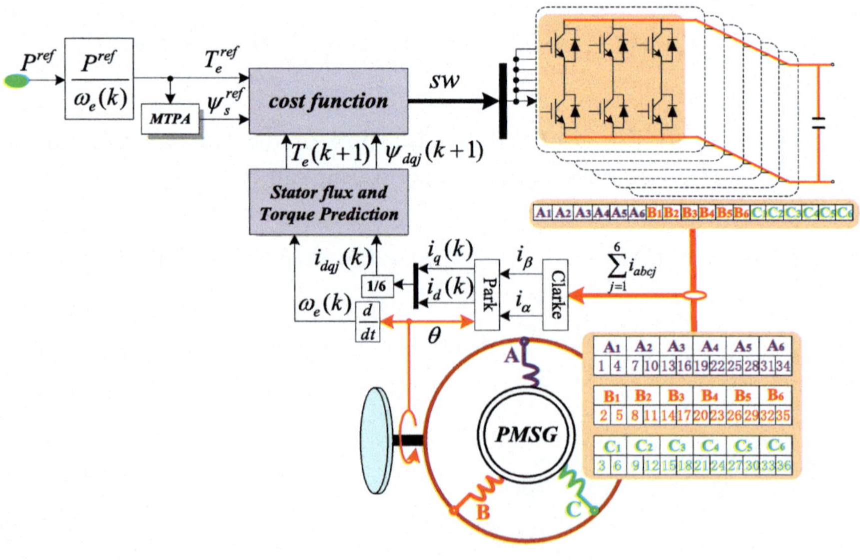

Fig. 14.5 Structural diagram of the modular multiphase PM energy storage motor system with the PPC strategy.

of windings of modular multiphase PM energy storage motor, controlling one of them can realize the control of the whole modular multiphase PM energy storage motor system. The proposed PPC strategy is used to obtain the excellent torque and stator flux control performance. The proposed PPC strategy can effectively eliminate the steady errors of the stator flux and torque while obtaining lower stator flux and torque ripple.

14.4 Simulation results

Taking the modular 6★3-phase PM energy storage motor as an example to verify the PPC algorithm, the modular 6★3-phase PM energy storage motor drive system is established using MATLAB/Simulink. During the charging and discharging process of the modular 6★3-phase PM energy storage motor, some simulation results are provided to verify the effectiveness of the proposed PPC strategy. The main parameters of the modular 6★3-phase PM energy storage motor used in the simulation are presented in Table 14.1. The simulation analysis of the charge and discharge states of the modular 6★3-phase PM energy storage motor is shown in Figs. 14.6–14.13.

Control performance of a modular multiphase PM energy storage motor under the charge state

To verify the operational performance of the modular 6★3-phase PM energy storage motor under the charge state, the power reference is set as

$$\begin{cases} P^{ref} = 2\times10^5 t + 1\times10^5 & 0 \le t \le 0.5s \\ P^{ref} = 2\times10^5 & 0.5s \le t \le 1s \end{cases} \tag{14.16}$$

Simulation results of the modular 6★3-phase PM energy storage motor under the charge state are shown in Figs. 14.6–14.9. Fig. 14.6 presents the simulation results of the torque and phase current of the modular 6★3-phase PM energy storage motor under the charge state. From Fig. 14.6, it can be seen that the peak-to-peak torque ripple of the modular 6★3-phase

Table 14.1 Main parameters of the modular multiphase PM energy storage motor.

Parameter	Value	Parameter	Value
Inductance (L_o)	5.572/mH	Stator phase resistance (R_o)	0.026/Ω
Flux linkage of PM (Ψ_{ro})	0.992/Wb	Number of pole pairs (n_p)	4

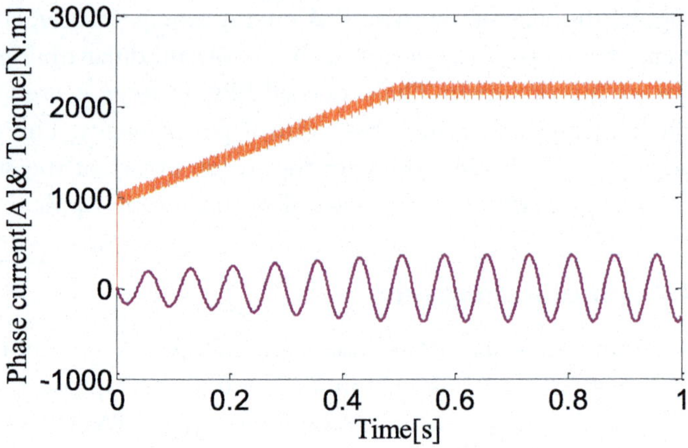

Fig. 14.6 Simulation results of the torque and phase current under the charge state.

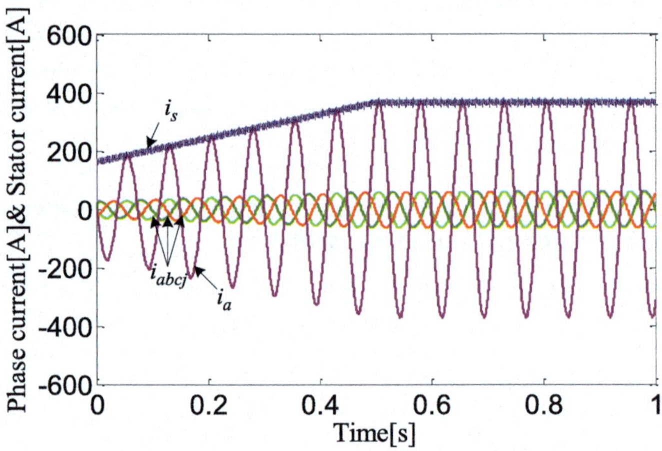

Fig. 14.7 Phase current of the modular 6*3-phase PM energy storage motor and unit motor under the charge state.

PM energy storage motor is small when using the proposed PPC strategy under the charge state. Furthermore, the stator current and electromagnetic torque of the modular 6⋆3-phase PM energy storage motor maintain a perfect sinusoidal waveform under the charge state, as clearly shown in Fig. 14.6. The simulation results demonstrate that the proposed PPC strategy can achieve excellent torque and stator current control performance under the charge state.

Fig. 14.7 presents the phase current of the modular 6⋆3-phase PM energy storage motor and the unit motor using the proposed PPC strategy

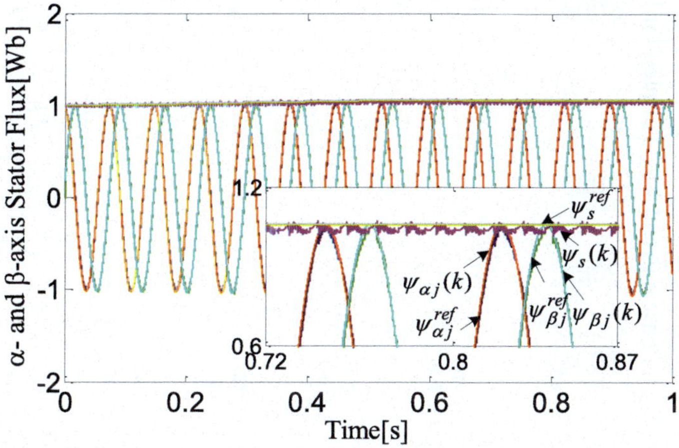

Fig. 14.8 The α-β stator flux linkage of the modular 6*3-phase PM energy storage motor under the charged state.

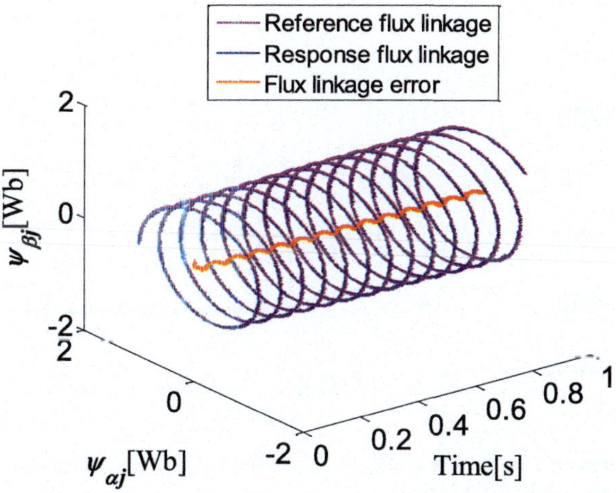

Fig. 14.9 The three-dimensional rotor flux trajectories of the modular 6*3-phase PM energy storage motor under the charge state.

under the charge state. From 0 to 0.5 s, the stator current of the modular 6⋆3-phase PM energy storage motor increases with the rise in the power reference value under the charge state. After 0.5 s, the stator current of the modular 6⋆3-phase PM energy storage motor remains constant when using the proposed PPC strategy, as shown in Fig. 14.7. The reason for this phenomenon is that the charging state of the modular 6⋆3-phase PM energy

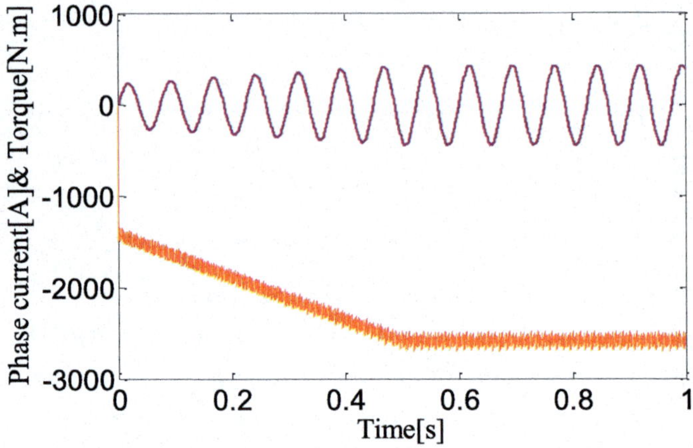

Fig. 14.10 Simulation results of the torque and phase current under the discharge state.

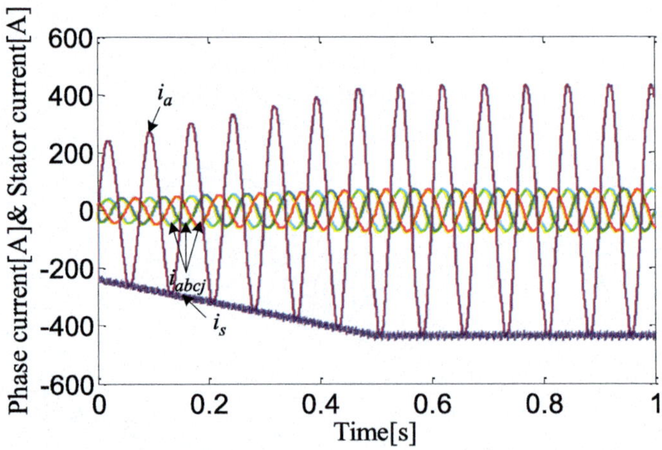

Fig. 14.11 Phase current of the modular 6*3-phase PM energy storage motor and unit motor under the discharge state.

storage motor remains unchanged. From Fig. 14.7, it can be observed that the stator current fundamental value of the modular 6⋆3-phase PM energy storage motor is six times that of the unit motor under the charge state. As shown in Fig. 14.1, due to the modular 6⋆3-phase PM energy storage motor consisting of six motor units with consistent characteristics, the three-phase stator current of the modular 6⋆3-phase PM energy storage motor is six times that of the unit motor. The reason for this mathematical relationship is that the modular 6⋆3-phase PM energy storage motor

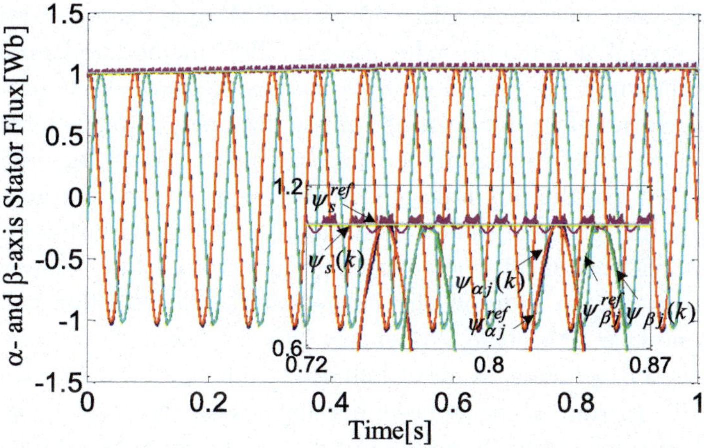

Fig. 14.12 The α-β stator flux linkage of the modular 6*3-phase PM energy storage motor under the discharge state.

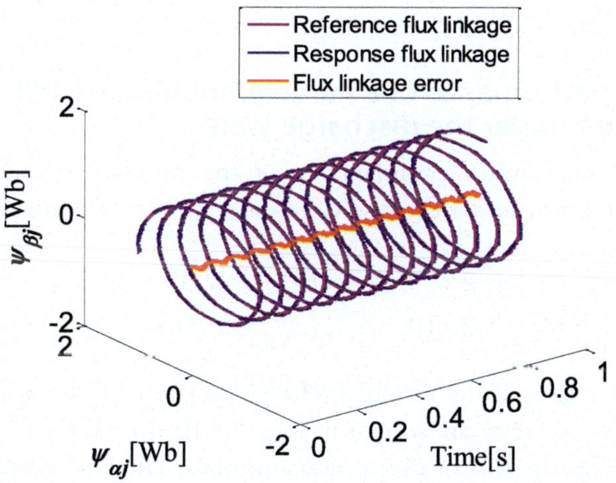

Fig. 14.13 The three-dimensional rotor flux trajectories of the modular 6*3-phase PM energy storage motor under the discharge state.

adheres to the motor superposition characteristics. The stator current control performance of the modular 6⋆3-phase PM energy storage motor is excellent when using the proposed PPC strategy under the charge state, as clearly shown in Fig. 14.7.

Fig. 14.8 presents the α-β stator flux linkage of the modular 6⋆3-phase PM energy storage motor using the proposed PPC strategy under the charge state. The simulation results show that the α-β stator flux linkage predicted value of the modular 6⋆3-phase PM energy storage motor can accurately track its reference value using the PPC method under the charge state, as shown in Fig. 14.8. Fig. 14.9 presents the three-dimensional rotor flux trajectories of the modular 6⋆3-phase PM energy storage motor using the proposed PPC strategy under the charge state. From Fig. 14.9, it can be observed that the flux linkage error between the flux linkage response value and the reference value of the modular 6⋆3-phase PM energy storage motor is very small when adopting the proposed PPC strategy under the charge state. In addition, the three-dimensional rotor flux trajectories of the modular 6⋆3-phase PM energy storage motor show little change under the charge state, as clearly illustrated in Fig. 14.9. The reason for the small variation in the three-dimensional rotor flux trajectories is that the α-β stator flux linkage response value of the modular 6⋆3-phase PM energy storage motor exhibits minimal variation under the charge state. In summary, the proposed PPC strategy can achieve excellent stator flux linkage and torque control performance for the modular 6⋆3-phase PM energy storage motor under the charge state.

Control the performance of a modular multiphase PM energy storage motor under the discharge state

To verify the operational performance of the modular 6⋆3-phase PM energy storage motor under the discharge state, the power reference is set as

$$\begin{cases} P^{ref} = -2\times10^5 t - 1\times10^5 & 0 \le t \le 0.5s \\ P^{ref} = -2\times10^5 & 0.5s \le t \le 1s \end{cases} \tag{14.17}$$

Simulation results of the modular 6⋆3-phase PM energy storage motor under the discharge state are shown in Figs. 14.10–14.13. Fig. 14.10 presents the simulation results of the torque and phase current of the modular 6⋆3-phase PM energy storage motor under the discharge state. The peak-to-peak torque ripple of the modular 6⋆3-phase PM energy storage motor is small when using the proposed PPC strategy under the discharge state, as shown in Fig. 14.10. From Fig. 14.10, it can be observed that the stator current and electromagnetic torque of the modular 6⋆3-phase PM energy storage motor maintain a perfect sinusoidal waveform when using the proposed PPC strategy. The simulation results demonstrate that the proposed PPC strategy achieves excellent torque and stator current control performance under the discharge state, as shown in Fig. 14.10. From 0 to 0.5 s, the stator current of the modular 6⋆3-phase PM energy storage motor increases with the rise in the power reference value when using the proposed PPC strategy. After 0.5 s, the stator current of the modular 6*3-phase PM energy storage motor remains unchanged under the discharge state, as shown in Fig. 14.10.

Fig. 14.11 presents the phase current of the modular 6*3-phase PM energy storage motor and unit motor using the proposed PPC strategy under the discharge state. After 0.5 s, the stator current of the modular 6*3-phase PM energy storage motor remains unchanged when using the proposed PPC strategy under the discharge state, as shown in Fig. 14.11. The reason for this phenomenon is that the discharge state of the modular 6*3-phase PM energy storage motor remains stable when using the proposed PPC strategy. From Fig. 14.11, it can be observed that the phase currents of the modular 6*3-phase PM energy storage motor and unit motor maintain a good sinusoidal waveform under the discharge state. In addition, the phase current amplitude of the modular 6*3-phase PM energy storage motor is approximately six times that of the unit motor, which verifies the torque superposition characteristic of the modular 6*3-phase PM energy storage motor. The mathematical relationship is due to the fact that the modular 6*3-phase PM energy storage motor adheres to the motor superposition characteristics. From Fig. 14.11, it can also be observed that the stator current control performance of the modular 6*3-phase PM energy storage motor is effective when using the proposed PPC strategy under the discharge state.

Fig. 14.12 presents the α-β stator flux linkage of the modular 6*3-phase PM energy storage motor using the proposed PPC strategy under the discharge state. The α-β stator flux linkage response value of the modular 6*3-phase PM energy storage motor can accurately track the reference value using the proposed PPC strategy under the discharge state, as clearly shown in Fig. 14.12. From Fig. 14.12, it can be observed that there is a small stator flux linkage error between the flux linkage response value and the reference value of the modular 6*3-phase PM energy storage motor under the discharge state. Fig. 14.13 presents the three-dimensional rotor flux trajectories of the modular 6*3-phase PM energy storage motor using the proposed PPC strategy under the discharge state. From Fig. 14.13, it can be observed that there is a small change in the three-dimensional rotor flux trajectories of the modular 6*3-phase PM energy storage motor under the discharge state. The reason for this phenomenon is that the variation in the α-β stator flux linkage response value of the modular 6*3-phase PM energy storage motor is very small when using the proposed PPC strategy under the discharge state.

14.5 Conclusion

This chapter begins by analyzing the output characteristics of the modular multiphase PM energy storage motor, based on which the PPC method was developed to achieve lower torque ripple and smaller stator winding currents under both charge and discharge states. For the first time, the proposed PPC algorithm is applied to control the modular multiphase PM energy storage motor in this chapter. The designed modular multiphase PM energy storage motor offers advantages such as a simple structure, strong robustness, and high power density. A multimodule parallel power electronics converter energy storage system, based on the modular multiphase PM energy storage motor, is designed to meet the demands of high-speed railway EMU systems. Furthermore, the novel PPC strategy is introduced to eliminate steady-state errors in torque and stator flux while achieving lower stator flux and torque ripple. Regardless of whether the modular multiphase PM energy storage motor is operating in the charge or discharge state, the proposed PPC method ensures excellent control performance. Simulation results confirm that the proposed PPC method performs effectively in both charge and discharge states, achieving lower torque and stator flux ripple, as well as smaller stator winding currents.

References

[1] Y. Huang, L. Yang, et al., Saving energy and improving service quality: bicriteria train scheduling in urban rail transit systems, IEEE Trans. Intell. Transp. Syst. 17 (12) (2016) 3364–3379.

[2] H. Sun, J. Wu, et al., A bi-objective timetable optimization model for urban rail transit based on the time-dependent passenger volume, IEEE Trans. Intell. Transp. Syst. 20 (02) (2018) 604–615.

[3] F. Faraji, A. Majazi, et al., A comprehensive review of flywheel energy storage system technology, Renew. Sustain. Energy Rev. 67 (2017) 477–490.

[4] X. Zhang, J. Yang, A DC-link voltage fast control strategy for high-speed PMSM/G in flywheel energy storage system, IEEE Trans. Ind. Appl. 54 (02) (2017) 1671–1679.

[5] W. Gengji, W. Ping, Rotor loss analysis of PMSM in flywheel energy storage system as uninterruptable power supply, IEEE Trans. Appl. Supercond. 26 (07) (2016) 1–5.

[6] W. Huang, W. Hua, et al., Model predictive torque control with SVM for five-phase PMSM under open-circuit fault condition, IEEE Trans. Power Electron. 35 (05) (2019) 5531–5540.

[7] W. Wang, J. Zhang, et al., Fault-tolerant control of dual three-phase permanent-magnet synchronous machine drives under open-phase faults, IEEE Trans. Power Electron. 32 (03) (2016) 2052–2063.

[8] W. Huang, W. Hua, et al., Enhanced model predictive torque control of fault-tolerant five-phase permanent magnet synchronous motor with harmonic restraint and voltage pre-selection, IEEE Trans. Ind. Electron. 67 (08) (2019) 6259–6269.

[9] Y. Zhou, X. Lin, et al., A fault-tolerant direct torque control for six-phase permanent magnet synchronous motor with arbitrary two opened phases based on modified variables, IEEE Trans. Energy Convers. 31 (02) (2015) 549–556.

[10] Z. Rao, Z. Zhang, et al., Characteristics and current harmonic control of N★ three-phase PMSG for HVDC transmission Based on MMC, Energies 13 (01) (2020) 178.

[11] Y. Demir, M. Aydin, A novel dual three-phase permanent magnet synchronous motor with asymmetric stator winding, IEEE Trans. Magn. 52 (07) (2016) 1–5.

[12] Y. Demir, M. Aydin, A novel asymmetric and unconventional stator winding configuration and placement for a dual three-phase surface PM motor, IEEE Trans. Magn. 53 (11) (2017) 1–5.

[13] M. Barcaro, N. Bianchi, et al., Analysis and tests of a dual three-phase 12-slot 10-pole permanent-magnet motor, IEEE Trans. Ind. Appl. 46 (06) (2010) 2355–2362.

[14] F. Luise, S. Pieri, et al., Regenerative testing of a concentrated-winding permanent-magnet synchronous machine for offshore wind generation—Part I: test concept and analysis, IEEE Trans. Ind. Appl. 48 (06) (2012) 1779–1790.

[15] F. Luise, S. Pieri, M. Mezzarobba, A. Tessarolo, Regenerative testing of a concentrated-winding permanent-magnet synchronous machine for offshore wind generation—Part II: test implementation and results, IEEE Trans. Ind. Appl. 48 (06) (2012) 1791–1796.

[16] M. Onsal, Y. Demir, et al., A new nine-phase permanent magnet synchronous motor with consequent pole rotor for high-power traction applications, IEEE Trans. Magn. 53 (11) (2017) 1–6.

[17] C. S.U, M. S.H, et al., Development of a 20-pole–24-slot SPMSM with consequent pole rotor for in-wheel direct drive, IEEE Trans. Ind. Electron. 63 (01) (2015) 302–309.

[18] E. Jung, H. Yoo, et al., A nine-phase permanent-magnet motor drive system for an ultrahigh-speed elevator, IEEE Trans. Ind. Appl. 48 (03) (2012) 987–995.

[19] B. G.S, Kazmierkowski, et al., Direct torque control of PWM inverter-fed AC motors—A survey, IEEE Trans. Ind. Electron. 51 (04) (2004) 744–757.

[20] W. Xie, X. Wang, et al., Finite-control-set model predictive torque control with a deadbeat solution for PMSM drives, IEEE Trans. Ind. Electron. 62 (09) (2015) 5402–5410.

[21] Z. Xiang, X. Zhu, et al., Multi-level design optimization and operation of a brushless double mechanical ports flux-switching permanent magnet motor, IEEE Trans. Ind. Electron. 63 (10) (2016) 6042–6054.

[22] C. Zhang, G. Wu, et al., Robust fault-tolerant predictive current control for permanent magnet synchronous motors considering demagnetization fault, IEEE Trans. Ind. Electron. 65 (07) (2017) 5324–5334.

[23] S. Huang, G. Wu, et al., Novel predictive stator flux control techniques for PMSM drives, IEEE Trans. Power Electron. 34 (09) (2018) 8916–8929.

[24] M. Lyu, G. Wu, et al., Robust nonlinear predictive current control techniques for PMSM, Energies 12 (2019) 443.

[25] F. Hu, D. Luo, et al., Cascaded robust fault-tolerant predictive control for PMSM drives, Energies 11 (2018) 3087.

CHAPTER 15

Robust predictive power control of modular multiphase PM energy storage motor with inductance parameter mismatch

15.1 Introduction

With the increasing demands for higher-power energy storage motor drives, multiphase PM motors, which have become the most ideal type of energy storage motor, are widely used in flywheel energy storage systems due to their strong fault tolerance and high operating efficiency [1–3]. Despite the many advantages of multiphase PM motors, achieving high-efficiency control of these motors is rather challenging because they are nonlinear, strongly coupled integrated control objects and, in addition, are subject to parameter disturbances. Aiming to obtain excellent servo control performance of multiphase PM motors, apart from conventional proportional-integral (PI) control [4], many advanced control strategies, such as model predictive control [5] and internal model control, have been proposed. The conventional PI control can achieve excellent dynamic and steady-state control performance under steady-state conditions. However, conventional PI control cannot guarantee the control performance of multiphase PM motors under dynamic conditions [6]. By comparison, the model predictive control method has great advantages because of its ability to account for both fast dynamic and static responses, and it tracks the reference value rapidly and accurately [5].

The model predictive control method is a type of control algorithm that relies strictly on the mathematical model of a multiphase PM motor. The model predictive control method can achieve excellent dynamic performance and outstanding static tracking control performance. However, model parameter mismatches can lead to inaccurate voltage control vectors because the model predictive control method depends entirely on an accurate multiphase PM motor model [7]. Furthermore, the one-beat delay

Robust Model Predictive Control of Permanent Magnet Synchronous Motor Under Parameter Mismatch
https://doi.org/10.1016/B978-0-443-44209-4.00005-1

of the digital control system can also seriously degrade the performance of the predictive control system [8].

In order to eliminate the influence of the one-beat delay, some advanced control methods have been proposed [9–14]. In [9], a novel fault-tolerant predictive current control method based on a composite observer was proposed, which can compensate for the influence of the one-step delay by using the current prediction value of the next time step. In [10], an improved deadbeat predictive current control method based on a stator current and disturbance observer was proposed, which can optimize the current control performance of the PM motor with one-step control delay and parameter mismatch. In [11], a novel deadbeat predictive current control algorithm was presented, which can eliminate the influence of the calculation time steps of the predictive control algorithm. In [12], a novel predictive stator flux control method based on a composite discrete-time sliding mode observer was proposed, which can eliminate the influence of parameter perturbation and one-step delay by adopting the stator flux linkage value of the next time step. In [13], a robust nonlinear predictive current control method based on the composite integral terminal sliding mode observer was presented, which can optimize the current control loop performance of the predictive control system with one-step delay. In [14], a new predictive control algorithm with a Luenberger observer was presented, which can offer robustness against the computational delay inherent in the digital implementation.

Model parameter mismatch is another important factor that affects the performance of predictive control systems. The cause of model parameter mismatch is often attributed to errors in parameter identification or high-temperature operation. This mismatch results in inconsistency between the parameters in the actual motor system and those in the predictive control system, leading to inaccurate predicted voltage vectors [15]. To enhance robustness against model parameter mismatch, several improved predictive control methods have been proposed [16–22]. In [16,17], a continuous-time model predictive control method was developed using a disturbance observer-based compensation technique, which effectively addresses model parameter uncertainties in practical engineering applications. In [18], a novel flux immunity robust predictive current control method based on an incremental model and extended state observer was introduced, improving inductance robustness to eliminate predicted errors caused by stator inductance mismatch. In [19], a robust model predictive current control method based on a disturbance observer was proposed, which minimizes instantaneous *d*- and *q*-axis current errors in every sampling period to eliminate the influence of unknown external disturbances and model uncertainties. In [20], a hierarchical model predictive control method based on integral sliding mode was analyzed, demonstrating its ability to reduce the impact of differences between the nominal predicted dynamic system and the actual system. In [21], a novel predictive functional control method with an extended state observer was proposed, enhancing the disturbance rejection performance of the predictive functional control method. In [22], a cascaded robust fault-tolerant predictive control method based on an integral terminal sliding mode observer was developed, achieving excellent dynamic response and strong robustness under parameter perturbation. However, these predictive control methods rely on disturbance observer compensation, which depends on the accurate detection of current and voltage.

In this chapter, a novel R-PPC method without the disturbance observer for a modular multiphase PM energy storage motor system is presented, which can effectively enhance robustness against inductance parameter mismatch and compensate for the influence of the one-beat delay. The modular multiphase PM energy storage motor system operates well in both the charge state and discharge state, while achieving excellent current tracking performance and smaller torque ripple by using robustness factors. Compared with the conventional PPC method, simulation and experimental results show that the proposed R-PPC strategy can achieve the merits of less torque ripple, lower *d*- and *q*-axis response current ripple, and excellent steady-state performance under inductance parameter mismatch. This chapter is organized as follows. The mathematical model of the novel modular multiphase PM energy storage motor system is established in Section 15.2. The topological structure of the modular multiphase PM energy storage motor system is illustrated in Section 15.3. The robust predictive power control of the novel modular multiphase PM energy storage motor system is proposed in Section 15.4. The simulation results are provided in Sections 15.5. Section 15.6 concludes this chapter.

15.2 The mathematical model of the modular multiphase PM energy storage motor

In this chapter, the modular multiphase PM energy storage motor is proposed to meet the capacity demand of high-power flywheel energy storage systems. The modular multiphase PM energy storage motor is composed of N motor units. The characteristics of these motor units are identical to each other, and each motor unit has magnetic and electrical isolation characteristics. Taking the modular 6★3 phase PM energy storage motor as an example, the structural diagram of the modular 6★3 phase PM energy storage motor is shown in Fig. 15.1.

Fig. 15.1 shows the star of slots for the modular 6★3-phase PM energy storage motor, with each phase sharing 12 identical slot vectors. For the proposed modular 6★3-phase PM energy storage motor, the six-segment winding is adopted by configuring the connection of the 36 coils. The no-load back EMF of the six three-phase PMSM units is identical and exhibits a sinusoidal waveform. The voltage state equation of the three-phase PMSM units is consistent with that of the conventional three-phase PMSM. The output electromagnetic torque produced by the modular 6★3-phase PM energy storage motor is the sum of the electromagnetic torque of each three-phase PMSM unit [2].

According to the characteristic analysis of the proposed modular multiphase PM energy storage motor [2], the mathematical model of the modular multiphase PM energy storage motor is described as

$$\begin{cases} \dfrac{di_{dj}}{dt} = -\dfrac{R_o}{L_o} i_{dj} + \omega_e i_{qj} + \dfrac{1}{L_o} u_{dj} \\ \dfrac{di_{qj}}{dt} = -\dfrac{R_o}{L_o} i_{qj} - \omega_e i_{dj} - \dfrac{\psi_{ro}}{L_o} \omega_e + \dfrac{1}{L_o} u_{qj} \\ T_e = \dfrac{3n_p}{2} \sum_{j=1}^{N} (\psi_{ro} i_{qj}) \\ T_e - T_L = \dfrac{J}{p} \dfrac{d\omega_e}{dt} \end{cases} \tag{15.1}$$

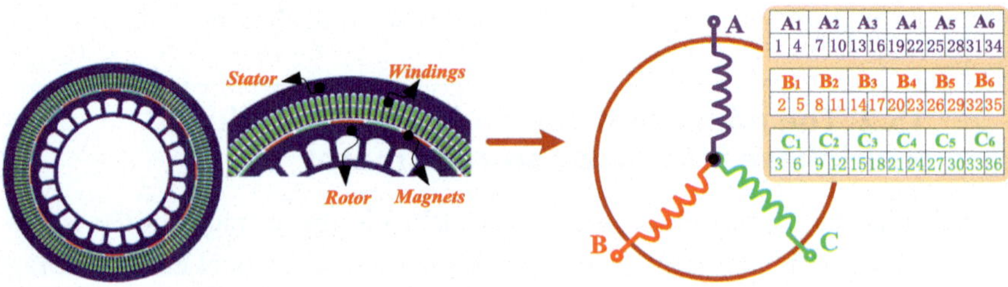

Fig. 15.1 The structural diagram of the modular 3*3-phase PM energy storage motor.

Where j stands for any unit of the motor; i_{dj} and i_{qj} are the d- and q-axis response current value, respectively; L_o and R_o are the stator inductance and stator resistance, respectively; ψ_{ro} is the flux linkage parameter; u_{dj} and u_{qj} are the d- and q-axis voltages, respectively; ω_e is the electrical rotor speed; n_p is the number of pole pairs, T_L and T_e represent the load torque and electromagnetic torque, respectively; J is the moment of inertia.

15.3 Topological structure of modular multiphase PM energy storage motor system

In order to expand the application of modular multiphase PM energy storage motor in rail transit, the multimodule parallel power electronics converter energy storage system is developed in this chapter. Fig. 15.2 illustrates the topological structure of modular multiphase PM energy storage motor system. During the braking state and starting state of the train, the modular multiphase PM energy storage motor is respectively used as generator and motor. When the train accelerates to leave the station, the modular multiphase PM energy storage motor is used as the motor for charging (i.e., $P > 0$). When the train decelerates to enter the station, the modular multiphase PM energy storage motor is used as the generator for discharging (i.e., $P < 0$). According to the above analysis, the discharging and charging control of modular multiphase PM energy storage motor

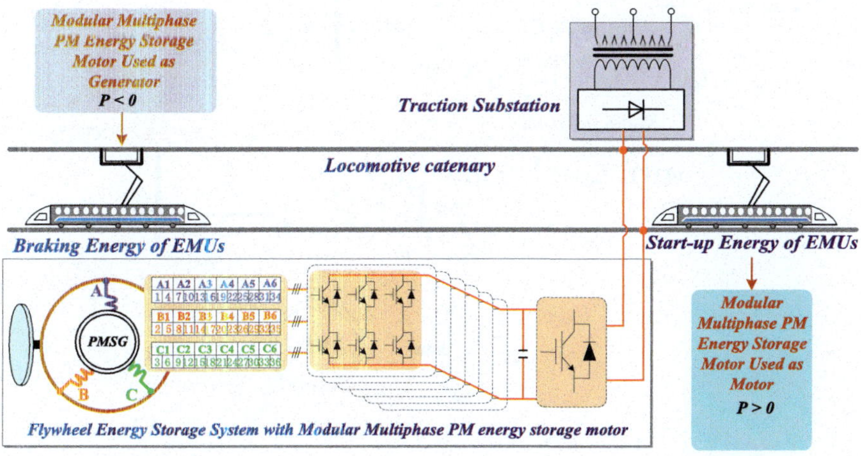

Fig. 15.2 Topological structure of a modular multiphase PM energy storage motor system for ground rail transit.

is essentially the power control, which can obtain excellent dynamic and static control performance by controlling the power.

15.4 Robust predictive power control of modular multiphase PM energy storage motor

Drive system of a modular multiphase PM energy storage motor

Taking a modular 6*3-phase PM energy storage motor as an example, the modular 6*3-phase PM energy storage motor is connected to six voltage source inverters (VSIs). The proposed R-PPC method for the modular multiphase PM energy storage motor system is illustrated in Fig. 15.3. According to the characteristic analyses of the modular multiphase PM energy storage motor, it is known that the control of VSIs can be analogous to that of a conventional VSI. In this chapter, a novel R-PPC method for the modular multiphase PM energy storage motor system is realized by using d-q axis current robust predictive control. The robust predictive current controller, utilizing robustness factors, is adopted to modify the current response values, which can achieve an optimal current response under inductance parameter mismatch. In addition, the proposed R-PPC

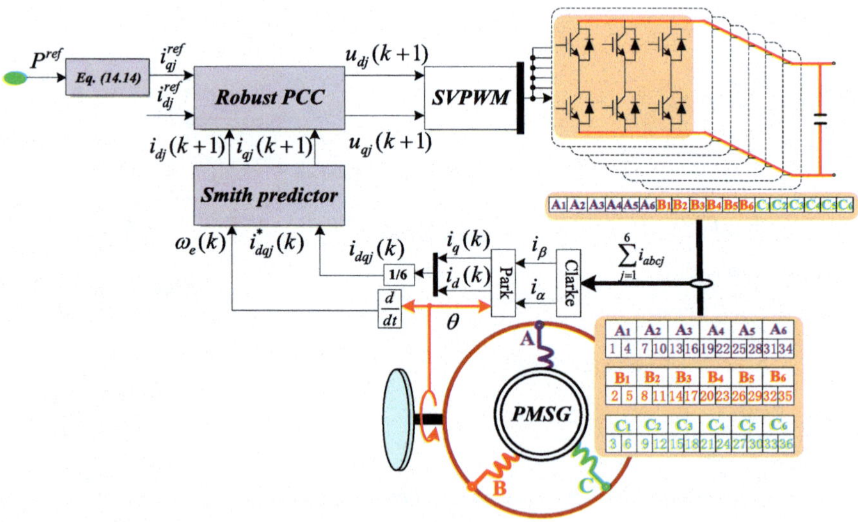

Fig. 15.3 Robust predictive power control method for a modular multiphase PM energy storage motor system.

method can compensate for the one-beat delay of the digital control system by using the Smith predictor.

Robust predictive power control of modular multiphase PM energy storage motor

According to Eq. (15.1), the general state equation of a modular multiphase PM energy storage motor can be described as [15]

$$\boldsymbol{x}_j(t) = e^{A(t-t_o)}\boldsymbol{x}_j(t_o) + \int_{t_o}^{t} e^{A(t-\tau)}(\boldsymbol{B}\boldsymbol{u}_j(\tau) + \boldsymbol{D})d\tau \tag{15.2}$$

where $\boldsymbol{x}_j(t) = \begin{bmatrix} i_{dj} \\ i_{qj} \end{bmatrix}$, $\boldsymbol{u}_j(t) = \begin{bmatrix} u_{dj} \\ u_{qj} \end{bmatrix}$, $\boldsymbol{A}(t) = \begin{bmatrix} -\dfrac{R_o}{L_o} & \omega_e \\ -\omega_e & -\dfrac{R_o}{L_o} \end{bmatrix}$, $\boldsymbol{B} = \begin{bmatrix} \dfrac{1}{L_o} & 0 \\ 0 & \dfrac{1}{L_o} \end{bmatrix}$,

$\boldsymbol{D} = \begin{bmatrix} 0 \\ -\dfrac{\omega_e}{L_o}\psi_{ro} \end{bmatrix}$.

When the sampling period T_s is set as very short, we can get

$$\begin{cases} e^{-\frac{R_o T_s}{L_{do}}} \approx 1 - \dfrac{R_o T_s}{L_{do}} \\ e^{-\frac{R_o T_s}{L_{qo}}} \approx 1 - \dfrac{R_o T_s}{L_{qo}} \\ \cos(\omega_e T_s) \approx 1 \\ \sin(\omega_e T_s) \approx \omega_e T_s \end{cases} \tag{15.3}$$

The discrete d- and q-axis voltage equation of a modular multiphase PM energy storage motor can be expressed as

$$\begin{cases} u_{dj}(k) = \dfrac{L_o}{T_s} i_{dj}(k+1) + \left(R_o - \dfrac{L_o}{T_s}\right) i_{dj}(k) - L_o\omega_e(k) i_{qj}(k) \\ u_{qj}(k) = \dfrac{L_o}{T_s} i_{qj}(k+1) + \left(R_o - \dfrac{L_o}{T_s}\right) i_{qj}(k) + L_o\omega_e(k) i_{dj}(k) + \psi_{ro}\omega_e(k) \end{cases} \tag{15.4}$$

where $u_{dj}(k+1)$ and $u_{qj}(k+1)$ are the d- and q-axis the output voltage of predictive current controller respectively; $i_{dj}(k)$ and $i_{qj}(k)$ are the d- and q-axis current values, respectively.

According to Eq. (15.4), the conventional current predictive controller of a modular multiphase PM energy storage motor can be designed as

$$\begin{cases} u_{dj}(k) = \dfrac{L_o}{T_s} i_{dj}^{ref} + \left(R_o - \dfrac{L_o}{T_s}\right) i_{dj}(k) - L_o \omega_e(k) i_{qj}(k) \\ u_{qj}(k) = \dfrac{L_o}{T_s} i_{qj}^{ref} + \left(R_o - \dfrac{L_o}{T_s}\right) i_{qj}(k) + L_o \omega_e(k) i_{dj}(k) + \psi_{ro} \omega_e(k) \end{cases} \tag{15.5}$$

where i_{dj}^{ref} and i_{qj}^{ref} are the d- and q-axis current reference values, respectively.

From Eq. (15.5), it can be known that the conventional current predictive controller relies heavily on the motor parameters. Especially when the inductance parameter are mismatched, the control performance of modular multiphase PM energy storage motor will be seriously deteriorated. In addition, there is one beat delay between the reference voltage value and the actual output voltage value in the digital control system. The one beat delay would also degrade the performance of the predictive current controller if the one beat delay is not eliminated in the proposed robust predictive controller. In order to eliminate the influence of one beat delay of the modular multiphase PM energy storage motor control system, the steps of one beat delay compensation are as follows: Firstly, the current response value (i.e. $i_{dj}(k+1)$, $i_{qj}(k+1)$) of $(k+1)T_s$ moment should be obtained at the kT_s moment. The ideal voltage vector (i.e. $u_{dj}(k+1)$, $u_{qj}(k+1)$) of the kT_s moment is then calculated by using the predicted current response value (i.e. $i_{dj}(k+1)$, $i_{qj}(k+1)$). Fig. 15.4 illustrates the principle of one beat delay compensation of predictive controller.

In order to eliminate the influence of a one-beat delay, Eq. (14.5) is modified to

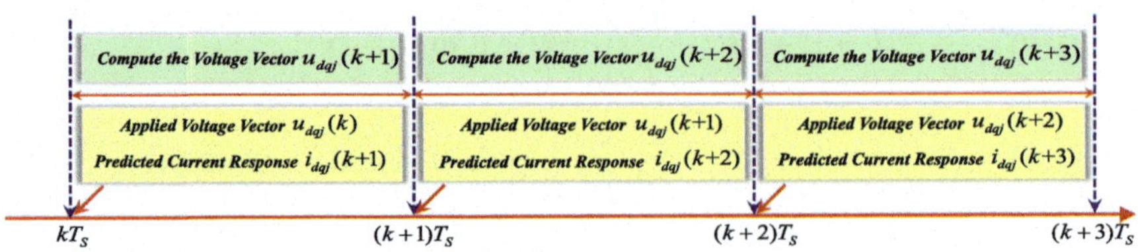

Fig. 15.4 Principle diagram of one-beat delay compensation.

$$\begin{cases} u_{dj}(k+1) = \dfrac{L_o}{T_s} i_{dj}^{ref} + \left(R_o - \dfrac{L_o}{T_s} \right) i_{dj}(k+1) - L_o \omega_e (k+1) i_{qj}(k+1) \\ u_{qj}(k+1) = \dfrac{L_o}{T_s} i_{qj}^{ref} + \left(R_o - \dfrac{L_o}{T_s} \right) i_{qj}(k+1) + L_o \omega_e (k+1) i_{dj}(k+1) + \psi_{ro} \omega_e (k+1) \end{cases} \tag{15.6}$$

At the kT_s moment, the current response value (i.e. $i_{dj}(k+1)$, $i_{qj}(k+1)$) of $(k+1)T_s$ moment can be calculated according to Eq. (15.6).

$$\begin{cases} i_{qj}(k+1) = \dfrac{1}{L_o} T_s u_{qj}(k) - \left(\dfrac{R_o}{L_o} T_s - 1 \right) i_{qj}(k) - T_s \omega_e (k) i_{dj}(k) - \dfrac{1}{L_o} T_s \psi_{ro} \omega_e (k) \\ i_{dj}(k+1) = \dfrac{1}{L_o} T_s u_{dj}(k) - \left(\dfrac{R_o}{L_o} T_s - 1 \right) i_{dj}(k) - T_s \omega_e (k) i_{qj}(k) \end{cases} \tag{15.7}$$

From Eq. (15.7), it can be known that the $i_{dj}(k)$, $i_{qj}(k)$, $u_{dj}(k)$ and $u_{qj}(k)$ are known quantities at the kT_s moment. Therefore the current response value (i.e. $i_{dj}(k+1)$, $i_{qj}(k+1)$) of $(k+1)T_s$ moment can be calculated by equation Eq. (15.7). The electrical speed can be considered as constant due to the electrical time is far less than the mechanical time (i.e. $\omega_e(k)=\omega_e(k+1)$). The ideal voltage vector (i.e. $u_{dj}(k+1)$, $u_{qj}(k+1)$) of the kT_s moment can be calculated by substituting Eq. (15.7) into Eq. (15.6).

$$\begin{bmatrix} u_{dj}(k+1) \\ u_{qj}(k+1) \end{bmatrix} = \frac{L_o}{T_s} \begin{bmatrix} i_{dj}^{ref} \\ i_{qj}^{ref} \end{bmatrix} + \frac{L_o}{T_s} \begin{bmatrix} -\left(1 - \dfrac{R_o}{L_o} T_s\right) & -T_s \omega_e (k) \\ T_s \omega_e (k) & -\left(1 - \dfrac{R_o}{L_o} T_s\right) \end{bmatrix}$$
$$\left\{ \frac{T_s}{L_o} \begin{bmatrix} u_{dj}(k) \\ u_{qj}(k) \end{bmatrix} + \begin{bmatrix} \left(1 - \dfrac{R_o}{L_o} T_s\right) & -T_s \omega_e (k) \\ -T_s \omega_e (k) & \left(1 - \dfrac{R_o}{L_o} T_s\right) \end{bmatrix} \begin{bmatrix} i_{dj}(k) \\ i_{qj}(k) \end{bmatrix} - \begin{bmatrix} \dfrac{T_s}{L_o} \psi_{ro} \omega_e (k) \\ 0 \end{bmatrix} \right\} + \begin{bmatrix} 0 \\ \psi_{ro} \omega_e (k) \end{bmatrix} \tag{15.8}$$

Since $1-\frac{R_o}{L_o}T_s \approx 1$, the Eq. (15.8) is then modified to

$$\begin{bmatrix} u_{dj}(k+1) \\ u_{qj}(k+1) \end{bmatrix} = \frac{L_o}{T_s}\begin{bmatrix} i_{dj}^{ref} \\ i_{qj}^{ref} \end{bmatrix} + \begin{bmatrix} -\frac{L_o}{T_s} & -L_o\omega_e(k) \\ L_o\omega_e(k) & -\frac{L_o}{T_s} \end{bmatrix}$$
$$\left\{\frac{T_s}{L_o}\begin{bmatrix} u_{dj}(k) \\ u_{qj}(k) \end{bmatrix} + \begin{bmatrix} 1 & -T_s\omega_e(k) \\ -T_s\omega_e(k) & 1 \end{bmatrix}\begin{bmatrix} i_{dj}(k) \\ i_{qj}(k) \end{bmatrix} - \begin{bmatrix} \frac{T_s}{L_o}\psi_{ro}\omega_e(k) \\ 0 \end{bmatrix}\right\} + \begin{bmatrix} 0 \\ \psi_{ro}\omega_e(k) \end{bmatrix} \tag{15.9}$$

Due to the response current ripples caused by the inductance parameter mismatch, the proposed R-PPC method using the robustness factors is developed to improve the system stability and avoid the response current ripple. After using the robustness factors, the current response values of kT_s moment can be expressed as

$$\begin{cases} i_{dj}^{*}(k) = \alpha i_{dj}^{ref} + \beta i_{dj}(k) \\ i_{qj}^{*}(k) = \alpha i_{qj}^{ref} + \beta i_{qj}(k) \end{cases} \tag{15.10}$$

where α and β are the robustness factors, and $\alpha + \beta = 1$.

According to Eq. (15.9) and Eq. (15.10), the ideal voltage vector (i.e. $u_{dj}(k+1)$, $u_{qj}(k+1)$) of the kT_s moment can be expressed as

$$\begin{bmatrix} u_{dj}(k+1) \\ u_{qj}(k+1) \end{bmatrix} = \frac{L_o}{T_s}\begin{bmatrix} i_{dj}^{ref} \\ i_{qj}^{ref} \end{bmatrix} + \begin{bmatrix} -\frac{L_o}{T_s} & -L_o\omega_e(k) \\ L_o\omega_e(k) & -\frac{L_o}{T_s} \end{bmatrix}$$
$$\left\{\frac{T_s}{L_o}\begin{bmatrix} u_{dj}(k) \\ u_{qj}(k) \end{bmatrix} + \begin{bmatrix} 1 & -T_s\omega_e(k) \\ -T_s\omega_e(k) & 1 \end{bmatrix}\begin{bmatrix} \alpha i_{dj}^{ref} + \beta i_{dj}(k) \\ \alpha i_{qj}^{ref} + \beta i_{qj}(k) \end{bmatrix} - \begin{bmatrix} \frac{T_s}{L_o}\psi_{ro}\omega_e(k) \\ 0 \end{bmatrix}\right\}$$
$$+ \begin{bmatrix} 0 \\ \psi_{ro}\omega_e(k) \end{bmatrix} \tag{15.11}$$

According to Eq. (15.1), the electromagnetic torque discrete equation of the modular multiphase PM energy storage motor can be obtained as

$$T_e(k) = \frac{3Nn_p}{2}\psi_{ro} i_{qj}(k) \tag{15.12}$$

The power discrete expression of the modular multiphase PM energy storage motor is

$$P(k) = T_e(k)\omega_e(k) \tag{15.13}$$

According to Eqs. (15.12) and (15.13), one yields

$$i_{qj}^{ref}(k) = \frac{2P^{ref}(k)}{3Nn_p\psi_{ro}\omega_e(k)} \tag{15.14}$$

where $P^{ref}(k)$ is the power reference value.

15.5 Simulation results

In order to verify the effectiveness of the proposed R-PPC algorithm, the drive system structure of modular 6*3-phase PM energy storage motor is designed by adopting the MATLAB/ Simulink. The main parameters of modular multiphase PM energy storage motor are given in Table 15.1. The robustness factors of the proposed R-PPC method are set as α=0.4 and β=0.6, respectively. During the discharging state and charging state, some simulation results of modular multiphase PM energy storage motor are illustrated to verify the control performance of the proposed R-PPC strategy.

Simulation results of the proposed R-PPC strategy under the state of charge

In order to verify the operational performance of the modular 6*3 phase PM energy storage motor under the charged state, the power reference is set as

$$\begin{cases} P^{ref} = 2\times10^5 t + 1\times10^5 & 0 \le t \le 0.5s \\ P^{ref} = 2\times10^5 & 0.5s \le t \le 1s \end{cases} \tag{15.15}$$

Table 15.1 Main parameters of the modular multiphase PM energy storage motor.

Parameter	Value	Parameter	Value
Inductance (L_o)	5.572/mH	Stator phase resistance(R_o)	0.026/Ω
Flux linkage of PM (Ψ_{ro})	0.992/Wb	Number of pole pairs (n_p)	4

Simulation results of the modular 6*3 phase PM energy storage motor using the proposed R-PPC strategy under inductance parameter mismatch are shown in Figs. 15.5–15.9. Fig. 15.5 presents the simulation results of the torque and phase current of the modular 6*3 phase PM energy storage motor in the case of inductance parameter mismatch. From Fig. 15.5, it can be seen that the modular 6*3 phase PM energy storage motor achieves good torque control performance using the proposed R-PPC strategy under inductance parameter mismatch. In addition, it is evident from Fig. 15.5 that the torque ripple of the modular 6*3 phase PM energy storage motor is significantly suppressed by employing the proposed R-PPC strategy under inductance parameter mismatch. The simulation results demonstrate that the proposed R-PPC strategy effectively suppresses the torque and stator current effects of inductance parameter mismatch on the modular 6*3 phase PM energy storage motor, as clearly shown in Fig. 15.5.

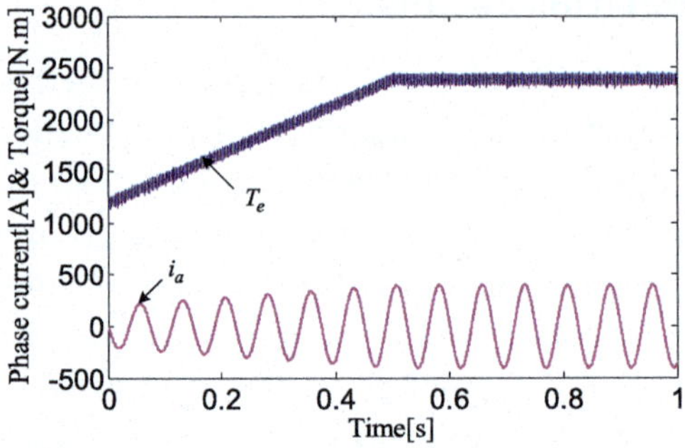

Fig. 15.5 Simulation results of the phase current and torque using the proposed R-PPC method under the charge state.

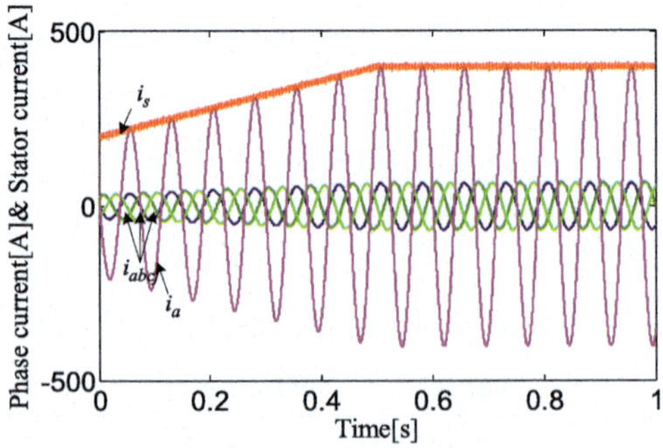

Fig. 15.6 Phase current of the modular 6*3-phase PM energy storage motor and unit motor under the charge state.

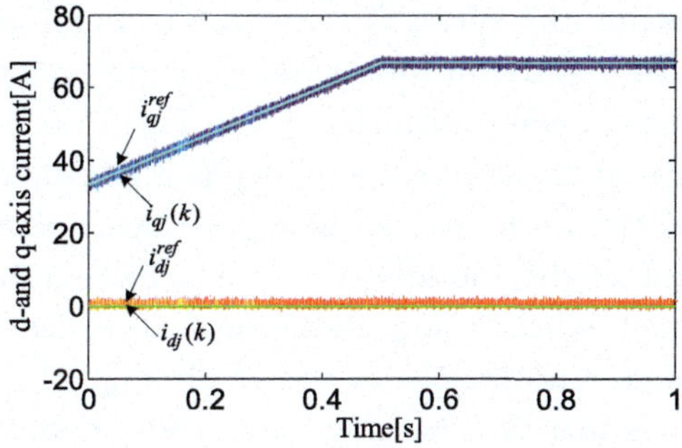

Fig. 15.7 Simulation results of the *d*- and *q*-axis currents using the proposed R-PPC method under the charge state.

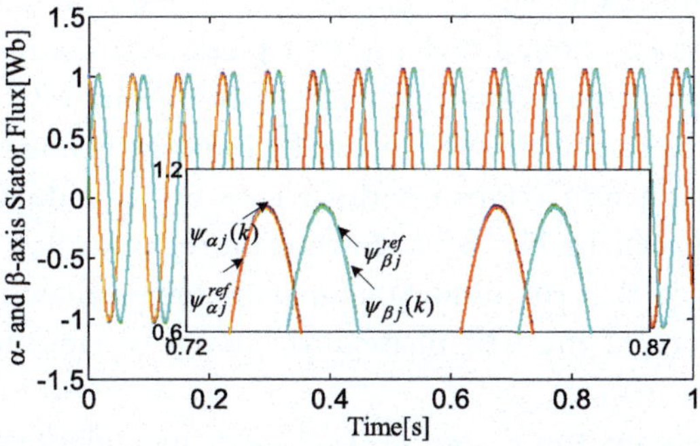

Fig. 15.8 The α-β stator flux linkage of the modular 6*3-phase PM energy storage motor under the charged state.

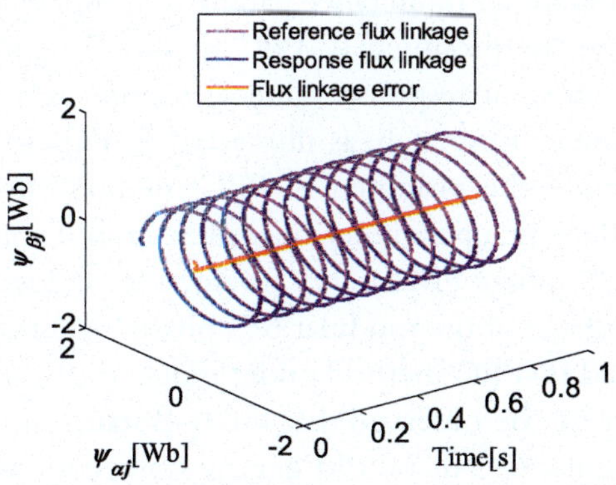

Fig. 15.9 The three-dimensional rotor flux trajectories of the modular 6*3-phase PM energy storage motor under the charge state.

Fig. 15.6 presents the phase current of the modular 6⋆3-phase PM energy storage motor and unit motor using the proposed R-PPC strategy under inductance parameter mismatch. From Fig. 15.6, it can be observed that the stator current of the modular 6⋆3-phase PM energy storage motor increases with the rise in the power reference value under the charge state. The stator current of the modular 6⋆3-phase PM energy storage motor remains unchanged when using the proposed R-PPC strategy under inductance parameter mismatch, as shown in Fig. 15.6. The reason for this phenomenon is that the charging state of the modular 6⋆3-phase PM energy storage motor remains constant. As illustrated in Fig. 15.6, due to the modular 6⋆3-phase PM energy storage motor consisting of six motor units with consistent characteristics, the three-phase stator current of the modular 6⋆3-phase PM energy storage motor is six times that of the unit motor. It can be concluded that the 6⋆3-phase PM energy storage motor meets the torque superposition characteristic. From Fig. 15.6, it can be observed that the modular 6×⋆3-phase PM energy storage motor achieves excellent stator current control performance under inductance parameter mismatch, as shown in Fig. 15.6. Furthermore, according to Fig. 15.6, it can be concluded that the inductance parameter mismatch does not affect the stator current of the 6⋆3-phase PM energy storage motor under the charge state.

Fig. 15.7 presents the *d*- and *q*-axis phase currents of the modular 6⋆3 phase PM energy storage motor using the proposed R-PPC strategy under inductance parameter mismatch. The *d*- and *q*-axis current response values of the modular 6⋆3 phase PM energy storage motor can accurately track the reference values under inductance parameter mismatch, as clearly shown in Fig. 15.7. When the proposed R-PPC method is applied, the ripple in the *d*- and *q*-axis current response values is significantly suppressed under inductance parameter mismatch, as illustrated in Fig. 15.7. This phenomenon occurs because the proposed R-PPC method effectively eliminates the impact of inductance parameter mismatch on the control performance of the modular 6⋆3 phase PM energy storage motor. Fig. 15.8 presents the α-β stator flux linkage of the modular 6⋆3 phase PM energy storage motor using the proposed R-PPC method under inductance parameter mismatch. From Fig. 15.8, it can be observed that the α-β stator flux linkage response values of the modular 6⋆3 phase PM energy storage motor accurately track their reference values when using the proposed R-PPC strategy under the charge state, as shown in Fig. 15.8. The simulation results demonstrate that the flux linkage error of the modular 6⋆3 phase PM energy storage motor, between the flux linkage response values and the reference values, is very small when using the proposed R-PPC strategy under inductance parameter mismatch, as clearly shown in Fig. 15.8.

Fig. 15.9 presents the three-dimensional rotor flux trajectories of the modular 6⋆3 phase PM energy storage motor using the proposed R-PPC method under inductance parameter mismatch. From Fig. 15.9, it can be observed that the three-dimensional rotor flux trajectories of the modular 6⋆3 phase PM energy storage motor show little change when using the proposed R-PPC method.The reason for this phenomenon is that the α-β stator flux linkage response value of the modular 6×⋆3 phase PM energy storage motor exhibits only a small variation under the charge state. In summary, the proposed R-PPC method can achieve excellent stator current and torque control performance for the modular 6×⋆3 phase PM energy storage motor under inductance parameter mismatch.

Simulation results of the proposed R-PPC strategy under the discharge state

In order to verify the operational performance of the modular 6⋆3 phase PM energy storage motor under the discharge state, the power reference is set as

$$\begin{cases} P^{ref} = -2\times 10^5 t - 1\times 10^5 & 0 \le t \le 0.5s \\ P^{ref} = -2\times 10^5 & 0.5s \le t \le 1s \end{cases} \tag{15.16}$$

Simulation results of the modular 6⋆3 phase PM energy storage motor using the proposed R-PPC strategy under the discharge state are shown in Figs. 15.10–15.14. Fig. 15.10 presents the simulation results of the torque

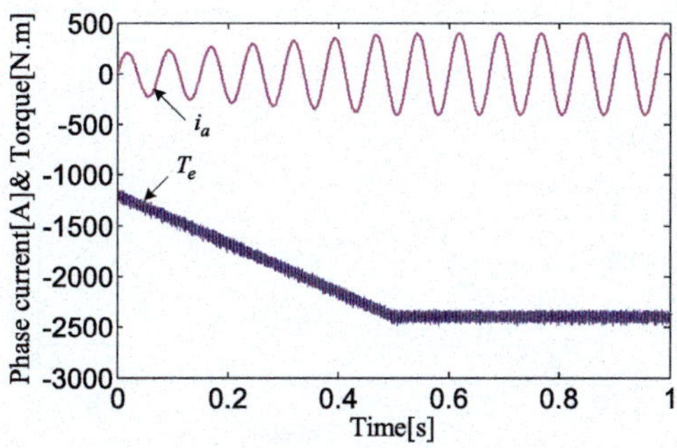

Fig. 15.10 Simulation results of the phase current and torque using the proposed R-PPC method under the discharge state.

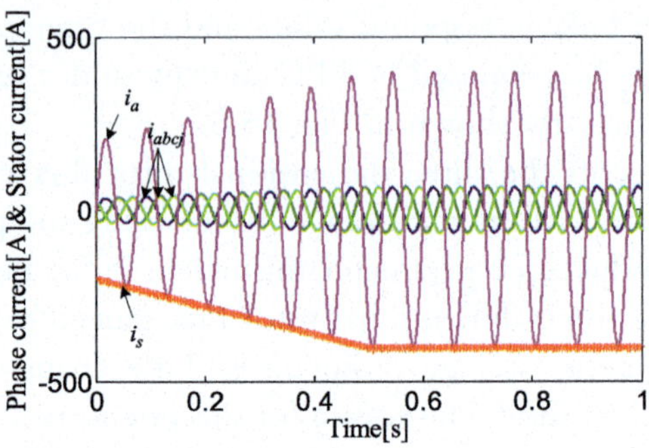

Fig. 15.11 Phase current of the modular 6*3-phase PM energy storage motor and unit motor under the discharge state.

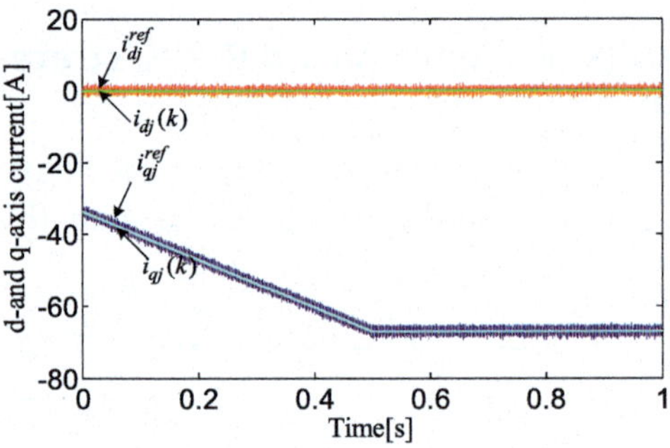

Fig. 15.12 Simulation results of the *d*- and *q*-axis currents using the proposed R-PPC method under the discharge state.

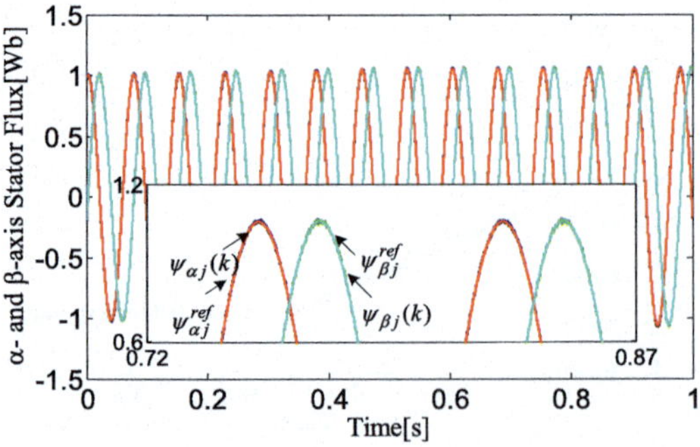

Fig. 15.13 The α-β stator flux linkage of the modular 6*3-phase PM energy storage motor under the discharge state.

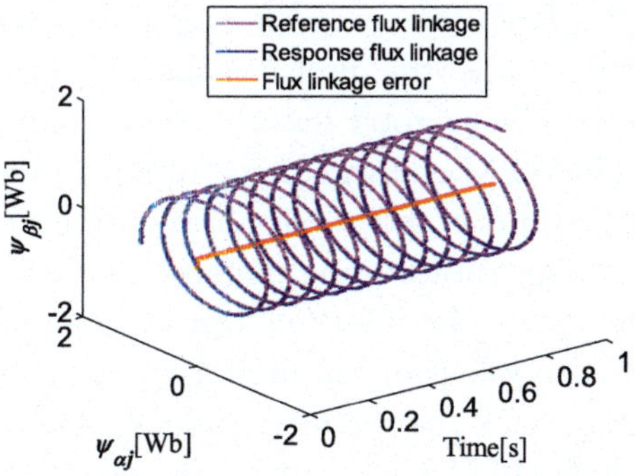

Fig. 15.14 The three-dimensional rotor flux trajectories of the modular 6×*3-phase PM energy storage motor under the charge state.

and phase current of the modular 6⋆3 phase PM energy storage motor under inductance parameter mismatch. From Fig. 15.10, it can be observed that the peak-to-peak torque ripple of the modular 6⋆3 phase PM energy storage motor is small when using the proposed R-PPC strategy under inductance parameter mismatch. The simulation results demonstrate that the proposed R-PPC strategy can achieve excellent torque and stator current control performance for the modular 6⋆3 phase PM energy storage motor under the discharge state, as shown in Fig. 15.10. Furthermore, Fig. 15.10 indicates that the torque ripple of the modular 6⋆3 phase PM energy storage motor can be significantly suppressed by using the proposed R-PPC strategy under inductance parameter mismatch. In summary, the proposed R-PPC strategy can effectively suppress the torque ripple and stator current effects caused by inductance parameter mismatch in the modular 6⋆3 phase PM energy storage motor under the discharge state, as clearly shown in Fig. 15.10.

Fig. 15.11 presents the phase current of the modular 6⋆3-phase PM energy storage motor and the unit motor using the proposed R-PPC strategy under the discharge state. It can be observed from Fig. 15.11 that the modular 6⋆3-phase PM energy storage motor with inductance parameter mismatch can achieve a perfect sinusoidal waveform using the proposed R-PPC strategy under the discharge state. From Fig. 15.11, it is evident that the stator current fundamental value of the modular 6⋆3-phase PM energy storage motor is six times that of the unit motor, which demonstrates the torque superposition characteristic of the modular 6⋆3-phase PM energy storage motor. According to the simulation analysis, it is evident that the inductance parameter mismatch does not affect the stator current of the modular 6⋆3-phase PM energy storage motor when using the proposed R-PPC strategy under the discharge state. In summary, the proposed R-PPC strategy can achieve good stator current control performance for the modular 6⋆3-phase PM energy storage motor in the discharge state.

Fig. 15.12 presents the *d*- and *q*-axis phase currents of the modular 6⋆3 phase PM energy storage motor using the proposed R-PPC strategy under the discharge state. From Fig. 15.12, it can be observed that the *d*- and *q*-axis current response values of the modular 6⋆3 phase PM energy storage motor accurately track the reference values when using the proposed R-PPC strategy under inductance parameter mismatch. The influence of inductance parameter mismatch is clearly suppressed by employing the proposed R-PPC strategy under the discharge state, as shown in Fig. 15.12. This phenomenon occurs because the proposed R-PPC method effectively eliminates the impact of inductance parameter mismatch on the modular 6⋆3 phase PM energy storage motor system under he discharge state.

Fig. 15.13 presents the α-β stator flux linkage of the modular 6×⋆3 phase PM energy storage motor using the proposed R-PPC method under the discharge state. From Fig. 15.13, it can be observed that the α-β stator flux linkage response value of the modular 6×⋆3 phase PM energy storage motor accurately tracks the reference value when using the proposed R-PPC method, even in the case of inductance parameter mismatch. The simulation results indicate that there is a small stator flux linkage error in the modular 6⋆3 phase PM energy storage motor between the flux linkage response value and the reference value when using the proposed R-PPC method, as shown in Fig. 15.13. Fig. 15.14 presents the three-dimensional rotor flux trajectories of the modular 6×⋆3 phase PM energy storage motor in the case of inductance parameter mismatch. From Fig. 15.14, it can be seen that there is a small change in the three-dimensional rotor flux trajectories of the modular 6⋆3 phase PM energy storage motor under inductance parameter mismatch. The reason for this phenomenon is that the variation in the α-β stator flux linkage response value is very small when using the proposed PPC strategy under the discharge state.

15.6 Conclusion

This chapter presents the R-PPC method to track its reference power for a modular multiphase PM energy storage motor with inductance parameter mismatch, which can improve robustness against inductance parameter mismatch and eliminate the one-beat delay in the digital control system.

The multimodule parallel power electronics converter energy storage system is introduced to expand the application of the modular multiphase PM energy storage motor in rail transit. In addition, a novel R-PPC technique using robustness factors is employed in the design to strengthen robustness against inductance parameter mismatch for the modular multiphase PM energy storage motor. The appropriate robustness factors are selected to achieve excellent power control performance under both the charge state and discharge state. Therefore the modular multiphase PM energy storage motor operates well under both the charge state and discharge state, while also achieving excellent dynamic and steady-state control performance. The simulation results show that the proposed R-PPC method provides the benefits of reduced torque ripple, lower *d*- and *q*-axis response current ripple, and excellent steady-state performance under inductance parameter mismatch.

References

[1] Y.A. Demir, A novel dual three-phase permanent magnet synchronous motor with asymmetric stator winding, IEEE Trans. Magn. 52 (12) (2016) 1–5.

[2] Z. Rao, Z.Z.H. S, et al., Characteristics and current harmonic control of N* three-phase PMSG for HVDC transmission Based on MMC, IEEE Trans. Ind. Electron. 13 (2) (2020) 178–186.

[3] S. Dwari, L. Parsa, An optimal control technique for multiphase PM machines under open-circuit faults, IEEE Trans. Ind. Electron. 55 (6) (2018) 1988–1995.

[4] X. Han, D. Jiang, T. Zou, R. Qu, Two-segment three-phase PMSM drive with carrier phase-shift PWM for torque ripple and vibration reduction, IEEE Trans. Ind. Electron. 34 (17) (2019) 588–599.

[5] F. Morel, X. Shi, J. Retif, A comparative study of predictive current control schemes for a permanent-magnet synchronous machine drive, IEEE Trans. Ind. Electron. 56 (4) (2009) 2715–2728.

[6] S. Li, Z. Liu, Adaptive speed control for permanent magnet synchronous motor system with variations of load inertia, IEEE Trans. Ind. Electron. 56 (13) (2009) 3050–3059.

[7] L. Tong, An SRF-PLL-based sensorless vector control using the predictive deadbeat algorithm for the direct-driven permanent magnet synchronous generator, IEEE Trans. Power Electron. 29 (2) (2014) 2837–2849.

[8] P. Cortes, J. Rodriguez, C. Silva, Delay compensation in model predictive current control of a three-phase inverter, IEEE Trans. Ind. Electron. 59 (12) (2012) 1323–1325.

[9] C. Zhang, G. Wu, F. Rong, et al., Robust fault-tolerant predictive current control for permanent magnet synchronous motors considering demagnetization fault, IEEE Trans. Ind. Electron. 65 (4) (2017) 5324–5334.

[10] X. Zhang, B. Hou, Y. Mei, Deadbeat predictive current control of permanent-magnet synchronous motors with stator current and disturbance observer, IEEE Trans. Power Electron. 32 (10) (2016) 3818–3834.

[11] Y. Sozer, T.D. A, E. Mese, Adaptive predictive current control technique for permanent magnet synchronous motors, IEEE Trans. Power Electron. 6 (13) (2013) 9–19.

[12] S. Huang, G. Wu, F. Rong, et al., Novel predictive stator flux control techniques for PMSM drives, IEEE Trans. Power Electron. 34 (4) (2018) 8916–8929.

[13] M. Lyu, G. Wu, D. Luo, et al., Robust nonlinear predictive current control techniques for PMSM, IEEE Trans. Ind. Electron. 12 (17) (2019) 443–451.
[14] J.C. Moreno, J.M. Huerta, R.G. Gill, et al., A robust predictive current control for three-phase grid-connected inverters, IEEE Trans. Ind. Electron. 56 (12) (2009) 1993–2004.
[15] Y.M. Lang, X. Long, et al., Flux immunity robust predictive current control with incremental model and extended state observer for PMSM drive, IEEE Trans. Power Electron. 32 (7) (2017) 9267–9279.
[16] J. Yang, Z.W. X, S. Li, et al., Design of a prediction-accuracy-enhanced continuous-time MPC for disturbed systems via a disturbance observer.IEEE, Trans. Ind. Electron. 62 (9) (2015) 5807–5816.
[17] J. Yang, S. Li, X. Yu, Sliding-mode control for systems with mismatched uncertainties via a disturbance observer, IEEE Trans. Ind. Electron. 60 (11) (2012) 160–169.
[18] H. Young, Analysis of finite-control-set model predictive current control with model parameter mismatch in a three-phase inverter, IEEE Trans. Ind. Electron. 63 (12) (2016) 3100–3107.
[19] C. Xia, M. Wang, Z. Song, et al., Robust model predictive current control of three-phase voltage source PWM rectifier with online disturbance observation, IEEE Trans. Ind. Electron. 8 (2) (2012) 459–471.
[20] M. Rubagotti, R.D. M, A. Ferrara, et al., Robust model predictive control with integral sliding mode in continuous-time sampled-data nonlinear systems, IEEE Trans. Ind. Electron. 56 (11) (2010) 556–570.
[21] H. Liu, S. Li, Speed control for PMSM servo system using predictive functional control and extended state observer, IEEE Trans. Ind. Electron. 59 (18) (2011) 1171–1183.
[22] F. Hu, D. Luo, C. Luo, et al., Cascaded robust fault-tolerant predictive control for PMSM drives, IEEE Trans. Power Electron. 11 (1) (2018) 3087–3094.

Index

Note: Page numbers followed by "*f*" indicate figures, "*t*" indicate tables.

R

S

T

U

V

Printed and bound by CPI Group (UK) Ltd, Croydon, CR0 4YY
14/09/2026
02216410-0003